Sustainable Development Goals Series

Industry, Innovation and Infrastructure

World leaders adopted Sustainable Development Goals (SDGs) as part of the 2030 Agenda for Sustainable Development. Providing in-depth knowledge, this series fosters comprehensive research on these global targets to end poverty, fight inequality and injustice, and tackle climate change.
The sustainability of our planet is currently a major concern for the global community and has been a central theme for a number of major global initiatives in recent years. Perceiving a dire need for concrete benchmarks toward sustainable development, the United Nations and world leaders formulated the targets that make up the seventeen goals. The SDGs call for action by all countries to promote prosperity while protecting Earth and its life support systems. This series on the Sustainable Development Goals aims to provide a comprehensive platform for scientific, teaching and research communities working on various global issues in the field of geography, earth sciences, environmental science, social sciences, engineering, policy, planning, and human geosciences in order to contribute knowledge towards achieving the current 17 Sustainable Development Goals.
This Series is organized into eighteen subseries: one based around each of the seventeen Sustainable Development Goals, and an eighteenth subseries, "Connecting the Goals," which serves as a home for volumes addressing multiple goals or studying the SDGs as a whole. Each subseries is guided by an expert Subseries Advisor.
Contributions are welcome from scientists, policy makers and researchers working in fields related to any of the SDGs. If you are interested in contributing to the series, please contact the Publisher: Zachary Romano [Zachary.Romano@springer.com].

More information about this series at http://www.springer.com/series/15486

Phillip B. Roös

Regenerative-Adaptive Design for Sustainable Development

A Pattern Language Approach

Springer

Phillip B. Roös
School of Architecture and Built Environment
Live+Smart Research Laboratory, Deakin University
Geelong, VIC, Australia

ISSN 2523-3084 ISSN 2523-3092 (electronic)
Sustainable Development Goals Series
ISBN 978-3-030-53233-8 ISBN 978-3-030-53234-5 (eBook)
https://doi.org/10.1007/978-3-030-53234-5

This Springer imprint is published by the registered company Springer Nature Switzerland AG
The registered company address is: Gewerbestrasse 11, 6330 Cham, Switzerland

To the Living Earth
For my dearest wife Pam and my children Arno, Ryan, Shalanda and Gemma

New Patterns for a Healthy Planet

I discovered Christopher Alexander's important idea of a Pattern Language more than 40 ago, when I was a young graduate student at the University of Oregon. It was a chance discovery as I was working part-time for a local planning consultant under contract to prepare a series of housing studies for the Warm Spring Consolidated Tribe, a Native American community located in the eastern part of the state. I had been tasked to prepare a housing options report and in the process of research found Alexander's books. Multiple volumes, beginning with *The Timeless Way of Building* and *A Pattern Language*, were like none I had ever read – more like sacred design scriptures, it seemed, than a typical book. The third volume in this remarkable sequence was one specifically applying these ideas to the University of Oregon (called *The Oregon Experiment*), where I was studying, and so to me it made the idea of patterns even more relevant and impactful (this document is apparently still the official master plan for the campus). To me the notion of timeless and ancient building and townscape patterns made considerable sense. Surely, we had learned much as a species over the several hundred thousand years of our evolutionary life (and out of 10,000 years of living in place in villages and settlements) about what we needed to build and about the physical and social relationships needed to flourish. Why reinvent what ancient builders and planners had known for millennia? The essential insight was an epiphany – that there were tested and timeless place-building patterns that could be tapped to guide our future design and planning work.

In the last few years there seems to have been an encouraging renaissance and rediscovery of the power of Alexander's pattern language idea and belief in the need to refresh and expand the repertoire of patterns. These efforts have included the *14 Patterns of Biophilic Design*, written by Bill Browning and his colleagues at Terrapin Bright Green; Michael Mehaffy's *A New Pattern Language for Growing Regions: Places, Networks, Processes;* and Phillip Tabb's work on biophilic patterns derived from the inspiring biophilic community of Serenbe, among others.[1]

Peter Kahn has argued for the need to collect and re-invigorate new patterns of nature experiences.[2] He believes – and rightly so, I think – that we have forgotten many of the ways we as a species have historically utilised, enjoyed, celebrated and closely experienced nature in our lives. We have even forgotten what we have lost: Kahn believes we suffer from a pernicious form of "shifting baselines," as fewer and fewer of us remember the many wondrous and life-affirming ways we tended to interact with nature in the past.

Against this backdrop of climate crisis, global biodiversity loss, and an unravelling of our planetary ecosystems, Roös's book adds immeasurably to a sense of what is possible and needed. Urban design

[1] E.g. see "14 Patterns of Biophilic Design," found here: https://www.terrapinbrightgreen.com/reports/14-patterns/; Mehaffy, et al, *A New Pattern Language for Growing Regions: Places, Networks, Processes*, found here: https://patterns.architexturez.net/doc/az-cf-193137; Phillip Tabb, *Serene Urbanism: A biophilic theory and practice of sustainable placemaking*, found here: https://www.taylorfrancis.com/books/e/9781315608433

[2] Peter Kahn, et al. "A Nature Language: An Agenda to Catalog, Save, and Recover Patterns of Human–Nature Interaction," *Ecopsychology,* 2010, found here: https://depts.washington.edu/hints/publications/A%20Nature%20Language%20(published).pdf

patterns – regenerative and biophilic – are more important than ever, and we need an even more ambitious system of new patterns, commensurate with the extraordinary challenges we face. The need to move beyond modest adjustments and small design tinkering is undeniable, and Roös's new patterns for regenerative-adaptive design are essential. They build on the new-found power of patterns but extend and expand their scope and ambition to fit the dire circumstances we are facing today. As this book shows, everything we build and design must be deeply biophilic; it must work in significant ways to repair and restore these lost connections, at the same time that species, landscapes and ecosystems are restored and regenerated.

What is evident here is a growing sense that our patterns of living and building must acknowledge the deep interconnectedness of our world. Modern cities source much of their food, materials, water and energy from distant hinterlands with little concern for, or transparency about, the long-term impacts, while externalising the waste and pollutants generated there. We don't care where the outputs go, as long as they go *away*. The good news is there is a growing movement which understands that we must make room for nature where we live and work and that we need and want nature around us – an especially important (finally) recognition that being truly healthy, happy, and human requires close contact with nature. Many of the patterns described in this book aim directly at bringing about these closer urban-nature connections. And there is also a growing sense that designing just to be a little less damaging or consumptive of resources and energy can't be enough – we need projects, neighbourhoods, and cities that are regenerative and adaptive, and that help to restore and repair a planet that is already heavily damaged.

We will want to take Roös's regenerative-adaptive patterns and find ways to amplify and accelerate their application. Part of this task will require us to re-think the design fields and what we are teaching in schools of architecture and design. How can the owners, occupants, users of buildings and spaces in our cities care about their regenerative qualities when architects and designers who work to create them don't have the passion, commitment and tools to aspire to higher goals themselves? I have been discouraged to see what we teach, or more importantly what we don't teach, in schools of architecture. There is not much about sharing space with other life, about building and living modestly, about the need to design spaces and places that restore and repair rather than destroy and deplete.

As we tell this story of profound interconnectedness we will need better and more effective methods of storytelling in an era of text messaging and short attention spans. The mystery and magic of nature will need to be part of the answer. I recently participated in a medicine walk in the forests of High Park in downtown Toronto that offers some hope. Led by Irish-Canadian forest ecologist and medical botanist Diana Beresford-Kroeger, it was a vivid and visceral demonstration of this principle of interconnection, but delivered with fun, beauty, and wonder. She explained how essential trees and forests were to human health and the source of a majority of our drugs. She spoke of even deeper connections, telling the group of rapt fellow walkers how terrestrial trees and forests sustained marine ecosystems, fallen leaves providing iron to support microorganisms at the bottom of the marine food chain.

Beresford-Kroeger's work (and she is not alone) reflects the critical power of awe and importance of love. At several points during the walk that day she lovingly embraced the trees she was discussing. The regenerative design responses we need and that are advocated in this book will require a merging of a deep understanding of interconnection with an even deeper sense of love and kinship with all other forms of life.

An appreciation for the value of this kind of ancient knowledge is gaining ground, which is encouraging and reflected strongly in Roös's work here. My brief experience in Oregon also reminds me of the essential role that native cultures can play in helping all of us renew our relationships to nature. They have, after all, managed to live wisely and sustainably for millennia and offer hopeful direction out of the climate and ecological disaster Western societies and economies have precipitated in.

I am especially excited to see in this book an attempt to learn from Indigenous cultures. Australian Aboriginals have lived sustainably on the land for more than 60,000 years, an astounding length of time. Some research conveyed the idea that this culture was mostly nomadic, with few efforts or innovations at creating settlements or actively managing their environments. New research and new books have shown this to be false. There are few silver linings to the horrific bushfires of the Australian summer, but in the emptiness of the devastation they did in many places show the lines and evidence of long-standing and ancient improvements on the land. Extensive and continuous land management, something careful scholars have known, was made evident in the aftermath of the bushfires when an ancient and elaborate system of aquaculture was discovered.

Hopeful as well are the ways that Indigenous cultures are merging ancient wisdom and modern legal tools, such as the efforts to attach legal and legally defensible rights to nature and natural systems. Ironically an idea expressed in the 1970s by another Christopher (Stone, not Alexander) in the important book *Should Trees Have Standing?*[3], it was the Maori of New Zealand who applied and tested this idea, notably through the Te Urewera Act of 2014, declaring this large ancient forest a legal person and as such entitled to rights of protection and the ability (through a special governance board) when threatened to sue in the courts.[4] More recently the idea is gaining hold and being applied in a variety of settings, including in ecosystems near cities (such as in the city of Toledo, Ohio, through the Lake Erie Bill of Rights[5]). The lessons may also include the need to shift our Western thinking more in the direction the Maori call "Mauri," or the spirit or life force that can be found in every element in nature, from fish to trees to rocks. This moves us closer to the notion of their inherent moral worth, and also to a sense of their pervasive and all-encompassing nature – it is all around us, even in cities, calling for acknowledgement and respect.

Several years ago I had the chance to participate in filming the story of how a group of citizens of Perth, Western Australia, organised to defeat a proposed highway expansion (the Row 8 highway) that was slated to destroy an ancient banksia forest and rare wetland. These areas are highly sacred to the *Noongar* people, and during one on-camera interview with *Noongar* Elder Noel Nannup, I learned more about how deeply embedded in nature these ancient Aboriginal societies saw themselves. The larger non-Indigenous world is not likely to embrace this worldview completely, but there are clear lessons to be learned and patterns (some contained here in this book) that can be followed. Most impressively, Nannup talked about the sense of unity between Aboriginals and nature, a totemic culture. Nannup described how at a young age every child is assigned one (or more) totems in the natural world. His was the Bronzewing Pigeon. He spoke eloquently and with remarkable detail about this bird and its biology. It was a creature that he was both clearly fascinated by and that he loved deeply. It was his totem and as such he always carried with him the duty to stand up for the pigeon.

This is just one example of the wonderful traditions that offer compelling lessons for how we cultivate the love, respect and active care for nature we need today. It is easy to succumb to despair, as official reports and projections come piling in and images of burning forests and the deadening of nature bombard us daily. The book before you is one positive response to the challenge of ecological grief experienced by many of us today. It is hard not to fall into a morass of hopelessness and despair, especially when contemplating the diminished world we are leaving the young. But books like this one offer a dose of hope that we might develop the tools and methods and framework to address these challenges and to find (hopefully) a more restorative path and future.

Philosophers sometimes challenge us to consider the ultimate meaning of life: is it primarily about seeking pleasure or happiness, or is it also (or primarily) about purpose and meaning? The rekindling

---

[3] Christopher Stone, *Should Trees Have Standing? Law, Morality and the Environment*, Oxford University Press, 1972.

[4] See "Te Urewera Act," Environment Guide, found here: http://www.environmentguide.org.nz/regional/te-urewera-act/

[5] E.g. See Tom Henry, "More at stake than water with Lake Erie Bill of Rights court decision," *The Blade,* January 25, 2020, found here: https://www.toledoblade.com/local/environment/2020/01/25/more-at-stake-than-water-with-lake-erie-bill-of-rights-court-decision-toledo/stories/20200124136

of connections with nature and the natural world provides, I believe, at least some of what makes up meaning and purpose, and some of the antidote to rising levels of anxiety and depression. Perhaps even more essential to purpose and meaning is the renewed commitment at personal and collective levels to begin to tackle the daunting challenges we face. That may be the most important role of the book to follow: as a reservoir of compelling, inspiring and immensely usable patterns that we must all begin to advocate and apply in our work. To follow are at least some of the seeds of hope and the antidotes to despair.

Timothy Beatley

University of Virginia
Charlottesville, VA, USA

The Made Order and the Given Order, Unfolding

You are about to read a radical and visionary work of immanent practicality. The author gets straight to the point: The critical issues we face as a global society require "a new manner of thinking if humanity wants to survive."

Sometimes it seems that when it comes to academic writing, the smaller the idea, the more one gets support for publishing in ever narrower journals and conferences. The successful scholar becomes adept at navigating the strict limitations of length, scope and methods insisted upon by peer reviewers who have built their own careers on tweaking small increments of an idea inside an existing paradigm. This is decidedly not that sort of book! It is, instead, a book of big ideas.

In a theoretical landscape of increasing fragmentation, a political and social realm of increasing polarisation and an economic terrain of widening income disparity, Phillip Roös takes a reconstructive integrating view. Nothing in academia or in professional practice rewards grand integrated thinking. If we live in an era when many promote the end of truth and facts, the end result is a depressing nihilism where nothing means anything and the only truth allowed by a pathological postmodernism is that there can be no metanarratives. Of course, that in itself *is* a metanarrative, as philosopher Ken Wilber has pointed out. Wholeness does not exist to the postmodern deconstruction-obsessed consciousness. Yet, wholeness *is* what is required, and Roös begins with the idea of *wholeness*, an unbroken vision of the potential for a synthesis of nature and culture in which "humans and nature work together in a developmental process that improves the value of the whole." Wholeness is both a quality one can feel and a structure of systems and relationships that can be mapped and ordered.

The problems Roös wants to address – ecological degradation and collapse, the threats to waterfronts, loss of indigenous knowledge, climate change, how design creates the feeling of life, the regeneration of industrial wastelands, the nature deficiency of cities – are not solved by the elimination of big ideas, knowledge systems or overarching narratives. Instead they require new stories of the Earth and better explanations for how individual design patterns fit together to build complex multiscalar settlements. To enter this realm takes a particular kind of courage combined with the humility of an integral eye and mind.

Behind the deep solutions offered in this book is the author's willingness to authentically see the world, to trust his own experience and those of others *as real* in the face of so many apologists for the failures of modern and postmodern architecture and urban design. Architects love to critique the work of other architects but also to defend the value of their profession. As a discipline, we live in our heads and tend to distance ourselves from the truths the whole human self, body included, has to teach us. Few of us navigate a contemporary city in Australia or in America and exclaim about how life-affirming the experience is. Instead, as Roös puts it, the built environment is "almost un-imaginably bad and unacceptable." The ideals and principles taught in design schools simply don't produce the quality products as claimed. Even in terms of sustainable development, we are "still grasping at straws, striving to achieve sustainable goals," he says.

The theoretical part of this book defines a series of "fundamental" patterns for a "regenerative-adaptive pattern language." These fundamental patterns are not spatial solutions but "idea patterns" that link together in an intellectual network an ecology of ideas. To unpack the plethora of new terms requires reading the book. I encourage the reader to dive deep and immerse yourself in this journey, as this is not another typical work of speculative design theory, most of which can hardly be called theories at all. A real design theory has to explain what good form is and how it comes into being; it needs its injunctions and its methods. It is the application of these tools in practice from which the paradigm emerges and not the other way around. Following the designation of the text, in the discussion that follows, these fundamental patterns are shown in small caps, such as THE WHOLE.

What thinking can support the generation of the wholeness described in the pattern, THE WHOLE? In addition to getting out of our heads (at least partly) and leaving behind cherished but dysfunctional precepts, the unworkability of contemporary paradigms (including sustainability), we also need the mind, a higher mind that is. To access the abstract nature of climate change as an invisible, complex-systems phenomenon that is progressive and statistically evolving requires an integral cognition. It requires trans-rational and trans-critical thought. Ecological design itself is a post-postmodern cognitive task, what Harvard's Robert Kegan calls 5th order consciousness, the "self-transforming mind" of the "inter-individual self." Taking such an integral prospect, an altitude if we may, it becomes apparent that the multiple perspectives of experiences, technology, ecology and cultures each have something powerful to offer a sustainable future, as articulated in an INTEGRAL SUSTAINABLE WORLDVIEW.

Any informed person knows that climate change is one of the grand challenges of our century. From the integral perspective, culture depends on human minds, which depend on life itself, which depends on base matter. Culture is built on nature, and therefore nature is in fact more fundamental, while culture includes nature in its depth. In this holarchic sense, nature is a part of culture. It is like a building where nature is the foundation. No building can stand when its foundation collapses. Any workable responses to curb or adapt to climate change call us to pay deep attention to both the success of the human species, but also to the diversity of species and ecosystems that support life on the planet. Roös therefore calls for CLIMATE CHANGE CO-ADAPTATION.

The example study area is a coastal town in Southern Australia. Embedded somewhere deep in the archetypal psyche is an innate AFFINITY TO WATER that humans seem to have. Perversely, our desire to settle near water, to look at it from our homes, to access it in our cars (and to use it as our sewer), ends up degrading the real degree of life in the very coastal places we seek out. It is a tension found from the California coast to the Aegean Islands to the shores of Victoria. In Christopher Alexander's thinking, a good pattern is the order of space that resolves the tensions inherent in a particular kind of recurring design situation. Tensions between human biophilic attraction to water and coastal ecological conservation, between a riparian diversity and the relative monoculture of human settlement, form the context of this critical pattern.

One of the characteristics of any integral theory is its embrace of the empirical realities revealed by developmental studies of individuals, cultures, socio-economic systems, etc. Wilber coined the phrase "transcend and include" to acknowledge that each new stage of human developmental awareness has the capacity to both include the workable aspects of a previous stage (its dignities) and to simultaneously transcend what is unworkable (its disasters). Modern thinkers, especially in the design fields, tend to ignore knowledge from traditional societies as primitive and undeveloped, while postmodernists tend to romanticise traditional people as noble Edenic non-modern ecologists. This book takes a more nuanced approach to ADVANCED INDIGENOUS KNOWLEDGE as a way to access, for design purposes, a deep knowledge of the land and how people have adapted to live with the rhythms of the days and seasons. In true integral fashion, the author practices taking the perspective of the land's Indigenous custodians for what it reveals about the place, knowledge that from other viewpoints may be occluded.

I have argued in other essays and books that the human experience of nature is as important as designing for how nature works. Giving people rich human experiences of nature and natural forces presents the opportunity to develop relationships with the natural world, and meaningful relationships beget care. Connecting people to nature is obviously a critical aspect of any design-with-nature approach. "Biophilic design" is the term of the day, although much of the current literature on the topic fails to distinguish "philia" as an interior human experience from the physiological, cognitive and health impacts and correlations of encounters with "bio." In integral terms, such writers collapse the subjective into the objective. In fact, there is a dearth of concrete biophilic design patterns with useful spatial guidance for designers to create such deep and affiliative experiences of nature. Thus, the LOVE FOR NATURE pattern and its consequent "nature language" – connecting patterns of space with patterns of ecopsychological experience – is well-timed.

A regenerative design approach will look to nature for its inspiration and model. Inevitably, one arrives at the question of how human systems can be ordered in ways analogous to and integrated with the order of nature's systems. The structure, processes and distribution of ecosystems have evolved over four billion years and constitute what works in the long term on Spaceship Earth. NATURE'S DESIGN helps us locate development in places with the resources and absorptive capacity for human life and also in places where sensitive and eco-productive systems are not harmed.

While nature can be a model for human design, biomimicry is far from enough to sustain regenerative design. John Lyle spoke of "human ecosystems" in which human activity and natural process are always happening co-spatially. We are short on language that transcends the dualistic terms of nature/culture, human/natural and design/nature. REGENERATIVE ECOSYSTEMS is such a transcending pattern that recognises all living systems as a manifestation of their underlying processes and flows of energy, information, materials and organisms. Such new order, arising from the integration of all the species, including humans and their processes, will give rise to authentic and novel form languages that are sophisticated enough to handle their underlying regenerative patterns.

If you ask a restoration ecologist how to go about recreating the pre-settlement diversity of a site, for example a degraded stream corridor, she will tell you that the sequence in which plants, animals, resources and landscape structures, such as bank grading or dechannelising, are introduced is critical to how the ecosystem develops and even what it ultimately becomes. The unfolding of the wholeness at a later stage has to be embedded in the patterns of process and structure at each earlier stage. Architects and urban planners mostly have no consciousness of this principle. Yet how can we expect a rich, diverse and healthy urban human ecosystem without the TRANSFORMATIONS OF WHOLENESS found even in a simple seed? This is fertile intellectual ground for the designer and connects the design patterns to design processes and methods. Roös here is recognising that wholeness is an unfolding incremental process, a necessary and, I will note here, difficult agenda.

Regeneration suggests the healing of places beyond the homeostasis implied by sustainability. It has a vector to return settled landscapes to the richness, diversity and complexity – the aliveness – that was present in earlier times when humans had little or no impact on ecosystem health. Yet, there is no singular end climax state in nature. Thus, the design patterns that catalyse regeneration also have to adapt to changes in the dynamic forces of nature and culture. Species evolve, weather systems change and new technologies are invented. From the integral perspective, these changes might be lateral-horizontal, such as changes of a relatively temporary "state" in a system, for instance the rhythm of seasons or the normal cycles of the economy. Changes can also be vertical-developmental in individuals, cultures and ecologies, as they move from one more or less stable "stage" to another, typically increasing in complexity as they do. The NOTION OF REGENERATIVE-ADAPTIVE PATTERNS represents a conceptual leap forward from all the major urban design and planning approaches, none of which are equipped to deal with adaptation to significant change.

As a real theory needs testing, the book also provides a detailed application of its concepts, tools and methods. The EVOLUTIONARY ADAPTATION pattern more directly addresses ideas such as ecological and settlement succession or development in complexity as the collective actions of residents over time towards continuous improvement and self-transcendence.

Humans are now "the architects of the environment" in this Anthropocene epoch, as biologist E. O. Wilson put it. The book's final chapter questions whether or not humans can achieve a symbiosis with Gaia. If so, will our responsive actions be the result of GAIA'S REVENGE, that is, rebalancing itself to adjust for the pesky species causing the inflammation? In that event, perhaps the pain of enduring the consequences of climate change and the social, health and psychological impacts of our degenerative forms of development will become greater than the price of organising to force-evolve a regenerative-adaptive culture in order to survive. Alternatively, Roös invites us to a celebration of Gaia and the planetary rules of design required of all living systems. Will we choose to endure the revenge and survive in a less diverse and less hospitable future or celebrate the good news of our aligning ourselves with THE WHOLE for long-term thriving?

What a great question! What a great time to be alive and engaged in the production of the built environment! It is perhaps, given the overwhelming influence that buildings and towns have on climate change, the most interesting generation for design in the short history of our species. It is certainly the most consequential. As Wendell Berry wrote in his poem *Healing*, "Seeing the work that is to be done, who can help wanting to be the one to do it?" A regenerative-adaptive pattern language offers designers and planners the mind tools for the work of making a new order that "seeks the given order and finds its place in it." Such a work cannot be done alone; it is a difficult work – without rest – that requires a learning community of practice. It is, however, the work that is to be done.

University of Tennessee
Knoxville, TN, USA

Mark DeKay

Acknowledgements

You carry Mother Earth within you. She is not outside of you. Mother Earth is not just your environment. In that insight of inter-being, it is possible to have real communication with the Earth, which is the highest form of prayer. - Zen Master Thich Nhat Hanh (Interview in *The Guardian*, 2012)

As a young child I always loved to play outside, explore the large expanse of grass fields and savannah bushland around our house, make believe that I could fly above the tree tops and join the flocking birds when the seasons changed – I felt connected, part of everything around me. I communicated with the Earth. As an adult in a fast-paced industrialised world, I lost this privilege during certain critical times in my life, but now again it is possible. In the first instance I want to thank Mother Earth to allow this re-connection between us.

This book has been years in the making. Many people have made contributions to my thinking and production. It is a result of a lifetime of professional practice and many years of research rather than just the time spent writing. From the perspective of THE WHOLE, one can see that this work is a framing and integration of knowledge by many others over decades. I am grateful to all those great thinkers upon whose work this book is built, with special mention to Christopher Alexander, Ian McHarg, John T. Lyle and Edward O. Wilson. Further, I acknowledge with sincere appreciation the assistance and encouragement given to me by many individuals during the course of preparing this book. The list is long, and I apologise in advance that I don't mention each individual, but you will know in your heart that I appreciate your support.

In particular I want to thank the following people:

The person who deserves the most thanks and acknowledgement is my dearest wife Pam, who gives love unconditionally. She supported me during times of doubt, frustration and many hours of writing sometimes throughout the night. Sometimes I wonder how she's had the patience to listen to me babbling along and talking endlessly about the ideas, thoughts, the working of the patterns and my philosophy of life. To my children Arno, Ryan, Gemma and Shalanda, thank you for your patience, support and encouragement. I also want to acknowledge the support and assistance from Professor David Jones, who inspired me to move into the world of academia: I am forever thankful to you. I am grateful for the contributions to the Foreword and Preface by two great minds, Professor Timothy Beatley and Professor Mark DeKay, respectively, as well as the wonderful discussions and collaborations the last few years, and I am sure our journey together has only just started. I pay my respects to the late *Aboriginal* Elder *Tandop* David Tournier; thank you for taking me to the special places in *Wadawurrung Country*, as well as the words of wisdom on how we need to live close to *Country*.

The production of the book would have been impossible without two brilliant help-mates: Dr Floriana Badalotti as copy editor, and Jesse Delmo as illustrator – a huge thank you to you both. Furthermore, I would like to acknowledge the administrative and financial support provided by the Live+Smart Research Laboratory, School of Architecture and Built Environment at Deakin University, which made the realisation of this book possible.

I hope this book meets the expectations of many people whom I shared this voyage with, and I must again thank you all for being part of sculpturing a new vision and path in the journey of my life.

Acknowledgement to Country

I acknowledge the Traditional Owners of the lands and waters that we live and work on across Australia and pay my respect to Elders past, present and future. I recognise that Aboriginal and Torres Strait Islander peoples have made and will continue to make extraordinary contributions to all aspects of Australian life including culture, economy and science. I specifically wish to respectfully acknowledge and pay my respects to the *Wadawurrung* peoples and their Elders, past, present and future, the Traditional Custodians of the lands and waters upon which my research and writing of this book focus and have been concluded. I duly acknowledge the rich cultural and intrinsic connections they have with their *Country*. I also recognise and acknowledge the contribution and interest of other Aboriginal peoples, Traditional Owners and organisations in the management of their lands, waters and natural resources in this region.

About This Book

This book originates from a PhD that researched regenerative-adaptive design and the application of a pattern language approach that considers our human-nature affiliations and adaptation responses to climate change impacts (Roös, 2017). The research was undertaken in a framework of exploration and critical thinking under an emancipatory paradigm. This book explains the narrative of the initial research, followed by an immersion in a journey of discovery, and the further analysis and engagement in writing and proposing a new *Regenerative-Adaptive Pattern Language* for sustainable development.

Taking us on a journey guided by patterns, this book follows a similar structure used in *A Pattern Language* (Alexander et al., 1977). A typical pattern language is a process of describing a set of patterns or good design practices of useful organisation within a field of expertise. In this context, *Regenerative-Adaptive Pattern Language* provides a more holistic process and framework by describing a set of patterns and adaptation practices in a useful, organised way for sustainable development, regeneration, and design and planning. Further, this new pattern language offers us a new guide towards re-establishing our deep innate human–nature connections.

The book begins with an introduction to the pattern language theory. It thereupon sets the scene by describing the critical issues we face as a global society and how a new manner of thinking needs to be adopted if humanity wants to survive. Connecting each chapter together, with upward and downward links within a hierarchical structure as part of a pattern language network, the book follows a sequence of *fundamental patterns*:

- The Whole – describing the state of the planet and how to progress from an Anthropocene dominant to a regenerative-adaptive future (Chap. 1)
- Integral Sustainable Worldview – supporting an argument that we need to move beyond sustainable development (Chap. 2)
- Climate Change Co-Adaptation – informing us that adaptation practice needs to include both humans and nature (Chap. 3)
- Affinity To Water – instructing us how to deal with the complexities living next to water (Chap. 4)
- Advanced Indigenous Knowledge – identifying the importance of including this knowledge in our design and planning for sustainable futures (Chap. 5)
- Love For Nature – stipulating an emphasis on the need to apply the deep human-nature affiliation to our design practice (Chap. 6)
- Nature's Design – how to apply this knowledge to the design and planning of our current and future human settlements (Chap. 7)
- Regenerative Ecosystems – representing the reasoning that an ecosystem thinking needs to be fundamentally embedded in the discourse of regenerative design (Chap. 8)
- Transformations Of Wholeness – providing guidance on an all-encompassing philosophy (Chap. 9)

- Notion Of Regenerative-Adaptive Patterns – providing instructions how to navigate through the complexities of the pattern language (Chap. 10)
- Evolutionary Adaptation – which includes practical applications of the regenerative-adaptive pattern language (Chap. 11)
- Gaia's Revenge – concludes the book with reminding us of the ultimate reality that we as humans are only a small part of this larger self-regulating system of Life on Earth (Chap. 12).

If the reader wishes to explore the detailed citation other than what is tabled in this book, I refer the reader to the PhD thesis (Roös, 2017).

Geelong, VIC, Australia Phillip B. Roös

References

Alexander, C., Ishikawa, S., Silverstein, M., Jacobson, M., Fiksdahl-King, I., & Angel, S. (1977). *A Pattern Language: Towns, Buildings, Construction.* New York: Oxford University Press.

Roös, P.B. (2017). *Regenerative-Adaptive Design for Coastal Settlements: A Pattern Language Approach to Future Resilience*. PhD Thesis, Deakin University, Geelong, Victoria, Australia.

Contents

Acknowledgements

You carry Mother Earth within you. She is not outside of you. Mother Earth is not just your environment. In that insight of inter-being, it is possible to have real communication with the Earth, which is the highest form of prayer. - Zen Master Thich Nhat Hanh (Interview in *The Guardian*, 2012)

As a young child I always loved to play outside, explore the large expanse of grass fields and savannah bushland around our house, make believe that I could fly above the tree tops and join the flocking birds when the seasons changed – I felt connected, part of everything around me. I communicated with the Earth. As an adult in a fast-paced industrialised world, I lost this privilege during certain critical times in my life, but now again it is possible. In the first instance I want to thank Mother Earth to allow this re-connection between us.

This book has been years in the making. Many people have made contributions to my thinking and production. It is a result of a lifetime of professional practice and many years of research rather than just the time spent writing. From the perspective of THE WHOLE, one can see that this work is a framing and integration of knowledge by many others over decades. I am grateful to all those great thinkers upon whose work this book is built, with special mention to Christopher Alexander, Ian McHarg, John T. Lyle and Edward O. Wilson. Further, I acknowledge with sincere appreciation the assistance and encouragement given to me by many individuals during the course of preparing this book. The list is long, and I apologise in advance that I don't mention each individual, but you will know in your heart that I appreciate your support.

In particular I want to thank the following people:

The person who deserves the most thanks and acknowledgement is my dearest wife Pam, who gives love unconditionally. She supported me during times of doubt, frustration and many hours of writing sometimes throughout the night. Sometimes I wonder how she's had the patience to listen to me babbling along and talking endlessly about the ideas, thoughts, the working of the patterns and my philosophy of life. To my children Arno, Ryan, Gemma and Shalanda, thank you for your patience, support and encouragement. I also want to acknowledge the support and assistance from Professor David Jones, who inspired me to move into the world of academia: I am forever thankful to you. I am grateful for the contributions to the Foreword and Preface by two great minds, Professor Timothy Beatley and Professor Mark DeKay, respectively, as well as the wonderful discussions and collaborations the last few years, and I am sure our journey together has only just started. I pay my respects to the late *Aboriginal* Elder *Tandop* David Tournier; thank you for taking me to the special places in *Wadawurrung Country*, as well as the words of wisdom on how we need to live close to *Country*.

The production of the book would have been impossible without two brilliant help-mates: Dr Floriana Badalotti as copy editor, and Jesse Delmo as illustrator – a huge thank you to you both. Furthermore, I would like to acknowledge the administrative and financial support provided by the Live+Smart Research Laboratory, School of Architecture and Built Environment at Deakin University, which made the realisation of this book possible.

I hope this book meets the expectations of many people whom I shared this voyage with, and I must again thank you all for being part of sculpturing a new vision and path in the journey of my life.

Acknowledgement to Country

I acknowledge the Traditional Owners of the lands and waters that we live and work on across Australia and pay my respect to Elders past, present and future. I recognise that Aboriginal and Torres Strait Islander peoples have made and will continue to make extraordinary contributions to all aspects of Australian life including culture, economy and science. I specifically wish to respectfully acknowledge and pay my respects to the *Wadawurrung* peoples and their Elders, past, present and future, the Traditional Custodians of the lands and waters upon which my research and writing of this book focus and have been concluded. I duly acknowledge the rich cultural and intrinsic connections they have with their *Country*. I also recognise and acknowledge the contribution and interest of other Aboriginal peoples, Traditional Owners and organisations in the management of their lands, waters and natural resources in this region.

Contents

About the Author

Phillip B. Roös is the Director of the Live+Smart Research Laboratory and the Associate Head of School – Industry Engagement at the School of Architecture and Built Environment, Deakin University. His teaching and supervision focus on architecture, environmental design and planning, ecological urbanism, and biophilic design. His work spans architecture, urban design and planning, landscape architecture, environmental design, teaching and research, as well as writings and art. Dr Roös is known internationally as a leader in environmental design and has been working as a design professional and architect for over 30 years on an extensive range of large-scale projects in Europe, Africa and Australasia. His work is influenced by whole systems thinking and his application of environmental design is closely related to the ordering of the large-scale aspects of the environment by means of architecture, engineering, landscape architecture, urban design and ecological planning. His teaching and research interests are centred on the human–nature relationship and the identification of optimised design processes based on a regenerative-adaptive pattern language theory. This approach incorporates the principles of biophilia and regenerative design as well as an adaptive pattern language that re-establishes our wholeness with nature, and considers the vulnerabilities of a changing landscape.

Chapter 1
The State of the Planet: From Anthropocene Dominant to Regenerative-Adaptive Futures

Abstract From the twentieth to the twenty first century humans have passed through a unique period; one in which the planning, design and construction of the built environment has been almost un-imaginably bad and unacceptable. The way we have designed and built our cities has resulted in a degeneration of the earth's natural systems, now eventuating in unprecedented impacts of a changing climate. The dominant approach to address this planetary issue is to apply sustainable development practices, focusing primarily on reducing damage to nature and aiming for a 'neutral' outcome using resources more efficiently. Ultimately, this practice is only resulting in a mitigation approach that is just slowing down the degradation of our living earth. Missing from the design, planning and adaptation practice discourse is a more deeply integrated approach to the design and planning of human settlements that considers the whole, which moves away from the current view that humans standing apart from nature, rather than participating, co-evolving and adapting with nature. Identifying key issues of concern to be dealt with – our human's affinity to water and unprecedented coastal development; the challenge of a changing climate with increased natural disasters; the degradation of ecosystems; and the half-hearted attempt by a global anthropogenic and utilitarianism society to achieve sustainability; this chapter introduces a *pattern language approach* to deal with these complexities. The narrative sets out the fundamentals for a more holistic, all encompassing, *integral method*, presenting a *regenerative-adaptive pattern language* for sustainable development, that re-establishes our wholeness with nature, and considers the vulnerabilities of a changing landscape. Setting in place the typical structure of *a pattern language*, the chapter concludes with the *fundamental pattern* THE WHOLE, [1].

Keywords The whole · Regenerative-adaptive · Earth system · Patterns · Biophysical environment · Patten language · Integral · Informed by nature · Sustainability · Climate change

Upward links:
We begin with the part of this [new] pattern language that cannot be simply put into words, but is rather inherently understood by all sentient beings as the non-verbal language of natural attraction, the fundamental truth, the existence of the *web of life*. This is a universal form of oneness: all things are connected and all things are part of THE WHOLE [1].

P. B. Roös, *Regenerative-Adaptive Design for Sustainable Development*, Sustainable Development Goals Series, https://doi.org/10.1007/978-3-030-53234-5_1

1.1 Introduction

> I think it would be a grave injustice to speak of the human species as in some sense evil, even though we are destroying the environment so efficiently at the present time. Basically that was not our intention, and it never was. It was very natural and it was necessary for the ancestral human being to throw everything they had against the wilderness in an attempt to conquer it and then utilise it. The nature of humankind is to expand its population, to gain security, to control, to alter. For millions of years that paid off without undue damage. But then what happened was, as we developed a modern industrial capacity, and then the techno-scientific capacity to eliminate entire habitats quickly and efficiently, we succeeded too well and at long last we broke nature. And now, almost too late, we are waking up to the fact that we have overdone it and that we are destroying the very foundation of the environment on which humanity was built (Edward Osborne Wilson, recorded in *The State of the Planet*; television program by David Attenborough, first transmitted November 2000).[1]

There is ample evidence that the human species is destroying the natural resources of earth, and that humanity's impact on the earth systems has become comparable to planetary-scale geological processes such as ice ages (Brito & Smith, 2012, p. 2). At the four-day convention *Planet Under Pressure: New Knowledge Towards Solutions,* held in March 2012 in London, over 3000 scientists and experts convened to evaluate the state of the planet and explored potential solutions to an impending global crisis. Consensus was reached by the attendees and recorded that humans have driven the planet into a new epoch, the Anthropocene,[2] in which many Earth-system processes and the living fabric of ecosystems are now dominated by humans (Brito & Smith, 2012, p. 10) (Fig. 1.1).

While many scientists debate this concept, recently a working group of scientists, put together by Zalasiewicz and tasked with confirming this epoch and its start date, came to conclusion that the Anthropocene is real, and a potential start date of this new epoch can be calculated around the early 1950s due to the industrial revolution (Davison, 2019).

As argued by Wilson, the very foundation of the environment on which humanity was built is in danger (Wilson, 2004). The human species has caused drivers of negative environmental changes which include the pollution of earth, oceans, waterways and atmosphere. These drivers are caused by industrial, agricultural, commercial and transportation activities of people and are now directly impacting the narrow band of climatic conditions that supports the on-going survival of the human species (MEA, 2005; Mgbemene, 2011; Parmesan, 2006; Pedersen Zari, 2012). The developing of environmental impact is rooted in the technological advancement of human habitation that resulted in developing populations, agricultural revolutions, urban and industrial revolutions, and the modern scene (Goudie & Viles, 2013). These impacts of human actions on the environment is as a result from the Anthropocene (Rockström et al., 2009; Steffen, Crutzen, & MCNeill, 2007). The *Anthropocene* is a name for describing the new epoch in Earth's history where human activities became a force so profound that it impacts the functioning of the Earth's system (Crutzen, 2002; Steffen, 2010). Human's impact on Nature came down to the fact that humans now have the greatest anthropo-mass of any living species on the planet and it is in the order of magnitude greater than the mass of all earthly mammals alive (Goudie, 2018, p. 27; Smil, 2011). This resulted in the biosphere transformed from a majority of terrestrial biosphere into intensively used 'anthromes' with predominantly anthropogenic processes with major negative impacts to nature (Ellis, 2011; Goudie, 2018).

Wilson further emphasise the fragile state of our planet in *Half-Earth: Our Planet's Fight For Life* (Wilson, 2016) by stating the 'world ends twice': the dawn of the Holocene replaced by the Anthropocene, the 'Epoch of Man', is finally leading this world into the peak of destruction called the Sixth Extinction. The human desire for enhanced and rapid technology will leave behind fossils of

[1] *The State of The Planet*, presented by David Attenborough is a three-part environmental documentary series, produced by Rupert Barrington for the BBC Natural History Unit, November 2000.

[2] The Anthropocene is an informal geologic chronological term that marks the evidence and extent of human activities that have had a significant global impact on the Earth's ecosystems. Eugene F. Stoermer first used the term in 1980 (Revki, 2011).

1.2 Affinity to Water: An Increased Risk

Our human species affinity to water in the landscape, more specifically the attraction to the sea and the resources it can offer, drew people to settle along the coastal areas of Australia more than 50,000 to 60,000 years ago (Nunn & Reid, 2016; Powell, Tournier, Jones, & Roös, 2019; Roberts, 1994). For centuries Aboriginal people have been confronted by climate, ecological and geological changes as well as annual seasonal variations. Weather patterns and climate change were gauged by the occurrence of natural events. Migrations of people along the Australian coast are captured in *Dreaming Stories*[4] of the *Sea Country*.[5] People lived in harmony with nature and practiced sustainable behaviours to adapt for survival. This was the scenario across the world, where Indigenous people settled along coastal areas to sustain generation upon generation living closely connected to their natural environments.

However, now in the twenty-first century people flock to coastal areas in ever-increasing numbers at a global scale, wanting a change in lifestyle from city life, seeking the pleasures, amenities and attributes that coastal settlements can offer. Contradictory to sustainable behaviours of the Indigenous people of the past, living with nature and understanding the patterns of the biophysical environment, today this predilection for the coast equates to an increased unsustainable development in the destination communities. The environmental changes associated with these land developments threaten the very qualities that make these places unique and attractive (Green, 2010).

Changes in the climate and rising sea levels make the coastal area more vulnerable and bring change to the coastal zone impacting beaches, marine environments, estuaries, wetlands and low-lying human settlements (DCC, 2009). Globally, the coastal areas are in peril. We know enough about how human actions are impacting the environment through the role we play. However, even with this knowledge, unsustainable behaviours of intense land development continue in the coastal areas. Graphic displays of the vulnerability of our species is right in front of us, demonstrated visually in events such as the 2005 Asian tsunami, the 2011 earthquake in Christchurch (Fig. 1.2), the large-scale disaster impacts of Hurricane Sandy in north-eastern United States (the largest Atlantic hurricane on record causing an estimated damage of around USD $ 50 billion), and in Australia the devastating 2010/11 Queensland floods, as well as the unprecedented floods in Queensland and Northern New South Wales in 2017 following the path of distraction by Cyclone Debbie.[6]

It seems that disasters due to a changing environment and climate are just getting worse and escalating, as it is especially noticeable in Indonesia and the Pacific. During 2018, Indonesia's National Disaster Mitigation Agency (BNPB) recorded approximately 2000 natural disasters that claimed nearly 4000 lives and displaced around 3 million people (Renaldi & Shelton, 2018). Undoubtedly, we are drawn to the coast, and human settlements continue to grow at a rapid rate in coastal areas. It is the coastal inhabited areas that are the most vulnerable to natural disasters, more so with the threats and risks as a result of a changing climate. For this reason, the case study area chosen for applying the *regenerative-adaptive pattern language* is the coastal town of Anglesea in a coastal area along the Great Ocean Road in Australia.

---

[4] *Dreamtime* or the *Dreaming* for Australian Indigenous people (sometimes referred to as the Dreamtime or Dreamtimes) is when the Ancestral Beings moved across the land and created life and significant geographic features. The Dreaming also means to have insight, and to see creation. *Dreaming Stories* pass on important knowledge and belief systems of the Aboriginal people from one generation to the next (Australian Government, 2008).

[5] *Sea Country*: Aboriginal people make no distinction between the sea and the land; the sea is part of *Country*. *Sea Country* is a term used to differentiate between land and sea. *Sea Country* includes the coastal land and the sea with everything connected to it. *Sea Country* includes the submerged lands that bear the footprints of the Aboriginal ancestors (Smyth, 2004).

[6] Cyclone Debbie made landfall along the Queensland coastline near Airlie Beach on Tuesday the 28 March 2017 (ABC News, 2017).

Fig. 1.2 Earthquake Christchurch 2011. (Photo by Koester 2019, CC BY 2.0)

1.3 The Challenge of Climate Change

The warming of the climate system with its related effects will continue. The modern industrial and techno-scientific capacity of the human species, developed to support and build mega cities to house the increased world population, is now estimated to reach a booming 9.8 billion people[7] on earth as early as the year 2050, and is continuously contributing to climate change. The built environment alone is contributing at least 30% of the total anthropogenic Global Greenhouse Gas (GHG) emissions, which come from the construction and the operations of buildings, and is consuming at least 40% of all energy (De la Rue Du Can & Price, 2008; Satterthwaite, 2008; UNEP, 2009). Estimates of how high the contribution to global GHG emissions from cities will increase vary from 30% to as high as 80% (Spiegelhalter & Arch, 2010). To put this in absolute terms, the IPCC (2014) estimated building-related GHG emissions have been around 8.6 million metric tons CO_2 equivalent in 2004, and under the high growth scenario (A2-ASF),[8] this figure could increase dramatically to 15.6 billion metric tons CO_2 equivalent in the year 2030 (Levine et al., 2007). Further, additional to GHG emissions, the building sector is also responsible for significant emissions of other harmful gases such as halocarbons, CFCs, HCFCs and hydrofluorocarbons (HFCs) through refrigeration and cooling applications and insulation materials (UNEP, 2009, p. 9).

[7] United Nations Population Division projects a world population growth with an estimated population between 8.3 and 9.8 billion by the year 2050 (UN, 2013; *World Population Prospects: The 2017 Revision*, UN, 2017)

[8] IPCC emission scenarios: The A2 scenario family includes slow improvements in the energy supply efficiency and a relatively slow convergence of end-use energy efficiency in the industrial, commercial, residential, and transportation sectors between regions. A combination of slow technological progress, more limited environmental concerns, and low land availability because of high population growth means that the energy needs of the A2 world are satisfied primarily by fossil (mostly coal) and nuclear energy. Source: IPCC Emission Scenarios, http://www.ipcc.ch/ipccreports/sres/emission/index.php?idp=98

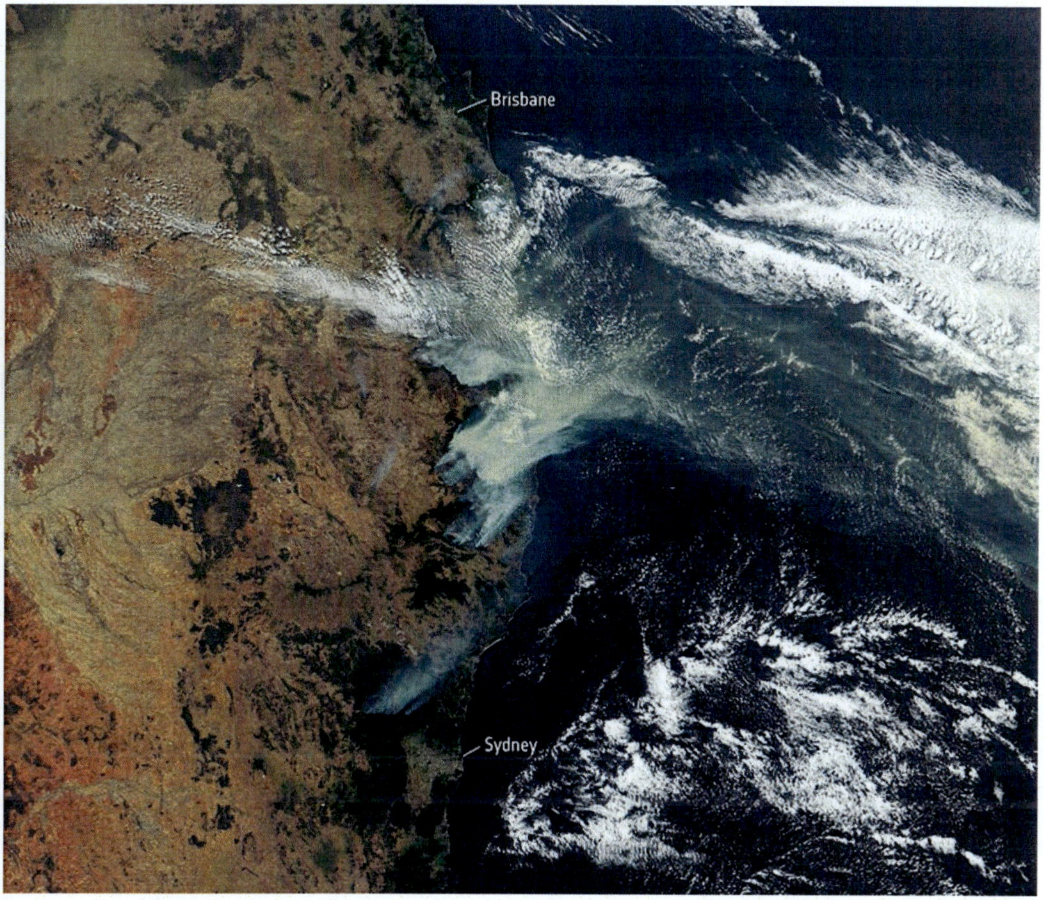

Fig. 1.3 Australian Bushfires 2019–2020. (Photo by European Space Agency 2019, CC BY-SA 2.5)

The human influence on the climate system is clear, and this is evident from the increase of global temperatures, greenhouse gas concentrations in the atmosphere, positive radiative forcing, ocean warming, loss of mass of the Greenland and Antarctic ice sheets, continued shrinking of glaciers almost worldwide, and sea levels continuing to rise (IPCC, 2013). Our human species is indeed facing a global challenge of unprecedented climate change proportions, recently evident in the 2019 Australian bushfires as never experienced before (Fig. 1.3). For the first time in the history the State of Victoria in Australia, the Premier Daniel Andrews declared a *State of Disaster* for 6 areas.[9] This demonstrates the increasing urgency for humans to address anthropogenic causes of climate change, as well as critically consider alternative methods of adaptation.

1.4 The Challenge of Ecosystems Degradation

Biodiversity and ecosystems and the services they support are fundamentally dependent on the climate; actually, on a stable climate with gradual variations and changes. Human impacts are one of the key factors for biodiversity loss, being the main stressor for the shift in habitats and alteration (Mooney

[9] *Victorian fires: state of disaster declared as evacuation ordered and 28 people missing*. The Age, Australian Associated Press, 3 January 2020. https://www.theguardian.com/australia-news/2020/jan/03/victoria-fires-state-of-disaster-declared-as-evacuation-ordered-and-second-man-found-dead. Accessed on 3 January 2020.

et al., 2009). Additionally, climate change interacts with the stresses on ecosystems in many ways, and is a main cause of concern particularly because climate change accelerates natural habitat destruction, overexploitation and the increase of invasive species (Staudinger, 2013; Brook, Sodhi, & Bradshaw, 2008).

In 1988, Wilson warned that global biodiversity, defined as the variation of all life on earth and the ecological complexes in which they occur, faced an unprecedented threat from habitat loss and other anthropogenic stressors (Leadley et al., 2010). Since then efforts were made to introduce conservation measures to classify and preserve global biodiversity before it was lost forever. Unfortunately, biodiversity change is of such a large and accelerating magnitude that despite worldwide conservation efforts, rates of species extinctions are currently up to 1000 times greater than natural background rates (Chapin et al., 2000; Diaz, Fargione, Chapin, & Tilman, 2006; Parmesan, 2006). Biodiversity is fundamental to ecosystem structure and function and underpins the broad spectrum of goods and services that humans derive from natural systems (Chapin et al., 1997; Naeem, 2009; Walther, 2010). As such it is in our own interest to consider alternative options for supporting a sustainable ecological-human integration system, where decisions are made in relation to the planning and design of the built environment that acknowledge that humans are part of nature, rather than above it.

1.5 Sustainability, Ecological Systems, and Climate Change

The word 'Sustainability' has become a 'buzz' word, applied to basically every aspect of our modern world—in economics, business, social structures, politics, and the environment. The attempt is to achieve the possibility of balance and permanence in the world. The concept of sustainability, as defined by the Brundtland Report, has been described by Van der Ryn as bias towards anthropogenic and utilitarianism that is based on a technical approach for the better management of the non-human world for its continued existence (Van der Ryn & Cowan, 1996, p. 5). Under various different definitions, sustainable development seeks to minimise pollution rather than achieve the regenerative outcomes of clean air, water and soil. It is clear that the focus of the sustainability discourse is on 'minimising'.

There is a global desire to achieve sustainable outcomes and provide a better world for society. Many governments and organisations worldwide have set in place policies, value statements and visions, trying to instil a sustainable future for their citizens. In 2015, all the United Nations Member States adopted the *17 Sustainable Development Goals* (SDGs), claiming that these are the world's best plan to build a better world for people and the planet by 2030 (UN, 2019) (Fig. 1.4).

However, we need to be aware of the mainstream status quo, where the standard approach for sustainability is merely superficial. Very disturbing assumptions are lurking behind the general vision of sustainability under a system of sustainable management (Van der Ryn & Cowan, 1996, p. 7). In his argument around the narrow-minded views on sustainability in this world, Van der Ryn and Cowan (1996) critiques this common approach of sustainable development and argues that it is in essence is a 'technological sustainability approach' with no substance. David Orr (1992) noted that the more urgent and best consideration is the 'ecological sustainability' approach. If ecological sustainability is applied to solutions for future outcomes, it is a very different vision of outcomes for society than the standard 'sustainability' approach:

> Ecological sustainability is the task of finding alternatives to the practices that got us into trouble in the first place; it is necessary to rethink agriculture, shelter, energy use, urban design, transportation, economics, community patterns, resource use, forestry, transportation, the importance of wilderness, and our cultural values (Orr, 1992, p. 24).

This approach by Orr provides the fundamentals for us to regenerate our environment, and as argued by Christopher Alexander, we need to consider the 'deep sustainability' that addresses the core of our culture, environment and our human connection to nature.

Fig. 1.4 Sustainable development goals. (By the United Nations, November 2019, public domain)

Deep sustainability[10] as defined by Alexander can be considered to be in contrast with the general understanding and current practice of sustainability (Alexander, 2004). In the general context of sustainability, along with the principles of the triple bottom line, the environment is seen as capital; in essence, in the decision-making process the key decision is how to maintain a stable pool of resources in the future for human benefits (O'Riordan, 2009). We can figuratively envision this standard discourse of sustainability as the branches and leaves of trees, the 'green' that we can see. In contrast, 'deep sustainability' is the roots of the tree that connect the whole ecological system together, described by Alexander as a whole integrated system of patterns illustrated in Fig. 1.5. This underground network of a complex ecosystem, connecting millions of species of fungi and bacteria as they swap nutrients between soil and the roots of the trees, forms a vast interconnected web (Popkin, 2019).

The standard and superficial context of sustainability (for that is what we can see) has now relocated into the wider context of the social response to climate change and global warming. Although sustainability was supposed to contribute to the classical green concerns by reinforcing environmental awareness, the climate change discourse operates on a different level, exposing the flaws of the traditional green approach of sustainability—or of sustainable development and design, for that matter. This raises the question of what we can do about the inevitable impacts of climate and environmental changes we are facing, and the relationship sustainable design and development has with this issue. Climate change is imposing a pragmatic turn on our approach to sustainability, and in regard to this narrative I investigate further the fundamentals that we need to pursue to be able to regenerate our environment, interconnect more with ecosystems, and adapt our human and natural environments for a resilient future. The discourse of this book is thus going far beyond the understanding and current practice of sustainability and sustainable development, seeking a *regenerative-adaptive pattern language* approach that considers *the whole*.

[10] Deep sustainability goes beyond the mere aspects of resource efficiency, energy reduction or sustainable growth; it requires the emotional, spiritual and cultural connections of people with their built and natural environment. Deep sustainability addresses the 'ends', rather than the means to an end (Alexander, 2004, p. 6).

Fig. 1.5 The visual and non-visual interconnectedness of trees. (Drawing by Jesse Delmo, synthesised by Roös)

1.6 A Pattern Language Approach

This topic of the regeneration of built and natural environments and forging human-nature interconnectedness with ecological systems, fundamentally initiating a different approach to sustainability, is indeed not an easy task. In the first instance this dynamic and major shift requires a *new world view*, the development of ecological thinking and the acknowledgement that we as humans are part of nature and not above it. As noted by DeKay (2011), this change is a shift to *integral awareness*, where cultural and psychological references are based on ecological thinking that sees the world in systems terms (DeKay, 2011, p. 60). To be able to do this is to add the perspective of 'wholeness', in essence the understanding of an ecosystems model where living systems are organised in patterns, and these patterns form networks and ultimately create a language.

As we are dealing with humans and their designed built environments and connections to the natural world, the structure and the narrative of this book adopted a similar approach to *A Pattern Language* as defined by Christopher Alexander (Alexander, 1979; Alexander et al., 1977). Further, the development of *a regenerative-adaptive pattern language* followed suit and structured the patterns in a similar fashion, as described in detail in Chaps. 10 and 11.

A typical pattern language is a process of describing a set of patterns or good design practices of useful organisation within a field of expertise. In this context, the *regenerative-adaptive pattern language* provides a more holistic process and framework by describing a set of patterns and adaptation practices in a useful, organised way for sustainable development, regeneration, design and planning.

Undeniably, the fundamental properties of a pattern language originate with Alexander's (1979) argument, whereby he questions the creation of living structures and the need of a code, like the genetic one, for the acts of building:

> So, I began to wonder if there was a code, like the genetic code, for human acts of building? Is there a fluid code, which generates the 'quality without a name' in buildings, and make things live? Is there some process which takes place inside a person's mind, when he allows himself to generate a building or a place which is alive? And is there indeed a process which is so simple too, that all the people of society can use it, and so generate not only individual buildings, but whole neighbourhoods and towns? It turns out that there is. It takes the form of a language (Alexander, 1979, p. 166).

A typical pattern language can also be an endeavour to articulate the wisdom of what brings 'aliveness' within a particular field of undertaking, through a set of interrelated patterns. 'Aliveness' is a term used to encapsulate the statement of 'quality without a name', in essence representing a sense of spirit, grace or of wholeness, which while of varying form, is precise and empirically verifiable (Alexander, 1979).

The core of the pattern language system is the solving of problems that occur over and over in our environment. To deal with this complexity, it is necessary to use a simple, straightforward structure to represent the pattern. Alexander's pattern (Alexander et al., 1977; Mehaffy, 2010, p. 103) consists of:

1. A *pattern title*, written in SMALL CAPS;
2. Main iconic and *representative image*, conveying the nature of the configuration and adding a qualitative dimension to the quantitative analysis;
3. *Upward links*, connecting the pattern to 'upper' patterns that may contain it and others;
4. *Problem statement* in bold, giving the configuration of the problem;
5. *Discussion*, assessing the associated issues to be discussed and addressed;
6. *Solution statement* in bold, describing the configuration of the solution;
7. *Diagram*, a sketch or cartoon of the solution with simple notes;
8. *Downward links*, connecting the pattern to 'lower' patterns that it may contain.

This system allows for nested hierarchical relationships, with overlaps, cross-linkages, and ambiguities. As the patterns combines into a system, into a 'language', these overlapping relationships creates a stronger, subtle and more complex kind of structure that is evident in the systems of nature.

Therefore, the structure of this book follows a similar structure to that of *A Pattern Language* (Alexander et al., 1977), and is set out as follows:

1. *Title* of the Chapter;
2. *Abstract*, which sets out the context;
3. Main *representative image*, which conveys the overall message from the chapter, or the fundamental patterns identified;
4. *Upward links*, which connect the fundamental pattern in the chapter to 'upper' fundamental patterns in previous chapters and key topics;
5. An *introduction*, giving the overview;
6. *Discussion*, the main body text of the chapter that provide details of the topic and detail context that supports the new fundamental pattern identified;
7. *Conclusion Statement*, providing a brief summary of the discussions and describes the configuration of a solution;
8. *Fundamental Pattern*, newly identified or developed as a result from the chapter discussion, with the pattern name written in SMALL CAPS;
9. *Diagram*, a sketch or cartoon with simple notes about the new identified fundamental pattern;
10. *Downward links*, which connect the fundamental pattern in the chapter to 'lower' fundamental patterns in forthcoming chapters and key topics.

1.7 Levels of Scale in a Pattern Language

Similar to words with grammatical relationships to each other to form a spoken language, so does patterns relate to each other in a pattern language. In a pattern language the patterns are part of a directed acyclic graph, each mode of which represents a pattern. Core to the workings of a pattern language is the hierarchy of these patterns (Borchers, 2008; in Neis, Brown, Gurr, & Schmidt, 2012, p. 92), linked to each other in levels of scale. This is clearly demonstrated in *A Pattern Language* (1977), where the language has a hierarchal structure within a network. This network has a sequence, and going through the collection of patterns, the sequence always moves from:

> ... the larger patterns to the smaller, always from the ones which create structures, to those ones which then embellish those structures, and then to those which embellish the embellishments... (Alexander et al., 1977, p. xviii).

In the *regenerative-adaptive pattern language*, levels of scale are represented by the higher level of *fundamental patterns* of the language, connected in the larger network with *regenerative-adaptive patterns*; these include:

- challenges of *design and adaptation*;
- *forces* that are continuously present;
- *centres* formed in the *dynamics of time*;
- *transformations* specific to place;
- *core patterns* embedded within communities;
- *regenerative attributes* that can regenerate environments; and
- the adaptive *evolutionary potential* as part of *the whole*.

The *fundamental patterns* identified in each chapter of this book are at the higher level of the hierarchy in the language; however, it must be noted that *fundamental patterns* can exist at all scales. In Chap. 11 the application of the *regenerative-adaptive pattern language* demonstrates the connections between *fundamental patterns* and the other entities in the larger network.

1.8 Use of the Regenerative-Adaptive Pattern Language

As described above, the *regenerative-adaptive pattern language* has a network structure. In essence, everything is connected together, linked in levels of scale. Each chapter in this book has a *fundamental pattern*, which is linked to other patterns and entities. Each pattern has a three-part rule, which expresses a relation between a certain context, a problem or issue, and a solution embedded in both the pattern and its entities (Alexander, 1979, p. 247). Since the language is in fact a *webstring*[11] *network*, there is no one sequence that perfectly captures it.

[11] In 1988, Michael J. Cohen (2008) developed an exercise in Ecopsychology in what he refers to as the *Webstring Natural Attraction Model*, where a *webstring network* in essence demonstrates the whole interconnected strings of attachment between all entities in Nature, including humans (Cohen, 2008, p. 12). When one string in this web of life is cut, it influences and has consequences for all other entities connected in this network. What is interesting about this *webstring network* is that there is no specific sequence of the connections, but they follow a line and form their own sequence upwards and downwards in the network.

However, when we use the network of the *regenerative-adaptive pattern language*, we always use it in a *sequence* moving from the higher patterns in the hierarchy towards the lower patterns and their connected entities in the language. The sequence in the network can flow both ways between the *upward* and the *downward links*, and interconnect between the other entities such as the *forces* that are continuously present, the *centres* formed in the *dynamics of time*, *transformations* specific to place, the *core patterns* embedded within communities, and so on. This sequence is further explained in Chap. 10, and applied and demonstrated in Chap. 11.

The reader will notice many simple sketches, diagrams and illustrations throughout this book, and in the use of the *regenerative-adaptive pattern language*. Their purpose is to visually represent the pattern or entity in a simple form while attempting to address the gaps in the traditional language of words and writing, which is inadequate to describe the complexities of the phenomena behind the pattern language. As Alexander (1964) puts it:

> The idea of a diagram [of a pattern] is very simple. It is an abstract pattern of physical relationships, which resolve a small system of interacting and conflicting forces, and is independent of all other forces, and of all other possible diagrams. The idea that is possible to create such abstract relationships one at a time, and to create designs, which are whole by fusing these relationships .... (Alexander, 1964, p. i)

1.9 Conclusion Statement

In this introductory chapter we begin with that part of the *regenerative-adaptive pattern language* that defines the consideration of THE WHOLE [1], describing the critical issues we face as a global society, and how a new manner of thinking needs to be adopted if humanity wants to survive. This requirement is clearly reflected in the words of Einstein:

> A human being is part of the whole called by us the universe, a part limited in time and space. We experience ourselves, our thoughts and feelings, as something separate from the rest, a kind of optical illusion of consciousness ... Our task must be to free ourselves from the prison by widening our circle of compassion to embrace all living creatures and the whole of nature in its beauty. We shall require a substantially new manner of thinking if humanity is to survive (Einstein, 1950, as quoted in *The New York Times*, 29 March 1972).

Therefore:

If we want to move beyond a damaging, anthropogenic-dominant, nature-destructive global society, we need to consider *the whole* and acknowledge that including the fundamentals of *wholeness* is a deeper, all-encompassing response to sustainable development.

1.10 Fundamental Pattern 1

THE WHOLE [1]

The universe, the earth, the biosphere, the biotic and abiotic matter, sentient beings along with all their systems and parts, constitute a complex unity of networks, an organised coherent system of many parts fitting and working together as one, to create wholeness and making up *the whole*. All living systems are part of a larger interconnected whole, and are part of an integral existence (Fig. 1.6).

Downward links:

The resilience of both humans and the natural environment in the future will depend on a new approach and thinking for sustainable development, where the consideration of THE WHOLE [1] will be core to a new global INTEGRAL SUSTAINABLE WORLDVIEW [2] ...

Fig. 1.6 Sketch of fundamental pattern – The Whole [1]. (Drawing by Jesse Delmo, synthesised by Roös)

References

ABC News. (2017). *Cyclone Debbie: Residents share stories as torrential rain and wind pummels region*. Retrieved 19 September 2019 from: https://www.abc.net.au/news/2017-03-28/cyclone-debbie-residents-recount-damage-as-storm-hits/8393540

Alexander, C. (1964). *Notes on the synthesis of form*. Cambridge, MA: Harverd University Press.

Alexander, C. (1979). *The timeless way of building*. New York: Oxford University Press.

Alexander, C. (2001–2005). *The nature of order – An essay on the art of building and the nature of the universe, book one: The phenomenon of life*. Berkeley, CA: The Center for Environmental Structure.

Alexander, C. (2003). *New concepts in complexity theory – Arising from studies in the field of architecture*. Berkeley, CA: Centre for Environmental Structure.

Alexander, C. (2004). *Sustainability and morphogenesis: The birth of a living world*. Berkeley, CA: Centre for Environmental Structure.

Alexander, C., Ishikawa, S., Silverstein, M., Jacobson, M., Fiksdahl-King, I., & Angel, S. (1977). *A pattern language: Towns, buildings, construction*. New York: Oxford University Press.

Australian Government. (2008, March 6). *The Dreaming*. Retrieved 14 January 2014 from Australian Government, Australian Stories, The dreaming: http://australia.gov.au/about-australia/australian-story/dreaming

Borchers, J. (2008). A pattern approach to interaction design. In P. G. Satinder (Ed.), *Cognition, communication and interaction: Transdisciplinary perspectives on interactive technology* (pp. 114–129). London: Springer.

Brito, L., & Smith, S. (2012). Planet under pressure: State of the planet declaration. In *International council for science, earth system science partnership*. London: ICS.

Brook, B. W., Sodhi, N. S., & Bradshaw, C. J. (2008). Synergies among extinction drivers under global change. *Trends in Ecology and Evolution, 23*(8), 453–460.

Chapin, F. S., Walker, B. H., Hobbs, R. J., Hooper, D. U., Lawton, J. H., Sala, O. E., et al. (1997). Biotic control over the functioning of ecosystems. *Science, 277*(5325), 500–504.

Chapin, F. S., Zavaleta, E. S., Evinger, V. J., Naylor, R. L., Vitousek, P. M., Reynolds, H. L., et al. (2000). Consequences of changing biodiversity. *Nature, 405*(6783), 234–242.

Cohen, M. J. (2008). *Educating, counseling and healing with nature*. Washington, DC: Project Nature Connect, Institute of Global Education.

Crutzen, P. J. (2002). Geology of mankind. *Nature, 415*, 23.

Davison, N. (2019). The anthropocene epoch: Have we entered a new phase of planetary history? *The Guardian*. Retrieved September 2019 from The Long Read, The Guardian: https://www.theguardian.com/environment/2019/may/30/anthropocene-epoch-have-we-entered-a-new-phase-of-planetary-history

DCC. (2009). *Climate change risks to Australia's coast: A first pass national assessment*. Canberra, ACT: Department of Climate Change, Commonwealth of Australia.

De la Rue Du Can, S., & Price, L. (2008). Sectoral trends in global energy use and greenhouse gas emissions. *Energy Policy, 36*(4), 1386.

DeKay, M. (2011). *Integral sustainable design: Transformative perspectives*. London/Washington, DC: Earthscan.

Diaz, S., Fargione, J., Chapin, F. S., & Tilman, P. (2006). Biodiversity loss threatens human well-being. *PLoS Biology, 4*(8), 1300–1305.

Ellis, E. C. (2011). Anthropogenic transformation of the terrestrial biospere. *Philosophical Transactions: Series A, Mathematical, Physical, and Engineering Sciences, 369*, 1010–1035.

Goudie, A. S. (2018). *Human impact on the natural environment*. Retrieved from https://ebookcentral.proquest.com

Goudie, A. S., & Viles, H. A. (2013). *The earth transformed: An introduction to human impacts on the environment*. New York: Wiley.

Green, R. J. (2010). *Coastal towns in transition – local perceptions of landscape change*. Melbourne, VIC: CSIRO Publishing.

IPCC. (2013). *Climate change 2013: The physical science basis. Contribution of Working Group I to the Fifth Assessment Report of the Intergovernmental Panel on Climate Change*. Cambridge/New York: Cambridge University Press.

IPCC. (2014). *Working Group II Contribution to the IPCC 5th Assessment Report: Climate Change 2014: Impacts, adaptation, and vulnerability AR5 Final Draft Report*. Geneva, Switzerland: Intergovernmental Panel on Climate Change, Working Group II (IPCC).

Irwin, A. (1995). *Citizen science: A study of people, expertise and sustainable development*. London: Routledge.

Koestler, A. (1967). *The ghost in the machine*. New York: Macmillan.

Leadley, P., Pereira, H. M., Alkemade, R., Fernandez-Manjarres, J. F., Proenca, V., Scharlemann, J. P. W., et al. (2010). *Biodiversity scenarios: Projections of 21st century change in biodiversity and associated ecosystem services*. Montreal, QC: The Convention on Biological Diversity.

Levine, M., Urge-Vorsatz, D., Blok, K., Geng, L., Harvey, D., Levermore, G., et al. (2007). *Residential and commercial buildings, climate change 2007: Mitigation, contribution of Working Group III to the Fourth Assessment Report of the Intergovernmental Panel on Climate Change. IPCC – Intergovernmental Panel on Climate Change*. New York/Cambridge: Cambridge University Press.

Lovelock, J. (2009). *The vanishing face of Gaia*. New York: Basic Books.

Lyle, J. T. (1994). *Regenerative design for sustainable development*. New York: Wiley.

McHarg, I. (1992 [1969]). *Design with nature*. Washington, DC: Wiley.

McHarg, I., & Steiner, F. (2006). *The essential Ian McHarg: Writings on design and nature*. Washington, DC: Island Press.

MEA. (2005). *Ecosystems and human well-being: Biodiversity synthesis. Millennium Ecosystem Assessment*. Washington, DC: World Resources Institute, Island Press.

Mehaffy, M. (2010). Horizons of pattern languages. In H. Neis & G. A. Brown (Eds.), *Current challenges for patterns, pattern languages and sustainability* (pp. 103–111). Portland, OR: PUARL Press.

Mgbemene, C. A. (2011, September 12–15). *The effects of industrialization on climate change*. Fulbright Alumni Association of Nigeria 10th Anniversary Conference Development, Environment and Climate Change: Challenges for Nigeria. University of Ibadan (pp. 103–111). Portland: PUARL Press.

Mooney, H., Larigauderie, A., Cesario, M., Elmquist, T., Hugh-Guldburg, O., Lavorel, S., et al. (2009). Biodiversity, climate change, and ecosystem services. *Current Opinion in Environmental Sustainability, 1*, 46–54.

Naeem, S. (2009). Gini in the bottle. *Nature, 458*, 579–580.

Neis, H. J., Brown, G. A., Gurr, J. M., & Schmidt, J. A. (2012). Generative process, patterns, and the urban challenge. In *Proceedings of the fall 2011 international PUARL conference*. Portland, OR: PUARL Press.

Nunn, P. D., & Reid, N. I. (2016). Aboriginal memories of inundation of the Australian coast dating from more than 7000 years ago. *Australian Geographer, 47*(1), 11–47. https://doi.org/10.1080/00049182.2015.1077539.

O'Riordan, T. (2009). Reflection on the pathways of sustainability. In W. N. Adger & A. Jordan (Eds.), *Governing sustainability* (pp. 307–328). Cambridge, UK: Cambridge University Press.

Orr, D. W. (1992). *Ecological literacy: education and the transition to a post-modern world*. New York: Albany State University of New York.

Parmesan, C. (2006). Ecological and evolutionary responses to recent climate change. *Annual Review of Ecology, Evolution, and Systematics, 37*(1), 637–669.

Pedersen Zari, M. P. (2012). *Ecosystem services analysis for the design of regenerative urban built environments*. Wellington, New Zealand: Victoria University of Wellington (VUW).

Popkin, G. (2019). 'Wood wide web' – The underground network of microbes that connects trees – Mapped for the first time. AAAS Science. Retrieved 19 September 2019 from: https://www.sciencemag.org/news/2019/05/wood-wide-web-underground-network-microbes-connects-trees-mapped-first-time

Powell, B., Tournier, D., Jones, D. S., & Roös, P. B. (2019). Welcome to *Wadawurrung* country. In D. S. Jones & P. B. Roös (Eds.), *Geelong's changing landscape: Ecology, development and conservation* (pp. 44–84). Melbourne, VIC: CSIRO Publishing.

Renaldi, E. & Shelton, T. (2018). Thousands killed as Indonesia devastated by nearly 2,000 natural disasters in 2018. *ABC News*. Retrieved on 19 September 2019 from: https://www.abc.net.au/news/2018-12-28/the-five-most-deadly-natural-disasters-in-indonesia-this-year/10668480.

Revki, A. C. (2011, May 11). Confronting the 'Anthropocene'. *The New York Times*. Retrieved from https://dotearth.blogs.nytimes.com/2011/05/11/confronting-the-anthropocene/?mtrref=www.google.com&gwh=E8C4BDB9C757DB1037E46D53CAB73A90&gwt=pay&assetType=REGIWALL

Roberts, R. J. (1994). The human colonisation of Australia: Optical dates of 53,00 and 60,000 years bracket human arrival at Deaf Adder Gorge, northern territory. *Quaternary Science Reviews, 13*, 575–583.

Rockström, J., Steffen, W., Noone, K., Persson, Å., Chapin, F. S., Lambin, E., et al. (2009). Planetary boundaries: Exploring the safe operating space for humanity. *Ecology and Society, 14*. http://www.ecologyandsociety.org/vol14/iss2/art32/. Accessed May 2020.

Roös, P., & Jones, D. (2017). Knowledge of making life: Design patterns for regenerative-adaptive design. In P. K. Collins (Ed.), *DesTech 2016: Proceedings of the international conference on design and technology* (pp. 203–210). Dubai, UAE: Knowledge E. https://doi.org/10.18502/keg.v2i2.616

Satterthwaite, D. (2008). Cities contribution to global warming: Notes on the allocation of greenhouse gas emissions. *Environment and Urbanization, 20*(2), 539–549.

Smil, V. (2011). Harvesting the biosphere: The human impact. *Population and Development Review, 37*, 613–636.

Smyth, D. (2004). *Kooyang sea country plan*. Heywood, VIC: Framlingham Aboriginal Trust/Winda Mara Aboriginal Corporation.

Spiegelhalter, T., & Arch, R. (2010). Biomimicry and circular metabolism for the cities of the future. In C. H. Brebbia, S. Hernández, & E. Tiezzi (Eds.), *The sustainable city VI: Urban regeneration and sustainability* (pp. 215–226). Southampton, UK: WIT Press.

Staudinger, M. G. (2013). *Impacts of climate change on biodiversity, ecosystems, and ecosystem services: Technical input to the 2013 National Climate Assessment*. Palo Alto, CA: United States Geological Survey.

Steffen, W. (2010). Observed trends in earth system behaviour. *Wiley Interdisciplinary Reviews: Climate Change, 1*, 428–449.

Steffen, W., Crutzen, P. J., & MCNeill, J. R. (2007). The Anthropocene: Are humans now overwelming the great forces of nature? *Ambio, 36*, 614–621.

UN. (2013). *World population projected to reach 9.6 billion by 2050 – UN News Centre*. Retrieved 10 March 2019 from UN – United Nations News: www.un.org/news

UN. (2017). *World population prospects: The 2017 revision*. Retrieved 19 March 2019 from UN Publications Department of Economic and Social Affairs: https://www.un.org/development/desa/publications/world-population-prospects-the-2017-revision.html

UN. (2019). *Helping governments and stakeholders make the SDGs reality*. United Nations Sustainable Development Goals Knowledge Platform. Retrieved 16 September 2019 from UN – SDGs Platform: https://sustainabledevelopment.un.org

UNEP. (2009). Buildings and climate change: Summary for decision-makers. In *United Nations Environment Programme (UNEP), Sustainable Buildings & Climate Initiative, Sustainable Consumption and Production Branch*. Paris: UNEP.

Van der Ryn, S., & Cowan, S. (1996). *Ecological design*. Washington, DC: Island Press.

Walther, G. R. (2010). Community and ecosystem responses to recent climate change. *Philosophical Transactions of the Royal Society B: Biological Sciences, 365*, 2019–2094.

Wilson, E. O. (2004). *On human nature*. Cambridge, MA: Harvard University Press.

Wilson, E. O. (2016). *Half earth: Our planet's fight for life*. New York/London: Liveright Publishing Corporation/W.W. Norton & Company.

Chapter 2
The Importance of a Sustainable Future

Abstract Globally, there is an acknowledgement that the inclusion of sustainability in our day-to-day operations is necessary. The United Nations has put forward the adoption of the 17 Sustainable Development Goals (SDGs), many governments have agreed to them, and are putting in place actions to achieve the targets. However, the mantra of sustainable development has been challenged by various scholars, scientists and philosophers, and there is a need for change to move beyond it. This chapter investigates the concept of sustainability, questions our linear throughput system of an industrial society that still prevails in the twenty-first century, and puts forward an implementation of a *new integral sustainable worldview*. With regenerative systems thinking at its core, this chapter introduces the fundamental thinking of *deep sustainability* by Christopher Alexander (Sustainability and morphogenesis: The birth of a living world. Berkeley, CA: Centre for Environmental Structure, 2004), and progresses further with the *Integral Sustainable Design theory* by Mark DeKay (Integral sustainable design: Transformative perspectives. London/Washington, DC: Earthscan, 2011), where an integral approach to sustainability is found in the consideration of multilevel complexity, and the intersecting of the domains of self, culture, and Nature. The chapter concludes with the *fundamental pattern* INTEGRAL SUSTAINABLE WORLDVIEW [2], supporting an argument that we need to move beyond sustainable development, towards a *regenerative-adaptive paradigm* which will consider sustainable design practices and that fundamentally forms a new *integral, sustainable worldview*.

Keywords Sustainable development · Sustainable development goals · Integral sustainable design · Regenerative design · Ecological design · New worldview · Resilient future · Integral · Sustainability

Upward links:
… when we want to consider a sustainable future where both humans and nature can co-exist in harmony, implement human settlement practices that are *to scale* with the surrounding environment, designing and planning according to *levels of scale* and *levels of complexity* that ultimately are part of THE WHOLE [1].

2.1 Introduction

The starting point is our crowded planet. It is estimated that our global human population will keep growing, rising from 7.7 billion in 2019 to 10.9 billion at the end of the century (UN, 2019). Further, nearly all of the global population growth is occurring in cities. The global trend is rapid urbanisation. The issue is that rapid urbanisation demands more resources to sustain our cities, but these resources are running out. Due to our global crisis, sustainable development became a central concept of our age

P. B. Roös, *Regenerative-Adaptive Design for Sustainable Development*, Sustainable Development Goals Series, https://doi.org/10.1007/978-3-030-53234-5_2

Fig. 2.1 Sustainable living at scale with the surrounding natural environment. (Photo by Author)

(Sachs, 2015). However, the concept of sustainable development is challenged. Brown (2015) claims that sustainable development is an oxymoron, stating that the population growth and economic development that continue to this day are a direct force against aspirations to achieve sustainability:

> Unfortunately, "sustainable development," as advocated by most natural, social, and environmental scientists, is an oxymoron. Continual population growth and economic development on a finite Earth are biophysically impossible. They violate the laws of physics, especially thermodynamics, and the fundamental principles of biology. Population growth requires the increased consumption of food, water, and other essentials for human life. Economic development requires the increased use of energy and material resources to provide goods, services, and information technology… The humans of the Anthropocene are changing the climate, decimating the biodiversity, and reducing the productivity of the biosphere (Brown, 2015, p. 1027).

From the perspective of considering *the whole*, and how we need to change our worldview that will consider the human-nature interconnectedness as fundamental to future sustainability, we need to further investigate the concept of sustainable development and its alignment with a more regenerative-adaptive future. Following from the previous one, this chapter raises the issues of the state of the planet, investigates the United Nations' 17 Sustainable Development Goals, and puts forward the proposition that we need to move beyond the current status quo of sustainability (Fig. 2.1).

2.2 A Rapid Urbanised World

It is evident that the future environment of human habitation is urban. The trend is that human settlements of villages become towns, towns turn into cities, and cities become megacities. Further, compact towns along coastlines (and inland) built of brick and stone sprawl into large megacities made of tarmac, concrete, steel and glass (Girardet, 2015, p. 3). Across many parts of the world large cities have sprung up on locations that were previously small villages and towns, and these cities spread like an invasive parasite along coastline and waterways. On the one hand these cities are an astonishing human achievement, where millions of people live in compact environments with rich cultures and advanced transport systems are built to move people around; moreover, the cities act as incubators for many new innovative technologies. On the other hand, the aggregated environmental impacts from these cities are a major concern, and the resources of Earth are depleted at a rapid rate to feed them. In 2012 our global civilisation used more natural resources in 8 months than what the Earth can produce, and under current trends we will need more than two Earths to supply the needy human society with biological resources by 2030 (Girardet, 2015) (Fig. 2.2).

Rapid urbanisation trends indicate that by 2050 more than 68% of the world's population will live in cities (UN, 2018). From the 1900s to 2013 the global human population increased 4.5-fold, from 1.5 to 7 billion. During the same time the global urban population expanded 16-fold. Nearly all the world's population growth occurs in cities. Cities uses resources directly, and do not replace the resources that they extract. Within cities 80% of the global GDP is produced, on just 2% of the earth's land surface. Urban areas currently account for up to 80% of global energy consumption, 75% of carbon emissions, and more than 75% of the world's natural resource consumption (IRP, 2013).

Fig. 2.2 A rapid urbanised world where cities are growing bigger than what their current infrastructure can support, encroaching on the natural environments. (Wellington, New Zealand, photo by Author)

The major contradiction that humanity is facing is that by building an urban future, the current form of urbanisation is the very threat of destruction for humanity and the natural world. The scale of this problem is clearly demonstrated by *The Millennium Project*, where their 2015–2016 report: *The State of The Future* (Glenn & Floresçu, 2015) raises *15 global challenges* that includes sustainable development and climate change, water and sanitation, population and resources, democratisation, global foresight and decision making, governance and ICT, rich-poor gap, health issues, education and learning, peace and conflict, status of woman, transnational organised crime, energy, science and technology, and global ethics. The report sketches a gloomy global scenario, whereby it reflecting back on the factors listed in the previous 2008 report, the new one identifies that the situation unfortunately has worsened over the years:

> Half the world is vulnerable to social instability and violence due to rising food and energy prices, failing states, falling water tables, climate change, decreasing water-food-energy supply per person, desertification, and increasing migrations due to political, environmental, and economic conditions – *2008 State of the Future* cited in *2015-2016 State of The Future* (Glenn & Floresçu, 2015, p. 2).

2.3 The Need for Change

In a modernised world we take a mechanical, industrial and linear approach to doing things. Our global cities are large engines that consume resources and dispose of what is not needed. The modern world has effectively become a disposable world, and humanity has been using natural resources as if there were no tomorrow (Girardet & Mendonça, 2009). Cities are in essence linear systems that produce waste. Lyle (1994) describe these as energy flows of the Industrial Society, where the entire process is a *throughput system,* from *inputs* of large quantities of non-renewable resources, *through processing*, to *outputs* as 'sinks' that include solid and liquid waste, smog, acid precipitation and large quantities of greenhouse gasses into the atmosphere (Lyle, 1994, p. 52). This linear system, which started in times of the industrial revolution and continues up to the current day, produces outputs that have resulted in the Anthropocene-induced global warming and rapid climate change. Lyle and other environmentalists acknowledged in the mid- to late 1990s that there needs to be a change. Lyle proposed that the linear system of resource use must be replaced by a regenerative, closed-loop system to be able to address this major global issue. Even though the awareness of 'sustainability' gains the majority of the interest, Lyle urges that we need to move beyond the concept of sustainable development practice, and he published his thinking in *Regenerative Design for Sustainable Development* (Lyle, 1994). Unfortunately, the focus on a *human-centric sustainability worldview* took centre stage, and 33 years onwards from the United Nations declaration for the need of *sustainable development* (1987), the world is still grasping at straws, striving to achieve sustainable goals. What is desperately needed is a renewable world, a world view that sustainable development is not the end goal, but only a small part of a *regenerative design paradigm* that aims to achieve a holistic, all-encompassing, renewable and resilient world.

2.4 Sustainable Development Goals

Globally there is an acknowledgement that the inclusion of sustainability in our day-to-day operations is necessary. Many governments as well as private and corporate organisations adopted and pledged to sustainability goals and values. Various frameworks are used to achieve sustainability in the building sector, including manufacturing, resource use, as well as day-to-day business processes. Voluntary organisations manage and promote rating tools for green buildings and sustainable developments; an example is the World Green Building Council, which represents about 70 members around the world, and including rating tools such as LEED (Leadership in Energy and Environmental Design), The Living Building Challenge, BREEAM, Green Star, and The Well Building Standard, amongst others. However, in the larger scale of our global society, the impacts and applications of these voluntary schemes are minimal.

Undeniably, the most globally reached-for and used framework for sustainable development is the United Nations' *17 Sustainable Development Goals* (SDGs), as described previously in Chap. 1. But what does this mean when an organisation like the United Nations issues such a requirement? It is worth further investigating these UN SDGs and identify how and if they align with a much-needed regenerative paradigm.

The *17 Sustainable Development Goals* (SDGs), which are at the core of *The 2030 Agenda for Sustainable Development* (2015), were adopted by all United Nations Member States in 2015. An urgent call to all countries was issued to participate in a global partnership to adopt the 17 SDGs (UN, 2015). This agenda for sustainable development is a plan of action that is applicable for people, the planet and prosperity. As part of this agenda, the *17 Sustainable Development Goals* included 169 targets to achieve to demonstrate an integrated balance within the three economic, social and environmental dimensions of sustainability.

Listed in *The 2030 Agenda for Sustainable Development* (UN, 2015), these ambitious sustainable development goals and targets include:

- Goal 1: End poverty in all its forms everywhere;
- Goal 2: End hunger, achieve food security and improved nutrition and promote sustainable agriculture;
- Goal 3: Ensure healthy lives and promote well-being for all at all ages;
- Goal 4: Ensure inclusive and equitable quality education and promote lifelong learning opportunities for all;
- Goal 5: Achieve gender equality and empower all women and girls;
- Goal 6: Ensure availability and sustainable management of water and sanitation for all;
- Goal 7: Ensure access to affordable, reliable, sustainable and modern energy for all;
- Goal 8: Promote sustained, inclusive and sustainable economic growth, full and productive employment and decent work for all;
- Goal 9: Build resilient infrastructure, promote inclusive and sustainable industrialization and foster innovation;
- Goal 10: Reduce inequality within and among countries;
- Goal 11: Make cities and human settlements inclusive, safe, resilient and sustainable;
- Goal 12: Ensure sustainable consumption and production patterns;
- Goal 13: Take urgent action to combat climate change and its impacts;
- Goal 14: Conserve and sustainably use the oceans, seas and marine resources for sustainable development;
- Goal 15: Protect, restore and promote sustainable use of terrestrial ecosystems, sustainably manage forests, combat desertification, and halt and reverse land degradation and halt biodiversity loss;
- Goal 16: Promote peaceful and inclusive societies for sustainable development, provide access to justice for all and build effective, accountable and inclusive institutions at all levels; and
- Goal 17: Strengthen the means of implementation and revitalize the global partnership for sustainable development.

According to the United Nations, these goals and targets will stimulate action over the next 15 years in areas of critical importance for humanity and the planet. However, in the recent *Global Sustainable Development Report 2016 Edition* (UN, 2016), in Chap. 3, which reports on the perspectives of scientists on technology and the SDGs, there is a large focus on technology, and the needs of a global society. Generally, the impression from the SDGs and targets is that the overall agenda still revolves around the issues and needs of humans to achieve a sustainable future. All though these attempts are plausible and will have potentially positive outcomes, the question remains: How will we address the desperate need of *regeneration and adaptation* in our current rapid changing climate, and move beyond the status quo of sustainability?

2.5 Moving Beyond Sustainable Development

The mantra of sustainable design and development has been challenged by various scholars. One of these scholars is Sim Van der Ryn, who challenges the Brundtland Commission's definition and effectiveness of '*Sustainable Development*' given its reliance on technological sustainability. The Brundtland Commission defined sustainable development in the United Nations World Commission on Environment and Development's report titled *Our Common Future* (1987) as:

> … development that meets the needs of the present without compromising the ability of future generations to meet their own needs (UN, 1987, p. 43).

The report recognises progressive changes in both societies and economies; it also recognises that development:

> … tends to simplify ecosystems and to reduce their diversity of species. And species, once extinct, are not renewable. The loss of plant and animal species can greatly limit the options of future generations; so sustainable development requires the conservation of plant species (UN, 1987; in Pojman, Pojman, & McShane, 2008, p. 420).

Since the Brundtland Commission's inquiry in 1987, sustainability has become a trending terminology for the 1990s, offering one perspective for balance and stability in a fragmented world (Van der Ryn & Cowan, 1996). This perspective was evident despite the continuing extraction of fossil fuels to provide energy and to support the ever-expanding industrial sector of the new modernised and mechanical world. The proponents of sustainability emphasised that minimising the impacts on the environment and being more efficient would be enough to achieve sustainability. Reed challenges the status quo of 'sustainability', and argues that the practice of sustainability is only just doing one 'hundred percent less bad' (Reed, 2007). Mang and Reed also contend that we need to move beyond sustainability to the application of regeneration systems (Mang & Reed, 2012). But what does it really mean to 'move beyond sustainability'? This is much more complex than we can comprehend, and both Alexander (2004), and DeKay (2011) are proponents respectively of a '*deep sustainability*', and an '*integral sustainability*' that is all-encompassing and part of *the whole*.

2.6 Integral Sustainable Design and Development

Moving beyond sustainable development needs a new worldview based on an 'integral' approach. The integral approach to sustainability is found in the consideration of multilevel complexity, the intersecting of the domains of self, culture, and Nature. When we want to achieve sustainable development, we need to start at the drawing board, with the fundamentals of sustainable design. In *Integral Sustainable Design: Transformative Perspectives* (DeKay, 2011), DeKay presents an integral theory of sustainable design that moves from a standard approach to an Integral Model. The Integral Theory as a meta-theory organises its fundamentals in two frameworks (DeKay, 2011, p. xxiii):

> *The four perspectives*, which arise from fundamental distinctions of value found in language (I, We, It/Its) and represent the methods of arts, humanities, basic sciences and the complex sciences… and

> *Levels of complexity*, which arise from the unfolding sequence of development in human individuals, cultures and physical systems, which manifest as developmental sequences such as those values, cognition, biological evolution, economic systems and worldviews (DeKay, 2011, p. xxiii).

Integral Sustainable Design goes beyond the technical aspect of sustainability, and investigates technological, ecological, experiential and cultural sustainability. DeKay's *integral* theory views sustainable design from four fundamental perspectives and includes (DeKay, 2011, p. xxiv):

- Behaviours perspective – the *what* of individual parts;
- Systems perspective – the *how* of complex wholes;
- Experiences perspective – the *who* that intends, thinks and feels; and
- Cultures perspective – the *why* of the collective we.

The Integral Theory (Wilber, 2000) combined with principles of sustainable design, is an approach that unites the art and the ethics of design with its true science. From the four quadrants perspective, Integral Sustainable Design must include:

- *Patterns of form* that order social and ecological relationships;
- With individual members and parts including their activities, functions and *performance*;
- *Experiences* of systemic members including perception, sentience and awareness; and
- Various *levels of complexity* from individual members acting with each other sharing meaning and understanding.

Taking this four-quadrant perspective to the application of sustainable design for sustainable development, a deeper sustainability worldview emerges. As indicated in Fig. 2.3, sustainability as part of our environment is viewed as *experiences* – self and conscious; *behaviours* – science and performance; *cultures* – meaning worldviews and symbolism; and *systems* – social and natural ecologies (DeKay, 2011, p. 17).

Integral sustainable design and development, supports the fundamental requirement that we as humans need to see our place in the system of life as *a whole*, collectively being part of Nature. As noted by DeKay, the overarching principle to design for sustainability is:

> … by considering multiple levels of developing complexity in the intersecting domains of self, culture and Nature (DeKay, 2011, p. 38).

This *integral sustainable worldview* supports the fundamental theories investigated in this book and which make up the fine grain elements of the *regenerative-adaptive pattern language*. The concerns and questions raised by *ecological design* (Van der Ryn & Cowan, 1996), progression beyond sustainability to *regenerative design* (Lyle, 1994), planning at ecological levels of scale to *design with nature* (McHarg, 1992 [1969]), and the deeper understanding of and design with living structures through the nature of order – *a pattern language* (Alexander et al., 1977; Alexander, 1964, 1979, 2001a, 2001b, 2004); all support a new worldview whose core philosophy is *integral*.

2.7 Conclusion Statement

We have identified that indeed there is a global desire to move towards a more sustainable world, recognising that sustainability needs to be at the core at everything we do as a global society. The efforts by many governments to achieve the SDGs set by the United Nations are encouraging, but as I have reasoned in Chap. 1 and in this chapter, this is not enough, and we need to move to a more holistic, all-encompassing method where we move beyond sustainable development towards a *regenerative-adaptive* paradigm that will include integral sustainable design practices. It is also important

Fig. 2.3 The four sustainable design perspectives integrated with the four perspectives of Integral Theory. (Drawing by Jesse Delmo, synthesised by Roös, derived from DeKay, 2011)

to acknowledge that in the face of a changing climate, unprecedented environmental degradation, and the declining of available resources we need to shift our thinking from *mitigation to adaptation*.

Therefore:

If we want to achieve a more sustainable and resilient future for both human-built environments and the natural environment, we need to apply the principles of *integral sustainable design* to our sustainable development practices. Further, new behaviours must be underpinned by the fundamentals of a *regenerative-adaptive paradigm*, enabling a new *integral sustainable worldview*.

2.8 Fundamental Pattern 2

INTEGRAL SUSTAINABLE WORLDVIEW [2]

Achieve a resilient and sustainable future by adopting integral sustainable design practices at multiple levels of complexity, including technological, ecological, experiential and cultural sustainability perspectives (Fig. 2.4).

Downward links:

However, we face an unprecedented change in the climate due to Anthropogenic causes and this will require a CLIMATE CHANGE CO-ADAPTATION [3] approach, where both humans and nature will be able to co-adapt to the effects and impacts of a new world …

Fig. 2.4 Sketch of Fundamental Pattern – Integral Sustainable Worldview [2]. (Drawing by Jesse Delmo, synthesised by Roös)

References

Alexander, C. (1964). *Notes on the synthesis of form*. Cambridge, MA: Harvard University Press.

Alexander, C. (1979). *The timeless way of building* (Vol. 1). New York: Oxford University Press.

Alexander, C. (2001a). *The nature of order – An essay on the art of building and the nature of the universe, book one: The phenomenon of life*. Berkeley, CA: The Center for Environmental Structure.

Alexander, C. (2001b). *The nature of order – An essay on the art of building and the nature of the universe, book two: The process of creating life*. Berkeley, CA: The Center for Environmental Structure.

Alexander, C. (2004). *Sustainability and morphogenesis: The birth of a living world*. Berkeley, CA: Centre for Environmental Structure.

Alexander, C., Ishikawa, S., Silverstein, M., Jacobson, M., Fiksdahl-King, I., & Angel, S. (1977). *A pattern language: Towns, buildings, construction*. New York: Oxford University Press.

Brown, J. H. (2015). The oxymoron of sustainable development. Book review of The Age of Sustainable Development by Jeffrey D. Sachs. *BioScience, 65*(10), 1027–1029. Retrieved on 6 February 2020 from https://academic.oup.com/bioscience/article/65/10/1027/1996049

DeKay, M. (2011). *Integral sustainable design: Transformative perspectives*. London/Washington, DC: Earthscan.

Girardet, H. (2015). *Creating regenerative cities*. Oxon, UK/New York: Routledge.

Girardet, H., & Mendonça, M. (2009). *A renewable world: Energy, ecology, equality. A report for the world future council*. Totnes, UK: Green Books.

Glenn, J. C., & Floresçu, E. (2015). *2015–2016 state of the future*. Washington, DC: The Millenium Project.

IRP. (2013). Urban resource flows and the governance of infrastructure transistions. *News/Environmental Development, 8*, 152–153.

Lyle, J. T. (1994). *Regenerative design for sustainable development*. New York: Wiley.

Mang, P., & Reed, B. (2012). Designing from place: A regenerative framework and methodology. *Building Research and Information, 40*(1), 23–28.

McHarg, I. (1992 [1969). *Design with nature*. Washington, DC: Wiley.

Pojman, L. P., Pojman, P., & McShane, K. (2008). *Environmental ethics: Readings in theory and application* (Vol. 1). Boston, MA: Cengage Learning.

Reed, B. (2007). Shifting from sustainability to regeneration. *Building Research & Information, 35*(6), 674–680.

Sachs, J. D. (2015). *The age of sustainable development*. New York: Columbia University Press.

UN. (2015). *Transforming our world: The 2030 agenda for sustainable development*. A/RES/70/1 United Nations. Retrieved on 18 January 2020 from https://sustainabledevelopment.un.org/content/documents/21252030%20Agenda%20for%20Sustainable%20Development%20web.pdf

UN. (2016). *Global sustainable development report 2016 edition*. New York: Department of Economic and Social Affairs. Retrieved on 18 January 2020 from https://sustainabledevelopment.un.org/content/documents/2328Global%20Sustainable%20development%20report%202016%20(final).pdf

UN. (2018). *News – 68% of the world population projected to live in urban areas by 2050, says UN*. United Nations Department of Economic and Social Affairs. Retrieved on 20 January 2020 from https://www.un.org/development/desa/en/news/population/2018-revision-of-world-urbanization-prospects.html

UN. (2019). *World population 2019: Population size and regional distribution*. United Nations Department of Economic and Social Affairs. Retrieved on 19 January 2020 from https://population.un.org/wpp/Publications/Files/WPP2019-Wallchart.pdf

UN – United Nations General Assembly. (1987). Chapter 2: Towards sustainable development, report of the World Commission on Environment and Development. In *Our common future*. New York: United Nations.

Van der Ryn, S., & Cowan, S. (1996). *Ecological design*. Washington, DC: Island Press.

Wilber, K. (2000). *A theory of everything: An integral vision for business, politics, science, and spirituality*. Boston, MA: Shambala.

Chapter 3
The Challenge of a Changing Environment

Abstract Climate change is one of the biggest challenges of our time. The impacts from a changing climate will affect both built environments and the natural world, requiring a new *holistic approach* to *adaptation* and the establishment ofc All decisions relating to the built environment, future planning of our cities, assessments of environmental impacts, identifying risks, setting standards and codes, are made based on data and knowledge of the past. The future will be very different and the challenges we face in a changing environment, from the degrading of ecosystems to a drastic change in the climate, will force us to consider a *new knowledge system*. Resilience comes from having the capacity to mitigate and diminish impacts or to adapt to change. It signifies the capacity of a system to absorb disturbances and surprises, achieving a state of *dynamic equilibrium*, that enables systems to grow and evolve while keeping their coherence. In the face of a rapid changing climate, this is indeed a complex, magnitude in scale, global issue. Due to the complexity of climate change and its related impacts, in this chapter we explore impacts and issues that will assist in framing the *overall narrative* in relation how we need to consider the application of *fundamental patterns* as part of the *regenerative-adaptive pattern language* approach, to adapt to a challenging and changing environment. The chapter concludes with the *fundamental pattern* CLIMATE CHANGE CO-ADAPTATION [3], informing us that adaptation practice needs to include both humans and nature.

Keywords Adaptation · Climte change · Regenerative-adaptive · Pattern language · Co-adaptation · Resilience · Built environment · Natural environment · Ecology

Upward links:
… undoubtedly it is recognised globally that we need to strive for more sustainable actions in our day-to-day lives as a society, and that we need to improve on the current status quo and adopt a new SUSTAINABLE WORLDVIEW [2]. However, various attempts for sustainable development are overturned due to the unforeseen impacts of a changing environment. This pattern, CLIMATE CHANGE CO-ADAPTATION [3], helps us to identify the key issues of a changing climate, its impacts on the built and natural environment, and what we need to consider to be able to adapt to these changes.

P. B. Roös, *Regenerative-Adaptive Design for Sustainable Development*, Sustainable Development Goals Series, https://doi.org/10.1007/978-3-030-53234-5_3

3.1 Introduction

> Anyone who says there is not a dramatic change in weather patterns,
> I think is denying reality.
> New York Governor Andrew Cuomo (Resnikoff, 2012).

Climate change is one of the greatest ecological, social and economic challenges facing society today (CSIRO, 2011) (Fig. 3.1). The Intergovernmental Panel on Climate Change (IPCC) concluded in their *Fourth Assessment Report* (2007) that there was unequivocal evidence of warming of the global climate system (IPCC, 2007a). This statement has been further supported and confirmed in the latest report from the IPCC, *Climate Change 2013: The Physical Science Basis. Contribution of Working Group I to the Fifth Assessment Report of the Intergovernmental Panel on Climate Change*, warning that 'Observational and model studies of temperature change, climate feedbacks and changes in the Earth's energy budget together provide confidence in the magnitude of global warming in response to past and future forcing' (IPCC, 2013, p. 16). Climate change is recognised as being indisputable, whereby plausible scenarios for global change and emissions of greenhouse gases conclude that temperatures would increase by between 2–6 °C (relative to 1980–1999) by the end of the century. The latest scientific research has concluded that after accounting for thermal expansion of ocean waters and ice cap retreat, that sea levels would rise from about 0.79 m to more than 1.0 m and even as high as 1.4 m by 2100 (IPCC, 2007b; IPCC, 2013).

Research findings presented at the Copenhagen Climate Congress in 2009 projected sea level to rise to 1.9 metres, and in some arguments these projections are considered as being conservative (DCC, 2009). These conclusions have raised some debate about whether the climate models used are reliable. Other reactions include support of these models, with some scientists arguing that the IPCC predictions are too conservative. Despite this debate, the IPCC fifth Assessment Report supports a scenario that 'positive climate change feedback'[1] may amplify the effects of climate change and force a sustained global warming greater than the threshold of 2 °C above pre-industrial times thereby leading to the near-complete loss of the Greenland ice sheet, causing an estimated global mean sea level rise of about up to 7 metres. The IPCC has offered this statement with a rating of 'medium confidence' (IPCC, 2014a, p. 1139).

All these predictions are subject to levels of uncertainty recognizing that the nature of the variables that influence sea level rise are uncertain and complex. For example, the predictions do not take into account cascading effects that could more rapidly result in a drastic increase in sea level rise. One example that could lead to cascading effects is if rapid melting of ice sheets occurs, it could result in a non-uniform rate of sea level rise due to adjustments in the Earth's gravitational field (IPCC, 2012). Another example that could cause cascading effects is the release of trapped methane hydrates gases under the Arctic sea ice sheets (Ryskin, 2003). In 2012 airborne study findings by NASA measured surprising high levels of greenhouse gas methane hydrate coming from the cracks in Arctic sea ice sheets, and partial sea ice cover (NASA, 2012).

Considering these uncertain scenarios, adaptation and approaches to climate change effects needs to be approached and managed in a flexible manner to avoid unnecessary early or late interventions, to manage risks from sea level rise and other related impacts.

[1] An interaction in which a perturbation in one climate quantity causes a change in a second, and the change in the second quantity ultimately leads to an additional change in the first. A negative feedback is one in which the initial perturbation is weakened by the changes it causes; a positive feedback is one in which the initial perturbation is enhanced. In the IPCC fifth Assessment Report, *Working Group III contribution to the IPCC fifth Assessment Report - Mitigation to Climate Change Technical Summary Final Draft*, a somewhat narrower definition is often used in which the climate quantity that is perturbed is the global mean surface temperature, which in turn causes changes in the global radiation budget. In either case, the initial perturbation can either be externally forced or arise as part of internal variability (IPCC, 2014a, p. 1450).

Fig. 3.1 The Challenge of a Changing Environment: Bushfires in Lithgow, Australia (Photo by Lithgowlights, 2013, CC BY-SA 3.0)

Further, the planning, design and construction of our built environment needs to consider new methods to cope with the future impacts of climate change, as well as those of the built environment on the natural environment and climate. Various innovative attempts are made by planning and design firms to address the impacts of climate change (AECOM & Droege, 2010). However, the question still remains how important is the considerations of the connections between the human and natural environments. This chapter explores the issues around climate change and the built environment – the impacts and risks – as well as offering critical considerations of the connections between the natural and built environments in relation to adaptation to a future changed climate scenario. What are the patterns that emerge, and what issues do we need to consider if we want to achieve a future environment that could secure ongoing resilience for humans, nature and the ecological systems?

3.2 Climate Change

Climate change and the degradation of our ecosystems is a challenge for the human race, and is at an unprecedented, unforeseen scale. Climate change will have many important and possible severe impacts upon human society as well as ecological systems (Staudinger, 2013; IPCC, 2013; , 2012;

Robertson, 2006; Alley et al., 2003). Because these impacts are numerous in scope, they are spatial, diverse and temporal in scale (Kenny, Warrick, Mitchell, Mullan, & Salinger, 1995). The diverse scale of the issues and consequences due to climate change are summarised in the IPCC fifth Assessment Report under 8 'key risks' as follows (Sharwood, 2014; IPCC, 2013):

1. Risk of death, injury, ill-health, or disrupted livelihoods in low-lying coastal zones and small island developing states and other small islands, due to storm surges, coastal flooding, and sea-level rise;
2. Risk of severe ill-health and disrupted livelihoods for large urban populations due to inland flooding in some regions;
3. Systemic risks due to extreme weather events leading to breakdown of infrastructure networks and critical services including electricity, water supply, and health and emergency services;
4. Risk of mortality and morbidity during periods of extreme heat, particularly for vulnerable urban populations and those working outdoors in urban or rural areas;
5. Risk of food insecurity and the breakdown of food systems linked to warming, drought, flooding, and precipitation variability and extremes, particularly for poorer populations in urban and rural settings;
6. Risk of loss of rural livelihoods and income due to insufficient access to drinking and irrigation water and reduced agricultural productivity, particularly for farmers and pastoralists with minimal capital in semi-arid regions;
7. Risk of loss of marine and coastal ecosystems, biodiversity, and the ecosystem goods, functions, and services they provide for coastal livelihoods, especially for fishing communities in the tropics and the Arctic (Fig. 3.2); and
8. Risk of loss of terrestrial and inland water ecosystems, biodiversity, ecosystem goods, functions, and services they provide for livelihoods.

Fig. 3.2 Retreat of the shoreline at Edwards Point, Port Phillip Bay, Australia. Loss of biodiversity amongst coastlines and coastal areas due to sea level rise and changes in weather systems is already evident in many places across the world. (Photo by Author)

This is indeed a complex and global issue. Due to the complexity of climate change and its related impacts, in this chapter we will explore only those climate change impacts and issues that will assist in framing the overall narrative in relation to how we need to consider the application of a *regenerative-adaptive pattern language* to solve issues for our settlements, one that needs to adapt to a new climate and environment. It is assumed that the reader accepts that major impacts on coastal environments in the near future due to sea level rise and climate change are undisputable, and that it is in the interests of our future generations to seek new methods of design and planning for sustainable development.

Methods are needed that result in offering possible resilience strategies for our communities, and providing a secure *regenerative-adaptive* future for humans and other species including the natural environment. Thus, both need capacity to adapt and evolve continuously in an integrated, holistic and resilient future. It is acknowledged that in a larger context, the environmental issues due to climate change need many solutions in order to adapt to the vast variety of climatic, environmental, economic, political and cultural conditions in which humans create their habitats. These cannot be solved by only one solution. However, I do propose as a potential strategy the application of a *regenerative-adaptive pattern language* to address this issue, which employ design and planning methods for the built environment that include the understanding of the interconnectedness of ecological and human systems, which considers THE WHOLE [1].

3.3 Climate Change and the Built Environment

As a starting point urban climates are different from rural climates, due to the hard surfaces and predominantly built structures. The more people move to and live in cities, the more there is an ever-increasing effect of the modification of theirclimates (Souch & Grimmond, 2006). Two major factors that impact the global climate as a result of urbanisation are the 'urban heat island' effect, and pollution that includes greenhouse gas (GHG) and sulfur dioxide emissions amongst others (Phelan, Kaloush, Miner, et al., 2015; Smith, Pitcher, & Wigley, 2001). The GHG emissions from buildings are projected to increase from 8.8 $GtCO_2$/yr. in 2010 to 17 or even 22 $GtCO_2$/yr. in 2050 according to IPCC baseline scenarios (IPCC, 2014a). Significant lock-in risks are the result of the increasing long-life spans of infrastructure and buildings, and their continued high demand for energy. The IPCC, however, does acknowledge that a change in lifestyle, culture, and behavioural change such as building usage, energy requirements, new technologies and a 'more green architecture' that supports sustainable development may assist in the reduction of GHG emissions (IPCC, 2014a, p. 60). This means that the built environment can serve as a potential medium where problems can be addressed, both at the cause point of climate change, and at the impact point by climate change levels (McGranahan et al., 2005). The climate change and built environment discourse can thus be divided in two sections; one is the impact on the climate as a result of the constant development and growth of the built environment, and the other is the impact upon the built environment due to climate change.

3.3.1 *Impact of the Built Environment on the Climate*

The built environment constitutes only 2.8% of the global land area of the earth (Ruth & Coelho, 2007, p. 207). This area hosts the main locations for human social, cultural and economic life, and more than half of the world's population lives in urban environments in and around cities (UN, 2007). The UN estimates that in the next thirty years the current 3.5 billion city dwellers will nearly double, at a rate of between 2.5 billion and 3 billion people (UN Habitat, 2013). Global economic growth and the urbanization of the built environment move in tandem, as economic growth and greenhouse gas emissions have for at least the last 100 years. Because most economic activity is concentrated in urban areas, the built environment has a key role in contributing to the global greenhouse effect. Affluence and lifestyle choices determine greenhouse gas emissions, and historically developed countries have had greater greenhouse gas emissions than developing countries. The world is urbanising quickly, and under the business-as-usual scenario it is likely that more than 70% of the world's population will live in cities by 2050; a scenario where greenhouse gas emissions will increase dramatically (Angel, Sheppard, & Civco, 2005).

Globally buildings contribute more than 30% of the total anthropogenic Global Greenhouse Gas (GHG) emissions, and are consuming at least 40% of all energy (De la Rue Du Can & Price, 2008; Satterthwaite, 2008; UNEP, 2009). In Australia the buildings sector accounts for 130 million tonnes (Mt) of GHG each year, contributing to about 23% of Australia's total emissions (AGO, 2006). Further, it must be noted that most of the power generated to supply energy to this massive growth of the built environment in Australia comes from non-renewable fossil fuels. Coal and gas account for about 80% of Australia's electricity generation. What is more disturbing, and having major impacts on the environment and communities, is that most of electricity is produced from brown and black coal at large power stations (DEE, 2020). More and more surrounding communities of these open cut power stations have been demanding their closure due to environmental impacts and impacts to the health of the population (Fig. 3.3). Undoubtedly, this contribution to greenhouse gas (GHG) emissions is adding to the discourse about climate change, and to impacts on climate.

Fig. 3.3 The Alcoa Anglesea Open Cut Coal Power Station was closed in August 2015 due to pressures from the community. (Photo of the Climate Guardians in Anglesea, courtesy of ClimActs, March 2015)

Therefore, directly and indirectly, the building sector globally and in Australia significantly impacts upon biodiversity loss, ecosystem degradation, and suburban sprawl, especially in coastal areas, as most of Australia's cities and regional towns are located on the coast. For this same reason the built environment in Australia is at high risk due to impacts of climate change.

3.3.2 *Impacts of Climate Change on the Built Environment*

The impacts of climate change upon the built environment are many and varied; these impacts are both indirect and direct. The built environment in larger urban areas and cities in coastal areas are particularly vulnerable in that they are immobile. Climate change poses serious threats to urban infrastructure, quality of life, and entire urban systems, especially settlements and cities that are located next to the coast (Roös, 2019, p. 203; Hoornweg et al., 2010). In Europe, 70% of the largest cities have areas that are particularly vulnerable to rising sea levels, and most of these cities are less than 10 metres above sea level. China alone has more than 78 million people living in vulnerable low-elevation settlements and cities; this number is increasing annually by 3% (McGranahan et al., 2005). In Australia 86% of the population lives in coastal zones, which is not surprising as two thirds of the continent is classified as arid. Up to 35% of buildings in the Australian coastal zone are at risk of inundation while many items of significant community infrastructure are also situated in vulnerable locations (DCC, 2009).

One very difficult issue, and one that is rarely mentioned in the literature, is the possibility that the built environment in these low-lying areas will be abandoned due to relocation of people threatened by inundation and flooding. This would represent one of the largest losses of value in land and infrastructure and the largest transfer of economic wealth in human history. This raises a daunting and challenging question: what will happen with the buildings, infrastructure and debris left behind? This is providing an enormous economic, as well as environmental challenge if the derelict built environment is left to its own devices. Tensions will grow as specific landowners and residents demand increasingly expensive infrastructure construction while others push for less costly shifts in habitation (Hoornweg et al., 2010). The complexity of solving future retreat and settlement establishment will provide major challenges. This is especially so when considering that ecological systems and natural environments are located next to settlements, and will be more so impacted by human activities. This provides an opportunity to reconsider current contemporary planning and design methods for sustainable urban development, and acknowledge the need to design and plan with nature, recognising the importance of ecological systems as an integral part of our species' future on this planet.

3.4 Climate Change and Ecology

Climate change alone is not the most serious environmental issue facing cities and the built environment today. The bigger issues include biodiversity loss and the degradation of ecosystems, escalating nitrogen pollution exceeding planetary limits, and population growth in cities resulting increased poverty (Hoornweg et al., 2010). However, climate change is a global issue, and the way our cities and built environment use resources in the ecological systems is an inherent part of the anthropogenic contribution to a global changing climate. The question we must ask ourselves is: How long will the earth's ecosystems still stay stable with such a large-scale climate change impact and resource depletion due to anthropogenic activities?

Rockström et al. (2009) identified the boundaries of resource use that can keep the earth's systems stable. In their 'planetary boundaries' model, three systems have already exceeded their stable state;

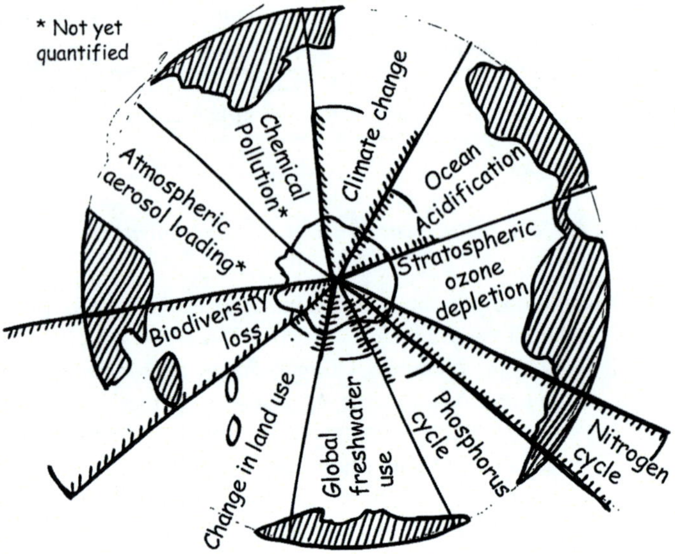

Fig. 3.4 The Planetary Boundaries of Sustaining the Earth's Systems (Drawing by Jesse Delmo, synthesised by Roös, from the model by Rockström et al., 2009)

these include the rate of biodiversity loss, climate change, and the human interference with the nitrogen cycle (Rockström et al., 2009). The 'planetary boundaries' model tries to identify an earth-system process and associated thresholds that, if exceeded and crossed, could generate unparalleled environmental impacts and change (Fig. 3.4). Rockström et al. (2009) have identified nine processes that they believe are necessary to define the planetary boundaries of sustaining the earth's systems, and these include:

- Climate change;
- Rate of biodiversity loss (terrestrial and marine);
- Interference with the nitrogen and phosphorus cycles;
- Stratospheric ozone depletion;
- Ocean acidification;
- Global fresh water use;
- Change in land use;
- Chemical pollution; and
- Atmospheric aerosol loading.

In the context of climate change, the proposed boundary for atmospheric CO_2 concentrations should not exceed 350_{ppm} (parts per million by volume), and radiative forcing should not exceed 1 watt per square meter above pre-industrial levels (Hansen et al., 2013). In December 2019 the average monthly level of CO_2 concentration in the Earth's atmosphere exceeded $413_{ppm,}$[2] and the annual average in 2018 was 407.4_{ppm} (Lindsey, 2019). The question that I raise is: Will we be able to reduce or stop the increase in CO_2 concentration levels?

The on-going growth of the built environment and global urbanisation will continue to have significant impacts on the global climate system, biodiversity and the different levels of organisation of ecosystems. On the other hand, the provision of ecosystem services to communities and society in general will be negatively affected (Campbell et al., 2009; McCarthy, 2001). There is also solid evi-

[2] CO_2 records provided by CO2.earth, compiled from two independent records and databases, NOAA-ESRL and SCRIPPS UCSD. Retrieved on 6 January 2020, from https://www.co2.earth

dence that warming of the climate will alter the known patterns of animal, plant and human diseases, and that such will affect ecological interactions, competition, predator-prey relationships, host parasite interactions, herbivory and pollination (Hellmann, Byers, Bierwagen, & Dukes, 2008; Harvell et al., 2002).

Actions are now starting to focus upon considerations of adaptation, in the form of developing adaptation policies to address the above-mentioned impacts on the built environment as well as to ecological systems and biodiversity loss. However, Campbell et al. (2009) identified that adaptation strategies tend to focus on technological, structural, social, and economic developments. The linkages between biodiversity and adaptation are often missed and the identification of vulnerability of systems (including human and ecological), function of character, magnitude, and sensitivity and adaptive capacity are not considered in a holistic sense (Campbell et al., 2009). In a holistic sense, when we take into account THE WHOLE [1], adaptation needs to consider the vulnerability issues as interconnected with both ecological and human environments.

In particular, the vulnerability of coastal environments is high. These landscapes are, in Australia and in various other parts across the world, directly exposed to rapid changes in a dynamic active environment where land meets ocean, and usually are intervened upon by human-based activities. Recent studies have shown a negative impact of many adaptation strategies on coastal ecological systems, especially in the case of 'hard defences built to prevent coastal and inland flooding' (Campbell et al., 2009). These hard defence systems result in so-called 'maladaptation' in the long term, because the ecological attributes that regulate the ecosystems are disturbed. It is thus important to understand the interconnectivity of humans as well as natural systems and the patterns of climate and nature, especially considering the tendency of humans to settle along the coast, resulting in the vulnerability of both.

3.5 Climate Change and Coastal Settlements

Human urban civilisation historically developed in fertile river valleys where fresh water was available for irrigation. Subsequently more and more settlements developed more densely where the rivers meet the sea, and resulted in most of the world's cities being located at river mouths with ports to be enable trade of their goods. The tendency to live close to rivers and the sea opens up risks of impact to these settlements due to the dynamics of the environment. It is estimated that in the United Kingdom about six million people (10% of the population, calculated in 2012) are at risk due to flooding from rivers and the sea. In addition, about 3000 kilometres of the UK coastline is eroding; floods and coastal erosion are already a serious risk in the UK, and these risks are projected to rapidly increase due to climate change (Ramsbottom, Sayers, & Panzeri, 2012). Similarly, the risks are high in other parts of the world, for example in the United States of America, where development along its coasts are particularly vulnerable to hurricanes and other kinds of severe weather incidents. It is now accepted and evidenced in various literatures including government policies internationally that there is a concern about the potential impacts of climate change in coastal areas (Macintosh, 2012).

In Australia, significant assets on the coast are already exposed, and of the estimated 711,000 existing homes in coastal zones up to 35% are at risk of inundation under a plausible sea level rise scenario of between 0.8 and 1.1 metres, and are located within a zone of less than 6 metres above sea level. Significant built infrastructure is situated in vulnerable locations (Gurran, Squire, & Blakely, 2011). As was recorded in *Visions of the Surf Coast - Changing Landscapes Under Future Climate Effects* (2013), most of the public infrastructure along the Victorian coast between Torquay and Lorne will be detrimentally affected by a sea level rise of 0.8 metres, and at the upper climate change impact scenario of 1.4 metres the damage will be inconceivable. Additional to public infrastructure, low-lying

Fig. 3.5 Increased Severity of Coastal Flooding due to Climate Change Impacts. Coastal flooding caused by increased tides from Cyclone Hamish along the Caloundra, Queensland coastline in 2009. (Photo by Jan Smith, March 2009, CC BY-SA 2.0)

properties, houses and built assets, as well as 700 hectares of native vegetation, wetlands, and heritage and cultural assets are at risk to inundation and coastal recession (Roös & Jones, 2013; GORCC, 2012).

Continued development and population growth in Australia's coastal areas will exacerbate risks from sea-level rise and increase the likely severity and frequency of coastal flooding caused by climate change (Fig. 3.5). A 2009 report by the Department of Climate Change titled *'Climate Change Risks to Australia's Coasts'* stated that:

> .... many coastal environments such as beaches, estuaries, coral reefs, wetlands and low-lying islands are closely linked to sea level. There is a lack of detailed knowledge as to how these environments will respond to sea-level rise, but the risk of beach loss, salinization of wetlands and inundation of low-lying areas and reefs beyond their capacity to keep pace must be recognised. (DCC, 2009, p. 32).

Notable is the 'lack of knowledge' of how these systems will respond to the impacts. A similar question can be raised about the extent of the adaptive capacity of low-lying built infrastructure on the coast. Nowhere is the adaptation response more crucial than in coastal zones, and there is an opportunity to adapt and enhance resilience of coastal settlements to climate change (Newton, 2009).

3.6 Climate Change Adaptation and Resilience

Adaptation refers to the adjustments in environmental, social and economic systems in response to actual or expected climatic stimuli and the effects or impacts from these (McCarthy, 2001). Adaptation involves changes in processes, practices, and structures to moderate potential damages or benefits associated with climate change (Smit, 2003). In the context of climate change, several definitions of adaptation are found in climate change literature, such as the following example:

> Adaptation is the process of adjustment to actual or expected climate and its effects. In human systems, adaptation seeks to moderate harm or exploit beneficial opportunities. In natural systems, human intervention may facilitate adjustment to expected climate and its effects. (IPCC, 2014b, p. 40).

Adaptation to climate change necessarily includes the adaptation to variability. Variability requires that adaptation include both the process of adapting as well as the condition being adapted to. Variability is evident in autonomous and spontaneous adaptation; it is argued that human and natural systems will, to some extent, adapt autonomously, and that planned adaptation can supplement autonomous adaptation. However, 'options and incentives are greater for adaptation of human systems than for adaptation to protect natural systems' (IPCC, 2001, pp. 6–8). In unmanaged natural systems, adaptation is reactive and is the means by which species and communities respond to changed conditions. In consideration of predicted accelerated climate changes, it is a recognised concern that many species will not be able to adapt to fast-changing environments (IPCC, 2012). Planned adaptation within managed ecosystems could be a possible solution for the changing future. The *patterns of nature* need to be considered in any process, implemented or managed by humans, so as to achieve sustainable outcomes. Unfortunately, this is not the case in the current practice of climate change risk assessments and adaption planning. The dominant approach for planned adaption is the 'risk management adaptation' approach.

Planned anticipatory adaptation based on the risk management approach is aimed at reducing a system's vulnerability by diminishing risk or improving adaptive capacity. Numerous reasons have been given for pursuing planned adaptation, and it can yield benefits regardless of the uncertainty and nature of climate change (Ali, 1999). Planned adaptation initiatives are regarded not as a substitute to the mitigation of climate change, but as a necessary strategy to manage its impacts. Risk assessment and adaption planning is best undertaken as a participative exercise that engages the community and draws on their knowledge of the 'system' (human systems) under consideration. Risk is usually defined as the effect of uncertainty on objectives (AGO, 2006). Thus, risk frameworks are developed to support planning and decision-making in the context of such uncertainty and the objectives to be achieved. Given the long-term nature of climate change and the considerable uncertainty about the timing and severity of some potential impacts, it is important that maladaptation does not occur, and actions are taken to avoid 'over or under responses' to adaptation (GORCC, 2012). This is known as the 'precautionary approach' to climate change adaptation (IPCC, 2007a).

Unfortunately, human adaptation considerations, including the use of the precautionary approach, are self-centred, and adaptation or mitigation considerations are framed and structured for human benefit only (Berke, 1995; Munasinghe, 2000; Campbell et al., 2009). A holistic integrated adaptation discourse, where ecosystems and human systems are in equilibrium, needs to be addressed. As argued by John Tillman Lyle in *Design for Human Ecosystems* (1991), we as humans are an inherent part of these ecological systems, and can enhance or destroy them (Lyle, 1991). Recognizing Lyle's argument, it is then important for humans as *part of the ecological system* to align our design and planning decisions with the inherent qualities of the natural system we belong to. In this alignment, to achieve persistence is to regenerate, and to be adaptable is to constantly acknowledge change and amend our built environment in parallel with the *changing natural environment*. This is further supported by the approach taken in the application of *A Pattern Language* (Alexander et al., 1977) to the development of the Eishin Campus project in Japan, where Alexander and his colleagues argued that local adaptation is crucial to the liveability of a place, as described in the following words:

> An environment or community will not come to life unless each place, each building, each room becomes unique, as a result of careful and piecemeal process of adaptation... We had to develop these tools so that adaptation could be a constant focus in each place, at many different scales, all over the Eishin Campus - (Alexander, Neis, & Moore Alexander, 2012, p. 19).

3.7 Adaptation and Sustainable Development

A strong conceptual understanding of sustainable development and its relationship to climate change has developed considerably over the past two decades, because the short and long-term implications of climate change and extreme events have become better understood. This is despite empirical evidence of progress with sustainable development being often elusive (IPCC, 2014b). An improved understanding of the short and long-term implications of climate change and extreme events has influenced conceptualisations of sustainable development and related objectives including poverty reduction, health, livelihood and food security, and other aspects of human welfare related to the idea of 'climate-resilient development' (IPCC, 2014b; Walther, 2010; Levine et al., 2007). One way that sustainable development can contribute to climate resilience is by pursuing consumption patterns that support social and economic development while reducing the use of natural resources and maintaining the available ecosystem services.

However, the links between climate change adaptation and sustainable development are cross-cutting and complex. The impacts of climate change and ill-designed responses to certain impacts may also derail current sustainable development policy actions and potentially offset already achieved gains. Impacts arising from 'maladaptation'[3], as well as continuing climate change impacts, are expected to affect numerous sectors of resource production and social structures, including agriculture, forestry and energy; they may threaten coastal zones and other vulnerable areas, and may pose critical challenges to governance and political systems (IPCC, 2012; Adger et al., 2011; World Bank, 2010). Adaptation to climate change and risks takes place in a dynamic social, economic, technological, biophysical, and political context. This adaptation process varies according to time, location, and sector, and this complex mix of conditions determines the capacity of systems to adapt. Ability to adapt (that is, adaptive capacity), in turn, depends on the state of development (Berke, 1995; Munasinghe, 2000). Under-development fundamentally constrains adaptive capacity, as illustrated by Ribot (1996), especially due to the lack of resources to cope with survival against extreme and expected events. Sustainable development of communities that have adaptive capacity for climate effects can result in resilient communities (Ribot, 1996). Resilience, however, requires more than just having adaptive capacity. Notably, most societies consider resilience as part of sustainable development in a social and economic context, whereas, in fact, resilience of society needs also to consider the adaptability and capacity of the ecological systems that it depends upon. When the ecological systems adaptability and capacity are integral to sustainable development planning – in other words, when nature is included in this paradigm – sustainable development becomes part of the INTEGRAL SUSTAINABLE WORLDVIEW [2].

[3] Maladaptation is defined in the IPCC *Climate Change 2001: Working Group II: Impacts, Adaptation and Vulnerability - Enhancing Adaptive Capacity* report as 'an adaptation that does not succeed in reducing vulnerability but increases it instead' (IPCC, 2001, p. 990). Barnett & O'Neill define maladaptation as: 'action taken ostensibly to avoid or reduce vulnerability to climate change that impacts adversely on, or increases the vulnerability of other systems, sectors or social groups' (Barnett & O'Neil, 2010).

3.8 Resilience – Ability to Mitigate and Adapt to Change

Resilience comes from having the capacity to mitigate and diminish impacts or to adapt to change. It signifies the capacity of a system to absorb disturbances and surprises, and it can mean the ability to reorganise, so as to retain the same essential function, structure and identity (Smit, 2003). According to the IPCC *Special Report on Managing the Risk of Extreme Events and Disasters to Advance Climate Change Adaptation* (SREX), 2012, resilience is 'the ability of a system and its component parts to anticipate, absorb, accommodate, or recover from the effects of a hazardous event in a timely and efficient manner, including through ensuring the preservation, restoration, or improvement of its essential basic structures and functions' (IPCC, 2012).

However, ecosystems change gradually over long periods of time in response to changes in average climatic conditions (Iverson, Schwartz, & Prasad, 2004). Seasonal shifts and extreme conditions due to changes in the climate system, will exert the greatest impacts, including droughts, frosts, and winter thaws that directly kill organisms or change their competitive balance, threatening their ability to adapt with changes. Further, disturbances associated with these extreme conditions (floods, wildfires, and hurricanes) strongly influence the dynamics of ecosystems over decades to centuries (Peters et al., 2011). When the resilience of an ecosystem or a socio-ecological system (defined as the capacity of a system to absorb disturbance and reorganize so as to retain essential function, structure, identity) and feedbacks is exceeded, transformation of the system to a new state is likely (Walker, Holding, Carpenter, & Kinzig, 2004). This is when the system regenerates or adapts to a new state.

Resilience is an inherent quality of all healthy living systems. It is a state of dynamic equilibrium that enables systems to grow and evolve while keeping their coherence (Berry, 2000). This process of achieving a *dynamic equilibrium* is indeed fundamental to a state of *wholeness*, supporting THE WHOLE [1]. This is true of ecological systems that regenerate. The features of ecological systems that make them distinct (diversity, memory, cross-scale interactions, sensitivity to environmental variability, and regeneration) are not just a passive effect of ecological interactions, but an important determinant of a system's persistence and adaptability in face of environmental change (Madhur, 2010). Regenerative systems in the natural environment can be considered to be resilient, and can be used to reduce the causes of climate change through increasing biomass and thus potentially increasing the storage or sequestration of carbon. This means the impacts of climate change can potentially be reduced, particularly over the long term. Increasing the health of ecosystems also increases the resilience of ecosystems, and potentially built environments if natural systems are integrated into these environments as an adaptation response to climate change (Gitay, Saurez, Watson, & Jon Dok, 2002; Chapin et al., 2000).

3.9 Conclusion Statement

The challenge of a changing environment due to climate change requires humans and nature to achieve resilience, meaning adaptive capacity to be able to evolve with a fast-changing environment. However, what is evident is that most approaches in the literature address the issue of human systems adaptation, whih includes consideration of social and economic vulnerability, and the services that ecological systems provide to the survival of humans as part of the adaptation discourse. Unfortunately, human adaptation considerations are self-centred, and as indicated, adaptation or mitigation considerations are framed and structured for human benefit only.

Therefore:

A holistic integrated adaptation discourse, where ecosystems and human systems are in equilibrium, needs to be implemented. This new way of responding to the changing environment at its core assumes that *humans and nature are equal*, and both should go through a process of *co-adaptation* to achieve ongoing resilience.

Fig. 3.6 Sketch of Fundamental Pattern – CLIMATE CHANGE CO-ADAPTATION [3] (Drawing by Jesse Delmo, synthesised by Roös)

3.10 Fundamental Pattern 3

CLIMATE CHANGE CO-ADAPTATION [3].

Changes on Earth operate at multiple levels of intricacy and takes place in dynamic social, economic, technological, biophysical, and political contexts. Due to anthropogenic influences, the climate and ecological systems are changing rapidly. Therefore, implement practices that considers both humans and nature as equal in the co-adaptation planning discourse (Fig. 3.6).

Downward links:
We have identified that coastal areas are highly susceptible to climate change impacts, and that human settlements, as well as natural environments that are vulnerable to these impacts, need to follow a process of CLIMATE CHANGE CO-ADAPTATION [3]. During this process we have to ensure that the context of THE WHOLE [1] and practices of an INTEGRAL SUSTAINABLE WORLDVIEW [2] are considered. However, our attraction to coastal environments – AFFINITY TO WATER [4] – is part of our innately affiliation to nature – LOVE OF NATURE [6] – and in order to keep this deep connection to place we need to apply ADVANCED INDIGENOUS KNOWLEDGE [5] to balance our adaptation responses with sustainable development decisions …

References

Adger, W., Brown, K., Nelson, D. R., Eakin, H., Folke, C., et al. (2011). Resilience implications and adaptation to climate change. *Wiley Interdisciplinary Reviews: Climate Change, 2*(5), 757–766.

AECOM & Droege, P. (2010). *Climate design: Design and planning for the age of climate change* (p. 2010). ORO Editions.

AGO. (2006). *Factors and methods workbook, December 2006: For use in Australian greenhouse gas emmissions reporting*. Canberra: AGO (Australian Greenhouse Office).

Alexander, C., Ishikawa, S., Silverstein, M., Jacobson, M., Fiksdahl-King, I., & Angel, S. (1977). *A pattern language: Towns, buildings, construction*. New York, NY: Oxford University Press.
Alexander, C., Neis, H. J., & Moore Alexander, M. (2012). *The Battle for the life and beauty of the earth: A struggle between two world-systems*. New York, NY: Oxford University Press.
Ali, A. (1999). Climate change impacts and adaptation assessment in Bangladesh. *Climate Research 12*(2–3):109–116. Special issue: National Assessment Results of Climate Change: Impacts and Responses (ed. N. Minura).
Alley, R. B., Marotzke, J., Nordhaus, W. D., Overpeck, J. T., Peteet, D. M., Pielke, R., et al. (2003). Abrupt climate. *Science, 299*(5615), 2005–2010.
Angel, S., Sheppard, S., & Civco, D. (2005). *The dynamics of global urban expansion*. Washington DC: The World Bank.
Barnett, J. & O'Neil, S. (2010). Maladaptation. Editorial, *Global Environmental Change 20*, 211–213.
Berke, P. R. (1995). Natural-hazard reduction and sustainable development: A global assessment. *Journal of Planning Literature, 9*(4), 370–382.
Berry, T. (2000). *The great work: Our way into the future*. New York, NY: Bell Tower Broadway Books.
Campbell, A.M., Kapos, V., Sharlemann, J., Bupp, P, Chenery, L., Coad, B., …, Rashid, M. (2009). Review of the Literature on the links between Biodiversity and Climate Change: Impacts, Adaptation and Mitigation. *Technical Series 42*, 124. Montreal, Canada.
Chapin, F. S., Zavaleta, E. S., Evinger, V. J., Naylor, R. L., Vitousek, P. M., et al. (2000). Consequences of changing biodiversity. *Nature, 405*(6783), 234–242.
CSIRO. (2011). *Climate change: Science and solutions for Australia*. Collingwood VIC: CSIRO Publishing.
DCC. (2009). *Climate change risks to Australia's coast: A first pass National Assessment*. Canberra: Department of Climate Change, Commonwealth of Australia.
De la Rue Du Can, S., & Price, L. (2008). Sectoral trends in global energy use and greenhouse gas emissions. *Energy Policy, 36*(4), 1386.
DEE. (2020). Energy Supply. Departmnent of Energy and Environment, Australian Government. Retrieved on 9 January 2020 from https://www.energy.gov.au/government-priorities/energy-supply.
Gitay, H., Saurez, A., Watson, R., & Jon Dok, R. (2002). *Climate change and biodiversity, IPCC technical paper*. Cambridge: IPCC.
GORCC. (2012). *Coastal climate change vulnerability and adaptation*. Melbourne: Great Ocean Road Coast Committee - GORCC.
Gurran, N., Squire, C., & Blakely, E. (2011). *Planning for climate change adaptation in coastal Australia: State of practice*. University of Sydney, Faculty of Architecture, Design and Planning. Sydney: University of Sydney, November 2011.
Hansen, J., Kharecha, P., Sato, M., Masson-Delmotte, V., Ackerman, F., Beerling, D. J., et al. (2013). Assessing "dangerous climate change": Required reduction of carbon emissions to protect young people, future generations and nature. *PLoS One, 8*(12), e81648. https://doi.org/10.1371/journal.pone.0081648
Harvell, C. D., Mitchell, C. E., Ward, J. R., Altize, S., Dobson, A. P., Ostfeld, R. S., et al. (2002). Climate warming and disease risks for terrestrial and marine biota. *Science, 296*, 2158–2162.
Hellmann, J. J., Byers, J. E., Bierwagen, D., & Dukes, J. S. (2008). Five potential consequences of climate change for invasive species. *Conservation Biology, 22*(3), 534–543.
Hoornweg, D., Sugar, l., Anderson, C., Brada, P., Trejos, C.C., Dave, R., & Lee, M. (2010). *Cities and climate change: an urgent agenda*. The international bank for reconstruction and development. Washington DC: The World Bank.
IPCC. (2001). *Climate change 2001:Working group II: Impacts, adaptation and vulnerability - enhancing adaptive capacity*. Geneva: IPCC (International Governmental Panel on Climate Change).
IPCC. (2007a). *Climate change 2007 impacts, adaptation and vulnerability. Contribution of working groups I,II and III to the fourth assessment report*. IPCC (international governmental panel on climate change). Cambridge: Cambridge University Press.
IPCC. (2007b). *Intergovernmental panel on climate change: Climate change 2007: The physical science basis. Contribution of working group I to the fourth assessment report of the intergovernmental panel on climate change*. London: Cambridge University Press.
IPCC (2012). *Managing the risks of extreme events and disasters to advance climate change adaptation: A special report of working groups I and II*. IPCC, SREX. Cambridge: Cambridge University Press.
IPCC. (2013). *Climate change 2013: The physical science basis. Contribution of working group I to the fifth assessment report of the intergovernmental panel on climate change*. Cambridge; New York: Cambridge University Press.
IPCC. (2014a). *Working Group III contribution to the IPCC 5th assessment report - Mitigation to climate change technical summary final draft*. Intergovernmental panel on climate change, working group III. Berlin: IPCC.
IPCC. (2014b). *Working Group II contribution to the IPCC 5th assessment report: Climate change 2014: Impacts, adaptation, and vulnerability AR5 final draft report*. Intergovernmental panel on climate change, working group II. Geneva: IPCC - Intergovernmental Panel on Climate Change.

Iverson, L. R., Schwartz, M. W., & Prasad, A. M. (2004). Potential colonization of new available tree species habitat under climate change: An analysis for five eastern U.S. species. *Landscape Ecology, 19*, 787–799.

Kenny, G. J., Warrick, R. A., Mitchell, N. D., Mullan, A. B., & Salinger, M. J. (1995). Climate impacts: An integrated model for assessment of the effects of climate change on the New Zealand environment. *Journal of Biogeography, 22*(4/5), 883–895.

Levine, M., Urge-Vorsatz, D., Blok, K., Geng, L., Harvey, D., Levermore, G., et al. (2007). Residential and commercial buildings, climate change 2007: Mitigation, contribution of Working Group III to the Fourth Assessment Report of the Intergovernmental Panel on Climate Change. IPCC – Intergovernmental Panel on Climate Change. New York/ Cambridge: Cambridge University Press.

Lindsey, R. (2019). *Climate Change: Atmospheric Carbon Dioxide*. NOAA (National Oceanic and Atmospheric Administration, Earth System Research laboratory), NOAAClimate.gov. Retrieved 7 January 2020 from https://www.climate.gov.

Lyle, J. T. (1991). *Design for human ecosystems - landscape, land use, and natural resources*. California: Island Press Washington DC.

Macintosh, A. (2012). *Coastal adaptation planning: A case study on Victoria, Australia.* Canberra: ANU centre for climate law and policy, ANU College of Law.

Madhur, A. G. (2010). Ecological systems as complex systems: Challenges for an emerging science. *Diversity, 2*, 395–410.

McCarthy, J. (2001). Climate change 2001: Impacts, adaptation and vulnerability. Working group II to the third assessment report to the intergovernmental panel of climate change. Cambridge, UK: Cambridge University Press.

McGranahan, B., Marcotullio, P., Bai, X., Balk, D., Bragga, T., Douglas, I., …, & Zlotnik, H. (2005). Urban systems. In R. S. Hassan, & R. S. Hassan (Ed.), Ecosystems and human well-being: Current State and Trends. 1*:* (pp. 795–825). Washington, DC: Island Press - Millenium Ecosystem.

Munasinghe, M. A. (2000). Climate change and its linkages with development, equity, and sustainability. *Proceedings of the IPCC Expert Meeting held in Columbo, Sri Lanka, April 1999.* (p. 319). Colombo: LIFE World Bank Washington DC.

NASA. (2012). Study finds surprising Arctic methane emission source. *ScienceDaily*, 24 April 2012. Retrieved on 20 June 2018 from www.sciencedaily.com/releases/2012/04/120424145145.htm.

Newton, G. (2009). Australia's environmental climate change challenge: Overview with reference to water resources. *Australian Journal of Environmental Management, 16*, 130–139.

Peters, D. P. C., Langley, C. M., Lugo, A. E., Collins, S. L., Driscoll, C. T., Groffman, P. M., et al. (2011). *Long term trends in climate and climate-related drivers*. Washington, DC: Agricultural Research Service Publication.

Phelan, P. E., Kaloush, K., Miner, M., et al. (2015). Urban heat island: Mechanisms, implications, and possible remedies. *Annual Review of Environment and Resources, 40*, 285–307.

Ramsbottom, D., Sayers, P., & Panzeri, M. (2012). *Climate Change Risk Assessment for the Floods and Coastal Erosion Sector: UK 2012 Climate Change Risk Assessment.* Department for Environment, Food and Rural Affairs (DEFRA). London: DEFRA.

Resnikoff, N. (2012, November 01). NY Gov. Cuomo: 'We have to accept the reality' of worsening climate change [Blog post]. *The rachel maddow show.* Retrieved 14 April 2014 from www.msnbc.com/rachel-maddow-show/ny-gov

Ribot, J. N. (1996). Climate variation, vulnerability and sustainable development in the semi-arid tropics. In J. M. Ribot (Ed.), *Climate variability, climate change and social vulnerability in the semi-arid tropics* (pp. 13–54). Cambridge; New York, NY: Cambridge University Press.

Robertson, D. (2006). Health effects of increase in concentration of carbon in the atmosphere. *Current Science, 90*(12), 1607–1609.

Rockström, J. W., Steffen, K., Noon, A., Person, F., Chapin, E., Lambin, T., et al. (2009). A safe operating space for humanity. *Nature, 461*, 472–475.

Roös, P. & Jones, D. (2013). *Visions of the Surf Coast - Changing Landscapes Under Future Climate Effects.* UPE10 symposium: International urban planning and environmental society. Sydney: ICMS Pty Ltd.

Roös, P. B. (2019). A landscape at risk. In D. S. Jones & P. B. Roös (Eds.), *Geelong's changing landscape: Ecology, development and conservation*. Melbourne, Australia: CSIRO Publishing.

Ruth, M., & Coelho, D. (2007). Understanding and managing the complexity of urban systems under climate change. *Climate Policy, 7*(4), 317–336.

Ryskin, G. (2003). Methane-driven oceanic eruptions and mass extinctions. *Geology, 31*(9), 741–744. https://doi.org/10.1130/G19518.1

Satterthwaite, D. (2008). Cities contribution to global warming: Notes on the allocation of greenhouse gas emissions. *Environment and Urbanization, 20*(2), 539–549.

Sharwood, S. (2014). New IPCC report: 8 ways climate change will throw world into peril. *The register.* Retrieved 28 April 2014 from http://www.theregister.co.uk/2014/03/31/intergovernmental_panel_on_climate_change_delivers_5th_report

Smit, B. a. (2003). *Adaptation to climate change in the context of sustainable development and equity*. Toronto, Canada: Quirk Press.

Smith, S. J., Pitcher, H., & Wigley, T. M. L. (2001). Global and regional anthropogenic sulfur dioxide emissions. *Global and Planetary Change, 29*, 99–119.

Souch, C., & Grimmond, S. (2006). Applied climatology: urban climate. *Progress in Physical Geography, 30*, 270–279.

Staudinger, M. G. (2013). *Impacts of climate change on biodiversity, ecosystems, and ecosystem services: Technical input to the 2013 National Climate Assessment*. Palo Alto, CA: United States Geological Survey.

UN. (2007). *World Urbinization prospects: The 2007 revision population database*. New York, NY: UN - United Nations Department of Economic and Social Affairs, Population Division.

UN Habitat. (2013). *The State of European Cities in Transition 2013*. United Nations Human Settlements Programme. Denmark: UN Habitat Publications.

UNEP. (2009). *Buildings and Climate Change: Summary for Decision-Makers*. United Nations environment Programme (UNEP), sustainable buildings & climate initiative, Sustainbale consumption and production branch. Paris: UNEP.

Walker, B., Holding, C. S., Carpenter, S. R., & Kinzig, A. (2004). Resilience, adaptability and transformability in social-ecological systems. *Ecology and Society, 9*, 2, art. 5.

Walther, G. R. (2010). Community and ecosystem responses to recent climate change. *Philosophical Transactions of the Royal Society B-Biological Sciences, 365*, 2019–2094.

World Bank. (2010). *World development report: Development and climate change*. Washington, DC: World Bank.

Chapter 4
Affinity to Water: The Coastal Zone and Coastal Settlements

Abstract Humans have *an affinity with water*, especially with the coast and the sea. This is not surprising because the coastal zone provides many resources for human use, including economic benefits that accrue from access to ocean navigation, coastal ports at river mouths, fisheries, tourism and recreation. Historically settlements have been established around these resources, and are scattered along coastlines world-wide. In the context of landscape, it is the *beauty and visual attributes* of coastal areas that attract humans, resulting in more and more people settling on the coast. This human-coastal interface sounds simplistic, but in fact it is an *interface* that is a *complex social-ecological system* characterized by *natural ecological processes* and *human-induced changes*. The discussions in this chapter investigate the beauty and visual attributes of coastal areas, the landscape and the coastal place character that attracts humans to the coast, and analyse the meaning of '*sense of place*', where settlements are inherently part of the surrounding natural environment. The chapter also further explores the *complex patterns* of human-ecological systems and conclude with the importance to manage and plan the *interface* between these *human and nature systems* towards a sustainable future. The chapter concludes with the *fundamental pattern* AFFINITY TO WATER [4], instructing us how to deal with the complexities of environments next to water.

Keywords Coastal settlements · Nature systems · Sense of place · Affinity to water · Coastal ecosystems · Sea level rise · Coastal change · Human systems · Climate change

Upward links:
… the coastal environment is a complex interface of natural ecological processes and human-induced changes, and is extremely vulnerable to climate change impacts. As a result, we need to follow a process of CLIMATE CHANGE CO-ADAPTATION [3]. However, we need to understand the complex attractions of humans to large bodies of water – AFFINITY TO WATER [4] – to be able to plan and develop sustainable settlements close to water.

4.1 Introduction

There is pleasure in the pathless woods, there is a rapture on the lonely shore, There is a society, where none intrudes, by the deep sea, and music in its roar – *Lord Byron, Childe Harold's Pilgrimage* (in Green, 2010, p. xi).

P. B. Roös, *Regenerative-Adaptive Design for Sustainable Development*, Sustainable Development Goals Series, https://doi.org/10.1007/978-3-030-53234-5_4

Humans have an *affinity with water*, especially with the coast and the sea. This is not surprising because the coastal zone[1] provides many resources for human use, including economic benefits that accrue from access to ocean navigation, coastal ports at river mouths, fisheries, tourism and recreation. Historically settlements have been established around these resources, and are scattered along coastlines worldwide. In the context of landscape, it is the *beauty and visual attributes* of coastal areas that attract humans, resulting in more and more people settling on the coast. As population density and economic activity in the coastal zone increases, so does pressure on *coastal ecosystems*. This human-coastal interface sounds simplistic, but in fact it is an *interface* that is a *complex social-ecological system* characterized by natural ecological processes and human-induced changes (Kittinger & Ayers, 2010). The cumulative impacts from human activities have made coastal ecosystems among the *most transformed* and *degraded environments in the world* (Halpern et al., 2008; Lotze, 2006). From a human built environment perspective, environmental degradation increases the vulnerability of coastal community settlements and threatens the supporting ecosystems and goods, services and values depended upon by coastal communities (Adger & Kelly, 2006; Cinner & McClanaham, 2006) (Fig. 4.1).

Fig. 4.1 The changing environment in coastal areas (Photo by Author)

[1]The Coastal Zone is defined as the interface where the land meets the ocean, encompassing shoreline environments as well as adjacent coastal waters. Its components can include river deltas, coastal plains, wetlands, beaches and dunes, reefs, mangrove forests, lagoons, and other coastal features. For the purpose of this study the coastal zone includes the features of the previous definition as well as the coastal hinterland, areas of land directly influenced by the sea or directly influencing the coastline and with critical impacts on the foreshore and near shore environment, and private and public land with built environment systems that engage and impact on the coastal environment (VCC, 2013).

In the concept of resilience of both ecological and human systems, the theme of sustainability research has emerged in literature. This literature identifies that patterns of climate disturbances occur along the coasts, especially exacerbating the vulnerability of settlements and the natural environment within a dynamic coastal landscape (Kittinger & Ayers, 2010; Levin, 1999; Connell, 1978). Understanding how both human and natural systems can potentially and successfully adapt to climate change is important for the resilience of both. The relationships between the systems are fundamental, and the complex pattern of human-ecological systems is important because systems can trend toward multiple stable states resulting in undesirable loss of critical ecosystem services and values.

Thus, this chapter investigates the nature of coastal settlements along the Australian coast, settlement patterns, place character, biophysical characteristics and their relationship to place and the natural environment, with regard to the vulnerability of sea level rise and climate change.

This chapter concludes with an investigation of current practices to manage and plan the interface of human and coastal ecological systems by reviewing coastal settlements in the state of Victoria, Australia, specifically along the Great Ocean Road. Even though the focus is on the Victorian Coast in Australia, the contexts and *vulnerability* of *the coast and human settlements* along this coast is similar to what other coasts and settlements across the globe experience. Therefore, the context can be matched with comparable attributes and issues for many coastal *landscapes across the world*. Further, I introduce the issues of coastal settlements and human interactions with the environment as a key part of the narrative of this book, due to the *high risk of impacts* from *sea level rise* and *climate change* we will face in the future.

4.2 Coastal Settlements and the Coast

More than 60% of the world's population live within 150 km of coastlines, with more and more densely populated settlements on just under 10% of the earth's land surface (Nicholls & Small, 2002), *sprawling all along the coastal zone*. It is predicted by 2050 that more than 75% of the world's population will be living on or near the coast (Green, 2010; UN., 2002). In Australia over 20 million people live within 50 km of the sea, and 20% of the population live in smaller coastal towns which are located right on the coast (Green, 2010). Rapid urban expansion in coastal settlements due to 'lifestyle migration' impacts negatively upon coastal amenities. The coast is under stress, with the growth pattern of settlements occurring in a sprawling linear fashion. This pattern is resulting in devastating effects on the natural local environment including biodiversity and habitat loss, degradation of water quality, damage to wetlands, coastal erosion, loss of native vegetation, introduction of pest animals and plants, and the loss of cultural heritage (VCC, 2008; DSE, 2004).

4.3 The Nature of Coastal Settlements

Globally, regional coastal settlements are similar in context and attributes. Coastal settlements and their attributes – the attractive natural environments and recreational opportunities that these places can offer – are the driving factors behind the migration of people to the coast. In Australia this phenomenon is referred to as '*sea change*'. This term refers to a variety of people moving to settlements along the coast, where the differences can be explained by understanding the location, settlement, and population variations that characterise coastal amenity communities. This migration is purely due to the human desire to seek a change in lifestyle (Burnley & Murphy, 2004). It is a form of amenity migration, whereby people seek the values and attributes that the place character of these coastal towns offer. These settlements were generally established due to local resources available such as fishing, timber, and fertile land close to the river inlets.

These coastal settlements differ in size and typology as well as character, and various factors influence the shaping of these towns. During my research for this book, it became evident that *similar typologies* between typical coastal settlements exists across countries on a global scale. The classification of these typologies by researchers at the University of Sydney represents the most fitting for the Australian context. They have distinguished a broad typology of coastal communities in Australia comprising five ideal types (Gurran, Norman, Gilbert, & Hamin, 2008; Gurran, 2005, p. 23–26):

- *Regional Coastal Cities* - smaller urban cities (populations above 100,000 people) situated beyond State capitals and located on the coast;
- *Coastal Commuters* - suburbanised satellite communities on the coast, located close enough to major centres to make commuting a feasible option;
- *Coastal Getaways* - small to medium coastal settlements within a three-hour drive of a capital city, located in predominately natural habitats with high ecological, visual and recreational values;
- *Coastal Lifestyle Destinations* - small to medium sized coastal settlements predominantly tourism and leisure focused, located three-hour drive or more from a capital city; and,
- *Coastal Hamlets* - small, rural coastal communities often surrounded by protected natural areas, situated more than three hours away from a capital city.

The case study town chosen for the application of the *regenerative-adaptive pattern language*, the town Anglesea, is unique according to the above classifications unique. Based on its population and setting, it can be classified as a *Coastal Hamlet*, but due to its distance from the capital city Melbourne of only approximately 115 km, it can be classified as a *Coastal Gateway*, or *Coastal Lifestyle Destination*. Most of the towns in this area of the Great Ocean Road region can be classified as *Coastal Hamlets*. This is unique in that they are within three hours of the capital city of Melbourne, but their 'remoteness' due to the *topography and natural environment* of the Otway Ranges, and accessibility of minor roads, results in small populations of around 1500 to 3000 permanent residents. Residents that migrate to these towns are attracted to the natural beauty of the coast, as well as the character of the towns and their '*sense of place*' experience.

4.4 Settlement Character in the Coastal Zone

In general terms, most coastal towns have a '*sense of place*', and settlements along the coast are generally referred to as having '*a lot of character*' (Fig. 4.2). The degree to which a place is said to have 'a lot of character' is mostly associated with how distinctive its environmental, cultural and social features are (Green, 2010). The uniqueness of a place usually is the result of the organic growth of a settlement; these qualities are mostly visible within old established coastal towns. Recent developments and new 'Greenfield developments' lack distinctiveness and uniqueness of place. They are perceived as 'faceless'; places that could be anywhere and yet nowhere (Hough, 1990). New land developments often are referred to as not having 'soul', and lacking a sense of '*wholeness*'. Alexander et al. (1977) has explored this phenomenon of 'wholeness' in architecture and settlement planning, and proposed in *A Pattern Language* (1977) a theory that the inherent patterns of the landscape, the morphological growth of a place, give it character and a sense of place resulting in wholeness (Alexander, 2004; Alexander et al., 1977). This theory and its application as part of the *regenerative-adaptive pattern language* are further explored and described in detail in Chaps. 10, 11 and 12, where we have identified that wholeness of place and place character only occurs when the development and design of the built form are integrated and connected with nature. Similarly, Green (2010) concludes that the place character of Victorian coastal towns is somehow unique, different, unusual, special and

Fig. 4.2 Barwon Heads is a typical example of a small coastal town as 'a place with lot of character' (Photo by Donna Squire, 2018)

are given their locality and engagement with the natural environment (Green, 2010). This wholeness is connected to '*deep sustainability*', where the integrated connections with nature and a place support resilience of local communities. *Deep sustainability* goes beyond the mere aspects of resource efficiency, energy reduction or sustainable growth; rather, it requires emotional, spiritual and cultural *connections of people with their built environment*, resulting in an environment that has a distinct '*sense of place*' and character.

Place character is the 'feel', 'atmosphere', 'ambience', and the 'soul' of a place. This is a difficult term to define, but a common description of place character is one that makes people feel at home and attracted to its inherent qualities. All places in the built environment as well as in the natural environment possess character (Green, 2010). Some places are considered to have recognizable qualities and have a unique character, and these are often referred to as having a 'sense of place'. 'Sense of place' is a term that developed from the term *genius loci,* a term initially used to describe the appreciation of landscapes (Jiven & Larkham, 2003). *Genius loci* can be acknowledged as a technical term, specific to the new landscape. However, because this term developed over time, the concept moved away and included its application to any landscapes, including the urban form (Jackson, 1994). *Genius loci*, in its application to perception of a place, evolved into describing the *quality of places* and in the terminology's transition to modern times it evolved into 'Sense of Place', which has been well described by landscape author J.B. Jackson:

> 'Sense of Place' is a much used expression, chiefly by architects but taken over by urban planners and interior decorators and the promoters of condominiums, so that now it means very little. It is an awkward and ambiguous translation of the Latin term *genius loci*. In classical times it means not so much the place itself as the guardian

divinity of that place. ... in the eighteenth century the Latin phrase was usually translated as 'the genius of a place', meaning its influence. ... We now use the current version to describe the atmosphere to a place, the quality of its environment. Nevertheless, we recognize that certain localities have an attraction which gives us a certain indefinable sense of well-being and which we want to return to, time and again (Jackson, 1994, p. 64).

When Jackson refers to the 'atmosphere' of a place, he indicates that this acknowledgement of *genius loci* is also linked to and has become allied to *place character* (Jiven & Larkham, 2003). This is the unique constellations of socio-physical characteristics that differentiate these environments from one another (Tuan, 1974). This place character – *the spirit of a place* – assumes high amenity value, and can be defined as the most important aspect that adds value to a place. Place character includes the values of that environment, i.e. the things that people experience that are special, including their emotional attachments and response to, and influence of, physical objects, including beaches, historical buildings, trees, and footpaths. It is the visual attributes and the interplays of these objects and spaces that contributes to a sense of place and the uniqueness of place-character, creating *distinct local values* in each coastal town (Roös & Jones, 2013). Communities on the coast in part exist due to their high visual and natural attributes they are situated within or adjacent to, which underpin the qualities of these values. Changes in these landscapes and built values may cause a dramatic effect upon their context, thereby influencing the overall place-character (Roös, 2014).

It can be argued that the principles of '*living centres*' of a settlement (as defined by Alexander in the *Nature of Order*; Alexander, 2001-2005, p. 80), which connect human built environments and nature together in a *wholeness*, add to the creation of place character and 'sense of place'. The living structures within a neighbourhood are crafted over many years by local residents, resulting in a unique character painted in the heritage and history of the place. This uniqueness is fundamental to the ongoing resilience of a place and results in the *sustainable futures of the communities* (Alexander, 2004). An important aspect of place-character is the heritage and cultural identity of place, shaped by the history of the settlement. In coastal towns it is this history that draws people to settle there, and results in increased numbers of overnight visits during holiday seasons. Heritage and history, culture as well as natural landscapes with its biophysical characteristics contribute to the appreciation of place (Roös, 2014). High visual and natural attributes of the coast are always in a transition stage, and the biophysical characteristics of the coastal zone are the foundation layer of patterns that influence the connection of humans to their natural environment. For centuries these *biophysical characteristics* have influenced society's decisions as to where to establish human settlements along the coast.

4.5 The Biophysical Characteristics of the Coast

Biophysical characteristics of the coast differ in various shapes and forms around the world, and this is a similar phenomenon in Australia. The Australian coast has four major coastal regions, each with their own geomorphology and climate characteristics. The dynamic interactions of land and sea over time shaped the coastline as it is today, with over 10,000 sandy beaches that make up half of the coastline, and the remainder comprising mostly rocky and cliff shorelines (CSIRO, 2009, p. 12). Most of the Victorian coast can generally be classified as a limestone rocky coastline, with sandy beaches predominantly located at river inlets. The calcareous beach sands are 'perched' on the Calcarenite ridges, as indicated in Fig. 4.3. This geomorphological arrangement is key in identifying which areas of the Victorian coast are vulnerable to climate change impacts, as well as to determine the *natural attributes* that contribute to the characteristics and values of coastal settlements in these sandy coastal zones. Further, the biophysical characteristics of the coastline provide fundamental information in considering the development of a pattern language to inform *regenerative-adaptive design and planning* decisions.

Fig. 4.3 Coastal Features of the Great Ocean Road Region, Victoria, Australia (Photo by Author)

The biophysical features of the Great Ocean Road Region Coast are, in most parts, highly resistant to coastal erosion due to the geological formations (GORCC, 2012). However, the inlets, beaches, sand dunes and low-lying areas are susceptible to coastal erosion. It is these areas that are more vulnerable to the rise in sea levels, and they are in easily accessible areas where the concentrations of human settlements are evident. These low-lying areas, inlets and sand dunes with estuaries are vulnerable to the impact of *coastal dynamics* including winds, flooding, storm surges and erosion.

4.6 Coastal Dynamics

The coastal zone is a dynamic place. It is the coast point where land interacts with the sea and storm systems gather energy from the ocean and intensify in conjunction with the forces of the climate. The most visual noticeable force are waves; wave energy has devastating impacts upon coastal attributes during storm surges, including erosion, inundation, sea bed disturbance and displacement of sand and sediment. Waves are part of a variety of geological, atmospheric, marine and terrestrial processes continually interacting to shape coastal landforms and conditions (McInnes, MacAdam, & O' Grady, J., 2009). These dynamics change in relation to the climate, weather patterns and the geological and landscape characteristics of the coast.

The following coastal processes of winds, waves, tides and storms are the primary driving forces of change upon coastal landforms and their environment (GORCC, 2012):

- *Wind* - is the result of horizontal gradients in air pressure. The intensity of pressures systems and their distribution and passage across the Southern Ocean and the Australian continent determine wind strength and direction. Winds generate waves and currents that directly shape coastal landforms;

- *Waves* - are the principal source of energy for most coastal systems and develop in response to wind. Waves colliding with the coast provide the energy needed to move sediment and reshape shorelines;
- *Tides* - are the vertical movements of the water level that result from the gravitational forces of the moon and sun acting on ocean waters. This movement is responsible for the inundation of low-lying coastal land at sub-daily to seasonal timescales. Tides are another major source of marine energy that shape coastal landforms and play an important role in the transport of coastal sediments;
- *Storms* - in Victoria, storms are generally experienced as cold fronts and intense low-pressure systems that pass across the Southern Ocean/Great Australian Bight. Their sharp pressure gradients generate strong winds and high waves that may lead to erosion of sensitive coastal landforms. Low air pressure and wind-generated waves may also lead to elevated water levels (storm surges) and inundation of low-lying coastal land as storm systems pass. Depending upon the timing of the storm, tides may exacerbate storm surge flooding. Rains generated by storms may also lead to flooding of land adjacent to coastal rivers and estuaries.

Climate change influences and change these dynamic coastal systems. Sea level rise is considered to be the biggest threat to coastal areas resulting in increased rates of coastal erosion, more extensive and frequent coastal flooding, increased intrusion of seawater into estuaries and coastal aquifers, changing water quality, changing groundwater characteristics and sedimentation, and increased sea-water temperatures that may affect ecosystems (DSE, 2012, p. 39).

4.7 Coastal Management & Planning

For many centuries, humans have deliberately exploited the resources of the coast and modified its environment. Throughout the world ancient civilisations built sea walls and ports, diverted rivers, use coastal soils for agriculture, and exploited marine life (Kay & Alder, 2005). Activities to preserve coastal character and its ecological integrity in the timescale of human interventions to the coast are very recent – as early as the 1930s, when the first protective coastal park was established under the auspices of ICM (*Integrated Coastal Management;* Wescott, 1998; Sorensen, 1997). Since then the concept of coastal management and planning has evolved, and now it is generally accepted that coastal resources can only be effectively evaluated and managed in the context of a total ecosystem associated with human cultural and social environments (Roös, 2019; Kay & Alder, 2005).

This concept is key to sustainability, and in coastal management the requirement for sustainable outcomes has become the dominant paradigm. *Sustainable coastal management* that includes an understanding of the social-ecological systems interface is closely linked to *sustainable development*. Effectively managing social-ecological systems requires an understanding of the complexity of human-environmental relationships, defining the pathways toward optimal planning outcomes, and implementing effective ecosystem-based management practices that support the goals of sustainable development.

Unfortunately, this is not always the case. Examples of the *opposite practice* to sustainable development for coastal management, is evident in the state of Victoria, Australia, where the decision-making and authority for approvals in the environmental scenario (where developments transverse public and private land) is a major problem in terms of effective implementation. Macintosh (2012) concludes in *Coastal adaptation planning: A case study on Victoria, Australia* that:

> …the Victorian framework is based on an awkward distribution of powers and responsibilities. It provides broad principles that are intended to guide decision- making by planning and responsible authorities, however it does

not dictate outcomes and is vague about what is expected in the treatment of coastal issues (Macintosh, 2012, p. 6).

Part of this problem is that the autonomy of local councils in the planning system is tightly constrained. This is due to the multiplicity of powers of other governmental bodies, such as the State Planning Minister, the Referral Authorities, the Victorian Coastal Council (including Regional Coastal Boards), and the Victorian Civil and Administrative Tribunal (VCAT). Another constraint to the implementation of a sustainability-focused policy is that an urban/growth boundary is not set, and the continued movement towards and across this notional boundary into the natural landscape of coastal areas results in a '*coastal suburban sprawl*' (Wescott, 1998; VCC, 2013). This is best demonstrated in the town of Torquay, with the approval of new greenfield developments on the edge of the urban boundary in the Spring Creek catchment[2] (Fig. 4.4). Coastal suburban sprawl have major impacts on the environment, and Smyth (2019) noted that increased land use changes to accommodate housing also result in major changes in the regions coastal and marine environments (Smyth in Jones & Roös, 2019, pp. 171–173). Notwithstanding all the issues around policy, even with good intentions, the main problem of future sustainable developments minimising human impacts on the coastal environment has not been resolved.

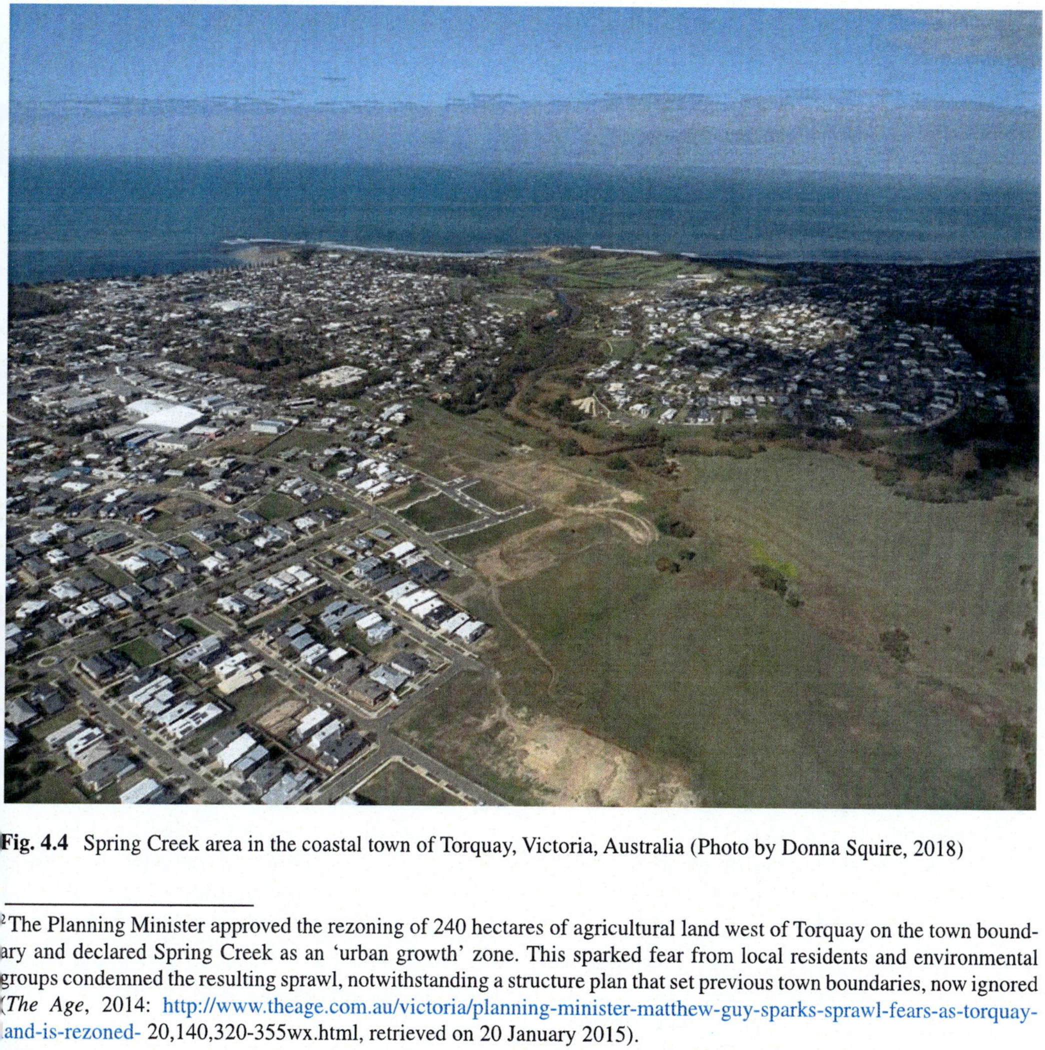

Fig. 4.4 Spring Creek area in the coastal town of Torquay, Victoria, Australia (Photo by Donna Squire, 2018)

[2]The Planning Minister approved the rezoning of 240 hectares of agricultural land west of Torquay on the town boundary and declared Spring Creek as an 'urban growth' zone. This sparked fear from local residents and environmental groups condemned the resulting sprawl, notwithstanding a structure plan that set previous town boundaries, now ignored (*The Age*, 2014: http://www.theage.com.au/victoria/planning-minister-matthew-guy-sparks-sprawl-fears-as-torquay-land-is-rezoned- 20,140,320-355wx.html, retrieved on 20 January 2015).

Effectively managing the *social-ecological systems* on our coasts across the globe requires an understanding of the *complexity* of *human-environmental relationships*, defining the pathways towards the best outcomes for environments, and implementing effective ecosystem-based planning and management. Existing literature suggests that this is best achieved through mechanisms of effective governance arrangements, adaptive management, institutional effectiveness, community participation, and policies that incentivise actions for achieving outcomes that consider the environment. This includes the consideration of natural hazards as well as the impacts of climate change (Olsson, Folke, & Hughes, 2008; Costello, Gaines, & Lynham, 2008; Christie & White, 2007). However, what is clearly lacking in various coastal management plans and strategies is the understanding of the complex patterns of human-ecological systems and patterns. These are important to inform efforts to manage and plan the interface between human and coastal environmental systems towards a sustainable future. I further conclude that the *deep connectivity* of *human systems with nature* is not evident in the discourse of coastal management, and the establishment, recognition and conserving of place character needs to include a deeper understanding of '*wholeness*'. This would provide a *connection with the coastal environment* that could result in resilience for the future of both humans and nature under a future changed climate.

4.8 Conclusion Statement

The discussions in this chapter indicate that it is the beauty and visual attributes of coastal areas, the landscape and the coastal place character that attracts human migrants to the coast, resulting in more and more people settling on the coast. The *biophysical characteristics* of the coast induce a '*sense of place*', where settlements are inherently part of the surrounding natural environment. The chapter also reveals that the relationships between *social-ecological systems* are fundamental, and understanding the complex patterns of human-ecological systems is important to inform efforts to manage and plan the *interface* between these *human and nature systems* towards a sustainable future. It is also evident that current policy and frameworks for coastal management and planning do not adequately address the requirements of sustainable practices for the integration of human and ecological systems along the coast. Further, the deep connectivity of human systems with nature is not evident, and the establishment, recognition and conserving of place character needs to include a deeper understanding of 'wholeness' to provide a connection with the coastal environment that could result in resilience for the future of both humans and nature under a future changed climate. This understanding of the '*wholeness*' and that of '*deep*' connectivity with the coastal environment is very important in social-ecological management. While this understanding is lacking in the current Western systems practice, it is very prominent in Indigenous belief systems, cultural heritage, values, and land management practices. Understanding of *Indigenous knowledge* about *Country* and *Sea Country* is paramount in developing planning solutions for our settlements along the coast, and this knowledge is further explored in Chap. 5, where *patterns of human-nature connection* inform considerations for the *regenerative-adaptive pattern language*. As noted by Jones and Roös (2019), decision-making for development and growth needs to better grasp and be accountable for a regions' environmental qualities, and needs to be ecologically responsive and consider regenerative design principles (Jones & Roös, 2019, p. 300).

Therefore:

To achieve a sustainable future for both human settlements and nature systems along the coast, develop and implement environmentally focused policy frameworks for coastal management and planning. These must adequately address the requirements of sustainable practices for the integration of human and ecological systems. Further acknowledge that the beauty of the coast is induced with a sense of place, where settlements are inherently part of the surrounding natural environment. This place character must be protected at all times.

Fig. 4.5 Sketch of Fundamental Pattern – AFFINITY TO WATER [4] (Drawing by Jesse Delmo, synthesised by Roös)

4.9 Fundamental Pattern 4

AFFINITY TO WATER [4].

In any new human settlement along the coast and next to water bodies should acknowledge the complex dynamics of coastal and water systems. Investigate the biophysical characteristics and social-ecological systems of place to understand the patterns of human-nature affiliations. Manage and plan the interface between these human and nature systems towards a rich, biodiverse enriched and sustainable future (Fig. 4.5).

Downward links:
When developing coastal management and planning solutions for our settlements along the coast, there is a need for the knowledge of the longer-term history of the land, the Indigenous knowledge that stretches over thousands of years. Our considerations for coastal settlement includes our AFFINITY TO WATER [4] and needs to embrace ADVANCED INDIGENOUS KNOWLEDGE [5] …

References

Adger, W. N., & Kelly, P. M. (2006). Vulnerability. *Global Environmental Change, 16*, 268–281.

Alexander, C., Ishikawa, S., Silverstein, M., Jacobson, M., Fiksdahl-King, I., & Angel, S. (1977). *A pattern language: Towns, buildings, construction*. New York, NY: Oxford University Press.

Alexander, C. (2004). *Sustainability and morphogenesis: The birth of a living world*. Berkeley, California: Centre for Environmental Structure.

Alexander, C. (2001-2005). *The nature of order - an essay on the art of building and the nature of the universe, book one: The phenomenon of life*. Berkeley, California: The Center for Environmental Structure.

Burnley, I., & Murphy, P. (2004). *Sea change: Movement from metropolitan to Arcadian Australia*. Sydney: University of New South Whales Press.

Christie, P., & White, A. T. (2007). Best practices for improved governance of coral reef marine protected areas. *Coral Reefs, 26*, 1047–1056.

Cinner, J., & McClanaham, T. R. (2006). Socioeconomic factors that lead to overfishing in small scale coral reef fisheries of Papua New Guinea. *Environmental Conservation, 33*, 73–80.

Connell, J. H. (1978). Diversity in tropical rain forests and coral reefs. *Science, 199*, 1302–1310.
Costello, C., Gaines, S. D., & Lynham, J. (2008). Can catch shares prevent fisheries collapse? *Science, 321*, 1678–1681.
CSIRO. (2009). *The climate change risks to Australia coast.* Commonwealth of Australia, Department of Climate Change. Canberra: Department of Climate Change.
DSE. (2004). *Great Ocean road region - land use and transport strategy 2004*. Melbourne, Australia: Department of Sustainability and Environment DSE.
DSE. (2012). *The victorian coastal hazard guide.* The Victorian Government, Department of Sustainability and Environment. Melbourne, Australia: DSE.
GORCC. (2012). *Coastal climate change vulnerability and adaptation*. Melbourne, Australia: Great Ocean Road Coast Committee - GORCC.
Green, R. J. (2010). *Coastal towns in transition - local perceptions of landscape change*. Melbourne, Australia: CSIRO Publishing.
Gurran, N., Norman, B., Gilbert, C., & Hamin, E. (2008). *Planning for climate change: Leading practice principles and models for sea change communities in coastal Australia.* Sydney: University of Sydney, Faculty of Architecture.
Gurran, N. S. (2005). *Meeting the sea change challenge: Sea change communities in coastal Australia*. The University of Sydney, planning and research Centre. Sydney: The University of Sydney.
Halpern, B. S., Walbridge, S., Selkoe, K. A., Kappel, C. V., Michelli, F., D'Agros, D., et al. (2008). A global map of human impact on ecosystems. *Science, 319*, 948–952.
Hough, M. (1990). *Out of place: Restoring identity to the regional landscape*. New Haven, CT: Yale University Press.
Jackson, J. (1994). *A sense of place, a sense of time*. New Haven, CT: Yale University Press.
Jiven, G., & Larkham, P. J. (2003). Sense of place, authenticity and character: A commentary. *Journal of Urban Design, 8*(1), 68–71.
Jones, D. S., & Roös, P. B. (Eds.). (2019). *Geelong's changing landscape: Ecology, development and conservation*. Melbourne, Australia: CSIRO Publishing.
Kay, R., & Alder, J. (2005). *A brief history of coastal management and planning*. London; New York: Taylor and Francis.
Kittinger, J. N., & Ayers, A. C. (2010). Shoreline armoring, risk management, and coastal resilience under rising seas. *Coastal Management, 38*(6), 634–653.
Levin, S. A. (1999). *Fragile dominion: Complexity and the commons*. New York, NY: Helix Books/Perseus Publishing.
Lotze, H. L. (2006). Depletion, degration and recovery potential of estuaries and coastal seas. *Science, 312*, 1806–1809.
Macintosh, A. (2012). *Coastal adaptation planning: A case study on Victoria, Australia*. Canberra: ANU Centre for Climate Law and Policy, ANU College of Law.
McInnes, K., MacAdam, I., & O' Grady, J. (2009). *The effect of climate change on Extreme Sea levels along Victorias coast. A project undertaken for the Department of Sustainbaility and Environment, Victoria as part of the 'Future Coasts' program*. Melbourne: CSIRO Marine Atmospheric Research.
Nicholls, R. J., & Small, C. (2002). Improved estimates of coastal population exposure to hazards released. *Eos, 83*, 303–305.
Olsson, P., Folke, C., & Hughes, T.P. (2008). Navigating the transition to ecosystem-based management of the great barrier reef, Australia. In *Proceedings of the National Academy of Sciences.* (Vol. 205, pp. 9489–9494). National Academy of Sciences.
Roös, P., & Jones, D. (2013). Visions of the surf coast - changing landscapes under future climate effects. In *UPE10 symposium: International urban planning and environmental society*. Sydney: ICMS Pty Ltd..
Roös, P. (2014). Landscapes in transition: Local perceptions of Indiginous and European settlement changes along the Victorian coast. In *UHPH 14: Urban history planning history conference proceedings*. Wellington, New Zealand: UHPH / Victoria University of Wellington.
Roös, P. B. (2019). A landscape at risk. In D. S. Jones & P. B. Roös (Eds.), *Geelong's changing landscape: Ecology, development and conservation* (pp. 202–215). Melbourne, Australia: CSIRO Publishing.
Smyth, C. (2019). Marine and coastal environments. In D. S. Jones & P. B. Roös (Eds.), *Geelong's changing landscape: Ecology, development and conservation* (pp. 171–184). Melbourne, Australia: CSIRO Publishing.
Sorensen, J. (1997). National and international efforts at integrated coastal managemenet: Definitions, achievements, and lessons. *Coastal Management, 25*, 3–41.
Tuan, Y. F. (1974). *Topophilia: A study of environmental perception, Atitudes, and values*. Englewood Cliffs, NJ: Prentice Hall.
UN. (2002). *Oceans: The source of life - United Nations convention on the law of the sea*. New York, NY: United Nations.
VCC. (2013). *Victorian coastal Startegy 2013 (draft report)*. Melbourne, Victoria, Australia: Victorian coastal council - VCC; Victorian Government Department of Environment and Primary Industries.
VCC. (2008). *Victorian coastal strategy*. Melbourne, Australia: Victorian Coastal Council, VCC.
Wescott, G. (1998). *Reforming coastal management to improve community participation and integration in Victoria, Australia*. London: Taylor and Francis.

Chapter 5
Origins of Advanced Knowledge

Abstract This chapter focuses on identifying the longer-term history of human and landscape interaction, and the knowledge of these landscapes, more specific those of Indigenous peoples. It is well known that Indigenous cultures have an intimate relationship and knowledge of the natural environment. It is thus imperative to explore the Indigenous knowledge of the land and nature, with an aim to identify the deeper *interconnections between people and their natural environments*. The identification of the longer-term histories of human and landscape interactions of a specific region can be discovered through investigating the Indigenous knowledge of a specific place, as well as the Indigenous living and continuing 'history' of the area. Relevant to the narrative of this book is the identification of the *deeper relationships* that were evident in the cultures and daily activities of the Indigenous people of the land. To understand the deeper forms of the land and waters, the intimate human-climate and nature relationship, and human settlement patterns of the past, this chapter further explores *advanced Indigenous knowledge* of the land through investigating specifically Aboriginal knowledge in Australia. Detailed accounts of the *Wadawurrung* people and their interconnections with the environment are presented. The chapter concludes with the *fundamental pattern* ADVANCED INDIGENOUS KNOWLEDGE [5], identifying the importance of including this knowledge in our design and planning for sustainable futures.

Keywords Indigenous knowledge · Aboriginal · Connection to the land · Six seasons · Country · Sea country · Human settlements · Advanced knowledge · Human-nature connection

Upward links:
… consider now the Indigenous knowledge of the land, before European settlement and global colonisation. The pattern AFFINITY TO WATER [4] helped us comprehend the desire of humans to live close to it. However, this human-nature connection needs to be supported by the ADVANCED INDIGENOUS KNOWLEDGE [5] of the land, so that we can work with nature to face the challenges of a future changing environment and achieve resilience with CLIMATE CHANGE CO-ADAPTATION [3].

Indigenous readers should note that the content of this chapter might include names of deceased persons. Where possible, attempts have been made to obtain permission to include these names. No offence is meant to any relatives of the deceased. The majority of the material in this chapter has been sourced from public domain publications, and previous published works. Most of the Wadawurrung information has been sourced from published works with additional content

P. B. Roös, *Regenerative-Adaptive Design for Sustainable Development*, Sustainable Development Goals Series, https://doi.org/10.1007/978-3-030-53234-5_5

Fig. 5.1 Origins of Advanced Knowledge – Aboriginal rock paintings of *Gwion Gwion* (Tassel Bradshaw) figures wearing ornate costumes (From Wikipedia Commons CC BY-SA)

provided through personal communications and recordings in accordance with consents and permissions provided and previously published in my PhD thesis work.[1]

5.1 Introduction

> According to Aboriginal belief, all life as it is today is part of one vast unchanging network of relationships, which can be traced back to the *Great Ancestors* of the *Dreamtime* (Veronica Brodie, Kaurna Elder, 1994, p. 4).

The accumulated Australian Aboriginal knowledge on the environment and Indigenous systems is over 50,000 to 60,000 years old (Powell & Tournier, 2019; Nunn & Reid, 2016; Roberts, 1994). During these ancient times hunter-gatherers initiated a consciousness of community (Fig. 5.1). Cycles and centres defined space and time and this period saw the beginning of conscious design and the building of shelters and villages. The people of the land built their own environments through an

[1] Roös, P.B. (2017). *Regenerative-Adaptive Design for Coastal Settlements: A Pattern Language Approach to Future Resilience*. PhD Thesis, Deakin University, Geelong, Victoria, Australia. Permission granted for the publication of Indigenous knowledge information and personal communications 2014–2016 by the late Aboriginal Elder - *Tandop* David Tournier. Additionally, this research entitled '*Coastal Climate Change & the Great Ocean Road Region: Community Workshops & Online Survey*', was subject to an approved human research ethics approval by Deakin University's Faculty of Science Engineering & Built Environment Human Research Ethics Advisory Group (HEAG), coded STEC-1-2013- ROOS dated 21 February 2013; and, this research entitled '*Regenerative-Adaptive Design for Coastal Settlements: A Pattern Language Approach to Future Resilience*', was subject to an approved Cultural Heritage Permit WAC-P0031 issued by the Wathaurung Aboriginal Corporation in accordance with s.36(1) of the Victorian *Aboriginal Heritage Act 2006* dated 28 August 2019.

interwoven fabric of long-standing cultural patterns that connected the patterns of nature. The building of shelters and the establishment of temporary to semi-permanent settlements was part of everyday learning and a critically important aspect for survival (Van der Ryn, 2005, p. 137). In Aboriginal culture this form-making was closely linked to the cycles of weather, climate and natural phenomena. As Aboriginal Elder Aunty Veronica Brodie explains, life is part of one vast network of relationships, which can be traced back to original Aboriginal wisdom (Brodie, 1994). Knowledge of the environment was very deep, and consciousness was formulated in the context of wholeness. Van der Ryn refers to this stage of human development as 'magical consciousness' where oral language and the graphic image and symbol were at the heart of awareness. It can be argued that the circle and spiral were representative of natural phenomena and oneness with the larger whole, reflecting their state of equilibrium with nature (Van der Ryn, 2005).

In *Notes on the Synthesis of Form*, Alexander (1964, p. 48) records that part of the bigger whole, subsystems that operate independently to create form in oneness, is fundamental to the process of 'unselfconscious cultures'[2] in building their structures and settlements. Patterns of tradition, myth and the use of materials are constant and link to nature (environment) from where these materials are used. Indigenous cultures will only introduce change to their form-producing systems if there are powerful and obvious interventions to the existing stage of equilibrium, and these changes accommodate a prolonged adaptive process. Ancient cultures and traditions, the 'unselfconscious cultures', were in essence formulated on the basis of adaptation linked to a cyclical process.

The aim of this chapter is thus to explore the role and values that Australian Aboriginal knowledge can bring to this narrative of the *regenerative-adaptive pattern language*, more specifically on a *deeper* understanding of the *interconnections of people and their natural environment*. Further, this will provide Aboriginal insights on historical adaption practices and the potential of including their chronicles as input to a *regenerative-adaptive pattern language*, and possibly providing answers for regeneration and adaptation strategies to assist sustainable development.

5.2 Indigenous Knowledge and Western Science

The understanding and potential application of Indigenous knowledge is different from the usual Western science discourse. This is because Indigenous knowledge is holistic, connected in storytelling with many parts that makes up the whole. Indigenous knowledge, as a way of knowing, is similar to western science in that it is based on an accumulation of observations. However, it is very different from western science in many fundamental ways. These two ways of knowing are parallel methods of acquiring knowledge about the earth and the universe, but distinct from each other. In both Western Scientific and Australian Aboriginal knowledge (Indigenous knowledge), information is organized to include experience as well as beliefs into 'knowledge'. Smylie et al. (2003) summarise the two systems as follows:

> In Western knowledge systems this process involves the organization of individual data into abstract theoretical systems, composed of multiple components, each of which requires a 'specialist' to be fully understood. In Indigenous knowledge systems, generation of knowledge starts with 'stories' as the base units of knowledge, proceeds to 'knowledge', an integration of the values and processes described in the stories, and then accumulates into 'wisdom', an experiential distillation of knowledge. This process can be viewed as cyclical, as "wisdom" keepers in turn generate new 'stories' as a way of disseminating what they know (Smylie et al., 2003, p. 141)

[2]Christopher Alexander identifies the *Traditional People* (Indigenous people) as 'unselfconscious cultures', and the modern-day western people as 'self-conscious cultures'. The distinction is made in the context of architecture and the built form, where the 'unselfconscious cultures' built form evolved due to functionality, use, and the direct connection to the natural environment, and on the other hand 'self-conscious cultures' built form evolved due to art, architecture and engineering, a trait that forces an egoistic view of the achievement of 'self' (Alexander, 1964).

The understanding and application of Indigenous knowledge is a difficult task. Firstly, the knowledge must be acknowledged and understood with a consideration of how this knowledge is translated by the respective Indigenous community themselves. The fundamental principle is to understand that Indigenous knowledge is 'holistic' in nature. As noted by Berkes, Indigenous knowledge is holistic in outlook and adaptive by nature, gathered by generations through observations, as their lives depended on this information for survival (Berkes, 1998). This knowledge accumulates incrementally and is tested by trial and error, told in narratives from generation to generation and shared in practical experiences (Berkes, 1998).

The sharing of this knowledge is also not straightforward, in a simple linear context, but it is more like a network between different members of families and clans, in a structured storytelling process tying everything together in a kinship system. The vast complexities in the Aboriginal kinship systems and *Country*, the way stories form intricate webs of networks across *Country*, the images and metaphors used to communicate across different cultural groups, and the ancestral practices are all required to share the knowledge (Yunkaporta, 2019, p. 199).

In Australia the Indigenous knowledge from Aboriginal Elders is only made available if the group of Elders (or in some instances, one Elder) grants permission. Collecting data and gaining and understanding of Indigenous knowledge occurs through listening to the story telling of the Elders at workshops, interviews and fieldwork assessments accommodating Elders by 'walking the land'. The 'Indigenous Knowledge Generation Process' by Smylie et al. (2003) has been adopted and used during interviews in 2014 by the author, and the below diagram (Fig. 5.2) demonstrates the process for capturing local Aboriginal knowledge. The diagram represents the overall fundamentals of an approach that is participatory. These findings are thus empirical and evidence-based.

However, it must be noted that in the reporting of stories by interviewees, their view and interpretation are reflected. According to Riessman (1993), attending and recording an experience suggests that the interviewee selects what they notice based on their own interests and biases. Thus, the interviewee will construct realities of the observations at this level in a way that is most relevant to that individual. The narrative is then inevitably a representation of the self in a socio-spatial context (Riessman, 1993). Finally, the reading of the story will be the final representation of the experience as an output of the participatory process, and the reader will interpret the narrative.

Examples of this participatory model trace back many years, as it has been used by early researchers of Australian Aboriginal culture and history, including A.W. Howitt's *The Native Tribes of South East Australia* (1904) and the preparation of the *Lake Condah Heritage Management and Strategy Plan* (1993). More recently, examples include Cultural Management Plans such as *Planning for the future: Yawuru Cultural Management Plan* (2011), the *Bonye Buru Booburrgan Ngmmunge: Bunya Mountains Aboriginal Aspirations and Caring for Country Plan* (2010), the *Ngarrindjeri Nation Yarluwar-Ruwe Plan, Caring for Ngarrindjeri Sea Country and Culture* (2006), and the *Kooyang Sea Country Plan* (2004). In *Geelong's Changing Landscape: Ecology, Development and Conversation* (2019), Jones and Roös captured knowledge of the *Wadawurrung* through this participatory process. Elders Uncle Bryan Powell's and *Tandop* David Tournier's storytelling of the history are recorded and written down in Chap. 4 – *Welcome to Wadawurrung Country* (Powell and Tournier, in Jones & Roös, 2019, pp. 44–84). All of these examples demonstrate how participation, interviews and local knowledge can be conceptualised and then recorded in the format of a Western report and publication.

The question is how this participative model can be applied in gathering Indigenous knowledge (Aboriginal knowledge), resulting in its integration into the discourse of the design and planning of future sustainable developments and regenerative-adaptive environments. It is my belief that the fundamentals of the *regenerative-adaptive pattern language* are embedded in an interconnected participatory process, similar to the arguments by Alexander that the people of the place have in-depth knowledge of it that can be captured in a similar framework to that of *A Pattern Language* (Alexander et al., 1977).

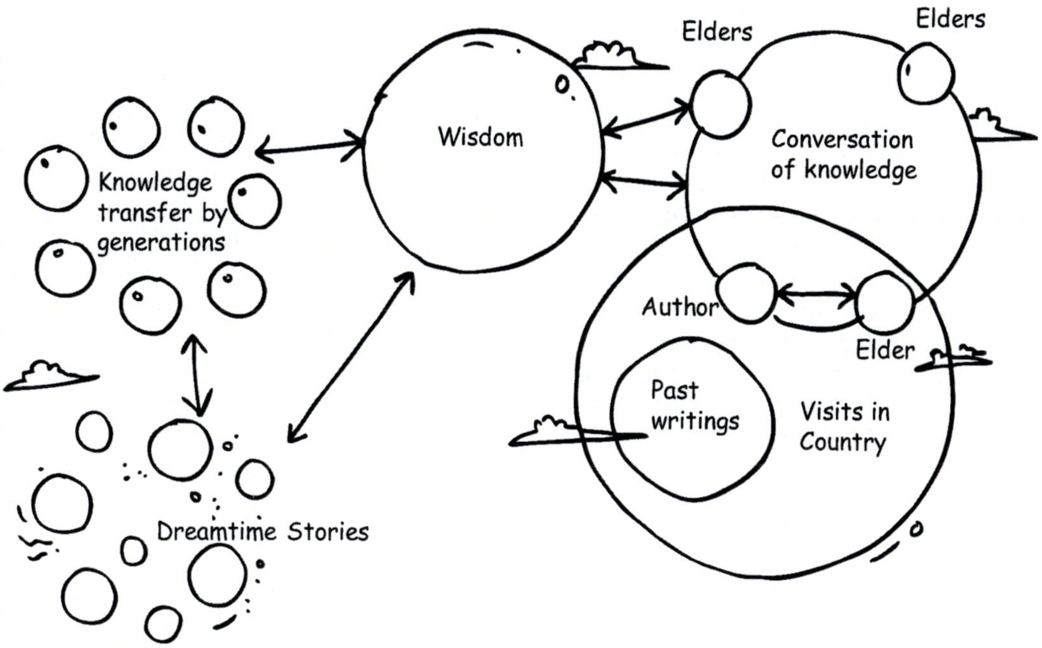

Fig. 5.2 Indigenous Knowledge Generation and Knowledge Capture Process (Drawing by Jesse Delmo, synthesised by Roös)

5.3 Connection to *Country*: Historical Context

Along the Australian coast, as well as in the harsh interior of this vast continent, for centuries Aboriginal people have been confronted with environmental, climate as well as annual seasonal weather variations. People living in settlements withstood harsh ecological, climate and geological changes; they believed that the natural phenomena of *Country*[3] are integrally connected with their spiritual and physical wellbeing (Roberts, 1975). '*Country*' is not only the physical landscape, but the term also embodies the tangible and intangible; it includes everything in time and space, the before, the now and after, below, here, the above and beyond, the realms of both day and night (Johnson & Jones, 2014). Changes in natural phenomena meant changes in a holistic context; changes to *Country* meant changes to the Aboriginal people themselves. Many of these changes have been captured in the cultural traditions of the people and the knowledge that has been transferred from generation to generation through *Dreaming* stories. Dreaming stories recount the forming of the main landscape *Country* and the coastal landscape *Sea Country*. The following example of one of these spiritual reflections of a Dreamtime story indicates the closeness to land and the importance of the changing environment. It relates the story of the creation of *Punjil* (also written *Bundjil*, *Bungil*, or *Bunjil*) and *Pallian* (Fig. 6.3), and the interconnection of nature and people:

> *Punjil* is the maker of earth, trees, animals, man and woman. *Punjil* had a wife named *Boi Boi*, but he never saw her face. She bore him two children, one a son named *Binbeal* and the other a daughter named *Kara-karook*. To *Binbeal* is committed the sovereignty of the heaven and to *Kara-karook* the incidental occurrences on earth; while great *Punjil* stalks like a big gentleman in the clouds, on the earth, and always carrying a big sword. *Pallian*, brother of *Punjil*, made all seas, rivers, creeks and waters, also all the fish in the oceans, seas and rivers. He governs the waters, was always in the waters, walking, bathing, and going over the seas. One day when our

[3] *Country* is a term used by the Aboriginal people of Australia to refer to the land to which they belong and the place of their dreaming (Smyth, 2004).

> ancestors awoke, *Punjil*, *Pallian* and *Karakarook* had gone up above. They had departed from *Deen Maar*, which remains sacred to our people to the present day (William Thomas [1969] cited in Smyth, 2004, p.6).

Additional to the spiritual connection to *Country*, Aboriginal people relied on their intimate knowledge of the environment, seasonal cycles and understanding of climatic events to ensure an adequate food supply, including natural medicines and protective resources (O'Connor & Prober, 2010). This knowledge was recorded over long-term and short-term cycles. Long term climate cycles were recorded in the two sequences; the *Mudong*, or life cycle that stretches for 12 years, and the *Garuwanga*, the Dreaming, a cycle that stretches between 12,000 to 20,000 years (Roös, 2014b; Kingsley, 2003) (Fig. 5.3).

The shorter-term cycles of annual changes were recorded in a 6- to 8-season cycle during the year. Various examples of this 6-season calendar are the recording of the seasons of the *Yawuru* Aboriginal people of north-west Australia in the Kimberly region, the *Wik Mungan* of north-east Australia in the Cape York area, the *Ngadju* of south-west Australia in the *Nullarbor* region, and the calendar of the *Gunditjmara* in south-west Victoria. All of these calendars reflect cultural rules and responsibilities about how they look after *Country*, animals and plants, and themselves. These rules align with their 6 seasons, and are very closely linked to the management and understanding of nature, as described in one example by Mati Gilbert:

> *Marrul* is after rain finish. *Wirralburu* is first southeast blow and the salmon start. *Barrgana* is proper cold time, when the *gardgu* tree is blooming. *Wirlburu* is coming hot. *Laja* is proper hot time. *Man-gala* is rain time, when we burn *gunggara* and bullock shit for mosquito. When the wattle flowers in *Barrgana* mean that the salmon and catfish are fat. When the wattle seeds, the mullet lay their eggs, and when the seeds open up, the eggs have hatched (Gilbert in Yawuru RNTBC, 2011, p. 81).

The use of the land and its resources, as well as the migration of the people and establishment of settlements, followed a specific climate cycle pattern, according to the six seasons as depicted in Table 5.1.

Fig. 5.3 *Bunjil* Shelter (Photo by Michael Barnett, 2012, CC BY-SA 3.0)

Table 5.1 Six Seasons of the *Yawuru, Wik Mungan, Ngadju, and Gunditjmara* (Derived from Roös, 2017, 2015)

Yawuru Seasons		
1	*Man-gala*	Wet season, Strong winds / cyclones. December to March.
2	*Marrul*	Changing season. Little wind and rain. April to May.
3	*Wirralburu*	Cooling season. Strong winds no rain. May to June.
4	*Barrgana*	Cold season. Dry winds and dust storms. June to August.
5	*Wirlburu*	Warming season. Westerly winds. September to October.
6	*Laja*	Hot. Built up to wet season. Late September to November.
Wik Mungan Seasons		
1	*Wurrgeng*	Dry and cold weather season. Early June to mid-August.
2	*Gurrung*	Hot dry weather season. August to late September.
3	*Gunumeleng*	Pre-monsoon storm season. Early October to December.
4	*Gudjewg*	Monsoon season. December to March.
5	*Bang Gerreng*	Storm season. Late March to end April.
6	*Yegge*	Humid cooler season. Late April to early June.
Ngadju Seasons		
1	*Ngarnngi*	Hot time, long season. September to March.
2	*Ngawu*	Egging season. September to October.
3	*Nganji*	Hot season. November to March.
4	*Kaluru*	Cold time, long season. April to August.
5	*Kupilya ngarrin*	Sleeping season. April to June.
6	*Karrikunja*	Courting and mating season. July to August.
Gunditjmara Seasons		
1	*Kooyong/Koorang*	Season of Eels. Late Jan to late March.
2	*Gwangal moron*	Season of honey bees. March to May.
3	*Chinnup*	Season of cockatoos. June to late July.
4	*Larneuk*	Season of nesting birds. Late July to August.
5	*Petyan*	Season of wildflowers. Sept to mid-November.
6	*Balalambar*	Season of butterflies. Mid November to late Jan.

5.3.1 Intimate Relationships with the Land

A typical reflection of living with nature and having a deep understanding and intimate relationship with the land is mentioned in the introduction to the *Yawuru* Cultural Management Plan:

> For thousands of years our ancestors have lived along the foreshores of Roebuck Bay, across the pindan plains, as far inland as the *Walan-garr*, the Edgar Ranges, and along the fringes of the Great Sandy Desert. Our *Country* is land and sea moulded by the cycle of seasonal change. We live by the seasons, reading the signs to know when and where we should go to harvest the riches of our country. When the first south-east winds start to blow across Roebuck Bay, bringing the colder weather, we head to the beaches and tidal creeks to fish for salmon. When the weather warms and the winds blow from the west, we head for the reefs to fish for bluebone and the myriads of other reef fish (Yawuru RNTBC, 2011, pp. 28-29).

The establishment of camps was based upon the knowledge and traditions passed down from generation to generation from ancestors (Elkin, 1977). Knowledge of the environment was passed on from Elders to young children when they went hunting, fishing, camping and engaged in discussion during gatherings. Understanding the subtle changes in *Country* and following the seasons was used as a guide to where to gather food, establish settlements and how and when to look after the resources of the land.

In *Yawuru Country* the migration of the Aboriginals between the *nagula* (coastal regions) and the *birra* (inland regions) is clearly aligned with the seasons and the changes in plants characteristics, migration of animals, birds and the seasons of different species of fish in the area. *Yawuru* peoples' habitats are aligned with the specific vegetation, geography and sources of food and shelter. The different groups of the *Yawuru* people are linked to the habitats of the *Yawuru* country, and can be identified with camps in the various *Yawuru* habitats, such as *Niyamarri* (Sand Dunes), *Bilarra* (Wetlands), *Warnangarri* (Rocky country), and *Jila* (Permanent freshwater source).

Another example of Indigenous people's intimate ties to land (to *Country*), and migration according to climatic phenomena and patterns are told in the Dreamtime story that relates to *Tjibruke*, a *Kaurna* ancestral being who taught lessons about life and respect and whose tears for his slain nephew created the fresh water springs along the Adelaide coastline from Kingston Park to Cape Jervis. The cultural landscape has been described in detail by Aboriginal Elder Aunty Veronica Brodie, who maps out Aboriginal representation of the landscape including human migration patterns according to climate. This includes sacred and significant *Kaurna* sites for which the *Kaurna* Elders are custodians, and according to belief should never have been revealed (Jones, 2010). Aunty Veronica Brodie describes the landscape and use of the landscape as follows:

> *Grandmother* told us of many, many camps that were sited from Glanville all the way to Outer Harbour, and that's quite a big stretch; and .... They wintered in the timbered foothill country, sheltering in the cover of gigantic and often hollow red gum trees along the riverbanks. Often as a basis for shelter they used a fallen tree, and hollowed out beneath it, a practice hazardous when floods and heavy rains destroyed their fires and flooded their sleeping places (Veronica Brodie, in Wood, 2007).

Both examples reflect the adaptive capacity and flexibility of the Indigenous people according to nature's cycles and environment, interconnectedness with the environment and the thorough understanding of the seasonal changes.

5.3.2 Climatic Seasons and Settlements

This knowledge of seasonal changes and longer-term climate cycles and events and the locations of settlements and migration patterns correspond to Memmott's hypothesis (2007) that climate influences Aboriginal settlements and the ethno-architecture of Aboriginal people across Australia. Settlements in different regions had specific shelter-type constructions, specifically created to adapt to the local weather conditions. Analysing the influence of climate upon shelter building and settlement types across the continent generated a classification of shelter type distribution (Memmott, 2007, pp. 7–12). This classification can broadly be summarised as follows:

- Open winter windbreaks – whole continent;
- Summer wet-weather enclosed shelters – central and northern;
- Cold wet-weather enclosed shelters – southern; and
- Shades for diurnal use – whole continent (Memmott, 2007).

This occurrence of connection to the land and seasonal relationships is evident through all Aboriginal tribes in Australia. It was initially recorded by anthropologist Thomson and considered a quantum leap in the theoretical understanding of Australian Aboriginals by anthropologists (Memmott, 2007). In his paper 'The Seasonal Factor in Human Culture Illustrated from the life of a Contemporary Nomadic Group', Thomson (1939) demonstrated the integrated role of hunter-gatherer lifestyles with seasons and coastal habitat changes as part of ethno-architecture. Thomson focused upon the coastal clan of the Cape Keerweer[4] region of the *Wik Mungan* people and considered the shelter types that had been constructed according to their cycle of six seasons. Seasonal shelter types and settlement patterns are described and summarized as what Thomson called 'The Seasonal Factor' (Thomson, 1939).

[4] Cape Keerweer is an area located on the Western coastal area of Cape York, the northernmost point of Australia.

In explaining 'The Seasonal Factor', Thomson was able to demonstrate that the seasonal year of the *Wik Mungan* and other groups was divided into specific and discrete numbers of climatic periods (Thomson, 1949; Thomson, 1939). These could be mapped against typical settlement patterns, foods, geographic focus, shelter types as well as domiciliary behaviour (Memmott, 2007). He concluded that migration movements and settlement patterns followed food sources, and shelter types corresponded to the seasons. In the selected case study area for this book that represents a more temperate climate of the Great Ocean Road region of Victoria, the Aboriginal knowledge of the *Wadawurrung* is further explored in the context of migration and settlement patterns in this chapter.

5.4 The *Gadubanud* and the *Wadawurrung*

Around 40,000 to 50,000 years before present (BP), most parts of the Australian continent had been settled by the Aboriginal people (Frankel, 1991). This settlement included the southern parts of the continent, including Victoria. The climatic changes during the *Last Glacial Maximum* (LGM) (around 26,500 BP to 19,000 BP), and in the late Pleistocene and the warmer Holocene (12,000 BP to Present) had an effect on the location, timing and reasons for Aboriginal people occupying different parts of the land in Victoria (Marshall & MacCulloch, 2012). In the case study area of the Great Ocean Road Region, the *Gadubanud* and *Wadawurrung* (Wathaurong) occupied both the coast and the hinterland during the Holocene era, and the most prevalent artefact type of the small tool tradition of the last 4000 years was identified in this southern region of Victoria (Coutts, 1978; Frankel, 1991).

Observations of the Aboriginal people of Victoria began as early as 1802, and the recording of ethno-historical texts increased after permanent European colonisation in 1835. The primary sources of Aboriginal history in Victoria have been recorded by Clark (1990) with his reconstruction of the Aboriginal traditional language boundaries in Victoria. The sources include journal entries and government correspondence produced by other explorers such as Matthew Flinders, as well as settlers and missionaries, including G.A. Robinson (the Chief Aboriginal Protector), as well as records by William Buckley (Marshall & MacCulloch, 2012).

The *Gadubanud* and the *Wadawurrung* resided, and still reside in the south-western regions of Victoria. Today, they still respect the land, and understanding the benefits of the coastal zone, as well as in an effort to protect the connection between human habitats and engagement with nature, they have developed the *Kooyang Sea Country Plan* (Smyth, 2004). As part of the *South-East Regional Marine Plan* (NOO, 2004), two *Sea Country Plans* initiated collaboration with coastal Indigenous groups to provide an opportunity for Indigenous involvement in natural resource use and planning of the coastal and marine environment, as well as supporting coastal Aboriginal cultural heritage conservation (Roös, 2014a, b, p. 668).

5.4.1 The Gadubanud People

The *Gadubanud* resided at various settlement nodes throughout the Otway region. Very few records exist of the *Gadubanud*, but although they were rarely recorded beyond the Otway region, the *Gadubanud* maintained complex relationships with other neighbouring Aboriginal groups, such as the *Gulidjan* of the Lake Colac region and the *Wadawurrung* of the Port Phillip Bay region. Through these relationships their history can be recorded (Mulvaney, 1961). Chief Protector Robinson first recorded their presence in 1842 when he met three *Gadubanud* at the mouth of the Hopkins River near Warrnambool, and received details of four clans that resided on the western edge of the Otway Ranges. Clark indicated (Fig. 5.4) that a pattern of semi-nomadic movements of the *Gadubanud* linked the

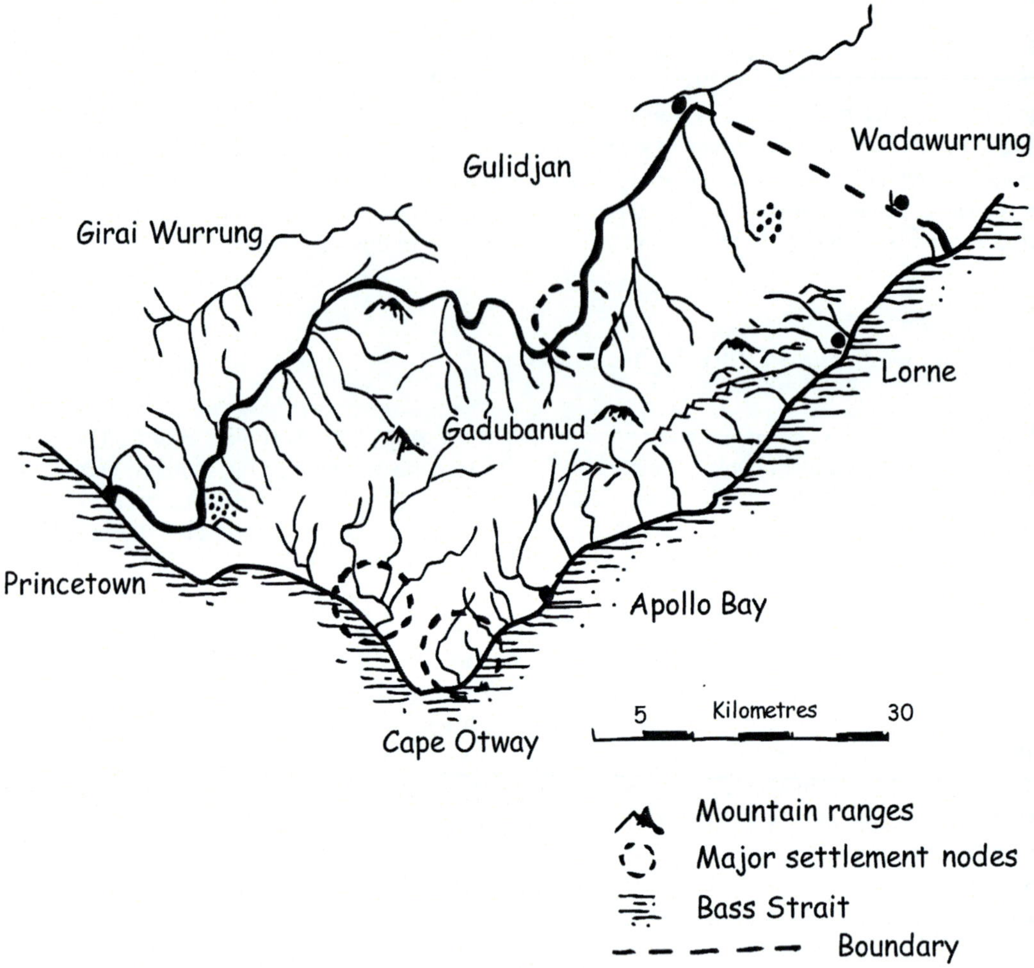

Fig. 5.4 The Country of the *Gadubanud* people indicating locations of clans and major settlement nodes (Drawing by Jesse Delmo, derived from Roös, 2017; adapted from Clark 1990)

midlands, estuarine and coastal settlement sites that were known for their high food yields and provided food supply throughout the year according to the seasons.

The land-based economy of the *Gadubanud* was predominantly focused toward the gathering and cultivating of food that would sustain their different clans. Settlement distribution of the *Gadubanud* was focused on the availability of regional food resources, and according to Niewójt, an investigation of specific locations of regional food sources helped to confirm the areas of settlement concentrations at the Aire River estuary, Cape Otway and the wetlands along the Barwon River (Niewójt, 2013). The recordings of *Gadubanud* camps by Robinson, Buckley and Armytage correlate with these concentrations, and significant archaeological deposits in the same nodes support these findings (Richards, 1998).

5.4.2 *The Wadawurrung People*

According to Clark (1990), the *Wadawurrung* (also referred to as *Wathaurong*) were a populous tribe with over 25 clans that occupied a territory of about 120 x 150 km, from the Otway ranges in the west to the northern inland woodlands of the central Victorian dividing range and Fiery and Mt. Emu

Creeks, down to the east Werribee plains and down south the coastal areas of the Bellarine Peninsula, along the Surf Coast down to Aireys Inlet. The *Wadawurrung* is closely related to the language-culture groups of the *Daung Wurrung*, *Bun Wurrung*, *Woi Wurrung*, *Djab Wurrung*, and *Dadja Wurrung*, which form the *Kulin* Nation[5] and consisted of more than 25 Clans (Marshall & MacCulloch, 2012). As with all *Kulin* peoples, the *Wadawurrung* had a patrilineal kinship system, in which a person's skin-group (otherwise known as a moiety) was inherited from their father. Clans belonged to either the *Bungil* (eagle-hawk) or the *Waa* (crow) moiety (CoGG, 2014; Richards, 1998; Clark, 1990).

The area of land over which the *Wadawurrung* hunted and gathered food has been described by Clark as 'the range':

> .... the tract or stretch of country identifiable as the economic range, normally included the estate and was thus owned by clans [of the Wathaurong]. The band seasonally occupied and utilised various parts of the range in a settlement pattern that was a response to the group's habitat (Clark, 1990, pp. 4-5).

5.5 The *Wadawurrung* and the Coast

> *Wadawurrung* travelled extensively in Country, moving with the seasons and following food resources. They moved from inland and the mountains across the plains to the sea, then back inland again. Temporary bark shelters were used for camping, but our people always returned to their permanent settlements with houses made of rock walls and mud roofs covered with bower spinach[6] and other herbs. (Aboriginal Elder *Tandop* David Tournier [dec.], personal communication, 2014, in Roös, 2017, p. 98)

The *Wadawurrung* had an affinity with the coast, and their cultural lands stretched along the coastline from the Werribee River to the Lorne peninsula and traversed inland in a north-westerly direction (WAC, 2014). *Wadawurrung Country* includes the '*Sea Country*', which is the sea and the coastal strip along the coast that includes adjoining estuaries, beaches, coastal waters (up to about 80 nautical miles offshore), rivers and ocean. The *Wadawurrung* make no distinction between land and sea. They see themselves as having responsibilities and custodianship across the land and sea boundaries that have been put in place over many thousands of years.

Sea Country is based on the relationship between the *Wadawurrung* and the *Sea Country* that goes back to Creation. The wetlands, estuaries, rivers, lakes, and the sea have sustained the people culturally and economically for tens of thousands of years. Ownership and rights of *Country* and *Sea Country* formed part of a complex structure that embraced all aspects of social, legal, economic and religious life. This structure was underpinned by the Aboriginal concept of Dreaming. The Dreaming stories of the Wadawurrung are similar to the other Aboriginal peoples that resided around Port Phillip Bay. Aboriginal Elder *Tandop* David Tournier noted that 'the Dreaming of *Bunjil* is very important for the *Wadawurrung*, as part of the larger *Kulin* Nation'.

[5] The *Kulin* Nation consisted of five language groups that resided around the Port Phillip Bay area and up to the Yarra River (Marshall & MacCulloch, 2012).

[6] Bower Spinach (*Tetragonia implexicoma*) is found mainly in the coastal regions of New Zealand and South Australia. It occupies habitats of coastal dunes, shrub land, coastal woodland and grassland. The succulent leaves and orange-red berries was a significant food source for the Aboriginal people of Tasmania and Victoria (DEPI, 2014).

5.5.1 Migrations to and from the Coast

Aboriginal people's migration from the coast to inland was evident due to past marine transgressions that have dramatically affected the landscape and Aboriginal settlement along Australian coastlines. These coastlines have varied significantly as a result of the broader climatic and sea level changes during the last glacial maximum and current sea levels (Marshall & McCulloch, 2012). The sea level changed dramatically during the last glacial period from about 75,000 years ago, and was at its lowest at the height of the last glacial maximum (LGM) at 18,000 years BP. The Bassian Plain linked the mainland and Tasmania during this time and the coastline followed the continental shelf, which was thousands of kilometres away (Marshall & MacCulloch, 2012, p. 69). As noted by Bird:

> ... about 18,000 years ago when the glaciers and ice sheets reached their maximum extent, the sea was 120 to 140 metres below its present level, and the continental shelves were exposed as wide coastal plains... The warming of the climate and the diminution of the glaciers and ice sheets caused a rise in the sea level which continued until the mid-Holocene about 6000 years ago; an event known as the late Quaternary marine transgression (Bird, 1993, pp. 8-9).

These changes in coastlines directly influenced the migration of Aboriginal people, including the peoples of the *Kulin* Nation. The migration of Aboriginal settlements inland due to rising seawaters can be traced and mapped through the occurrence of midden sites and their locations (Marshall, 1997).

The Dreaming stories of the neighbouring *Boon Wurrung* people reflect knowledge of changing sea levels and climatic events along the Victorian coastline as per the progressive changes indicated in the records of Marshall & McCulloch (2012). This knowledge is represented in the following extract of the account from *Boon Wurrung* Elder, Ms. Carolyn Briggs:

> Many years ago this land that we now call Melbourne extended right out to the ocean. Port Phillip Bay was then a large flat plain where *Boon Wurrung* hunted kangaroos and cultivated their yam daisy. But one day there came a time of chaos and crises. The *Boon Wurrung* and the other *Kulin* nations were in conflict. They argued and fought. They neglected their children. They neglected their land. The native yam was neglected. The animals were killed but not always eaten. The fish were caught during their spawning season. As this chaos grew the sea became angry and began to rise until it covered their plain and threatened to flood the whole of their country.
>
> The people went to *Bunjil,* their creator and spiritual leader. They asked *Bunjil to* stop the sea from rising. *Bunjil* told his people that they would have to change their ways if they wanted to save their land. The people thought about what they had been doing and made a promise to follow *Bunjil. Bunjil* walked out to the sea, raised his spear and directed the sea to stop rising. *Bunjil* then made the *Boon Wurrung* promise that they would respect the laws (Briggs, cited in Low Choy & Jones, 2013, p. 183)

5.5.2 *Wadawurrung* Migrations

The *Wadawurrung* settlement patterns followed food sources, and clans occupied a specific range over which they moved for gathering food, trading and fulfilling their social obligations. As indicated by Aboriginal Elder *Tandop* David Tournier at the introduction of this section, movement of the clans were very much connected to the seasons, and the food associated with these. *Wadawurrung Country* was also a very important area that included places of gathering of various Aboriginal groups from the Victoria region and beyond. Special gatherings and meetings occurred at special sites near the towns now called Queenscliff and Geelong. Clans from the *Yari Yari*, *Bangerang*, *Baraba Baraba*, *Wiradjuri*, as well as the other tribes of the *Kulin* Nation would have navigated with canoes along the rivers down to Port Phillip Bay, crossing the Bay and eventually meeting for weeks at the special sites to settle disputes, trade, arrange marriages, heal the sick and conclude with song and dance.[7]

[7] Information recorded about the settlement patterns, customs and social aspects of the *Waduwurrung* was done through a workshop, one on one discussions and special site visits with accompanying late Aboriginal Elder *Tandop* David Tournier from the Wathaurong Aboriginal Co-operative in 2014.

There were several semi-permanent settlement nodes within *Wadawurrung Country*, especially where food sources were available, trade could happen between tribes, and special places for ceremonies. People would go back to these nodes every second or third year, following the seasons in a cycle. The many layers of middens on top of each other at these sites indicate that people were there for a very long time, for thousands of years coming back and using the same place. More permanent campsites were established inland in areas such as the You Yangs.[8] These structures were made of stone and mud, with sticks on top woven together, closed with bark and mud on top. The mud roofs allowed for planting of herbs and plants to provide food, such as the Bower Spinach (*Tetragonia implexicoma*). On the coast, the shelters were not permanent, and small campsites included shelters made of bark, sticks, and in many cases just a placement in a half circle of brushes from the close by bush and shrubs, to protect against the wind. Movement of the people along the coast happened during summer months, and because it was warm with very little rain, it was not necessary to build shelters.

Main settlement nodes and coastal migration followed the seasons and food sources (bush tucker) and included the following as indicated in Table 5.2:

5.6 Patterns of Nature and *Wadawurrung* Settlements

By all accounts it is evident that the *Wadawurrung* moved across *Country* in specific patterns and settled with the changes according to the cycles of nature. Their seasonal cycles and calendar included various changes that signalled the construction of temporary and semi-permanent shelters, the management of the landscape through fire, the harvesting of plant roots, and migration along coastal areas and inland according to the patterns of the seasonal cycles. Figure 5.5 indicates the potential migration of the *Wadawurrung* across *Wadawurrung Country*, following the seasons in a cultural landscape.

Following a description of the seasons, the six-season cycle calendar offers a possible insight into *Wadawurrung* migration patterns across *Country* and can be described as follows[9] (Tournier, personal communication, 2014).

- **Early and Mid-Summer (November, December, January)**
  The Black Wattle tree and Swamp Paperbark flower in late November, and it is a sign for the beginning of the dry season, setting up camps around reliable water sources at creeks, rivers and billabongs. Due to the dry season, water bodies attract wildlife and Kangaroos, Emus, Wombats and Possums that come to drink and are caught. Grasses flower, fruits ripen, and manna is collected from the Coast Manna-gum trees. Fish move up the rivers, and fish traps are used. During this time of abundance, large gatherings of the clans and tribes take place. With permission from the Elders of the coastal clans, people move down to the sea and coastal wetlands and river mouths to gather shellfish, to catch Flatheads fish and Black fish. As the male eels start to move down the rivers to the sea, it signals the start of the next season - Late Summer.

- **Late Summer (February to March)**
  The early autumn rains arrive and the days become cooler. Various parts of the land are densely grown with scrub and grass, and by planned and carefully managed burning, areas are cleared to

[8] The You Yangs are a series of granite ridges that stretch for about 24 kilometres and rise to 364 m high at the Flinders Peak. Much of the ranges are protected and located in the You Yangs Regional Park (Parks Victoria, 2014).

[9] The *Wadawurrung* six-season cycle descriptions are combinations adopted from the *Boon Wurrung* seasons (Gott, 2011), and the *Gunditjmara* seasons (Jones, 2010), as well as information provided during the discussions directly with the late Aboriginal Elder *Tandop* David Tournier in 2014.

Table 5.2 *Wadawurrung* Settlement Areas and Bush Tucker (Food Sources) (Tournier, personal communication, 2014)

Open Areas and Plains	Season	High Ground and Mountains	Season
1. Kangaroo (*Goim*)	All year	1. Wombat (*Gnor-Gnor*)	All year
2. Fungi/mushrooms (*Tubar*)	Autumn/winter	2. Witchety grubs (*Wooeekam*)	Late summer
3. Grass flowers	Spring/all year	3. Tree exudates	All year
4. Lerp and manna (*Laap*)	Late summer	4. Fungi/mushrooms (*Tubar*)	Autumn/winter
5. Honey ants	After rain	5. Manna (*Laap*) - gum tree	Autumn/winter
6. Salt bush	All year		
Waterways, riverbanks, wetlands	**Season**	**Coast and river mouths, Sandhills**	**Season**
1. Frogs (*Djerrm*)	Summer	1. Lizards (*Gaan*)	Summer
2. Snakes (*Kangalang*)	Summer	2. Snakes (*Kangalang*)	Summer
3. Lizards (*Gaan*)	Spring/summer	3. Witchety grubs (*Wooeekam*)	Spring/summer
4. Ducks and swans	All year	4. Eggs (including turtles)	Early spring
5. Witchety grubs (*Wooeekam*)	Spring/summer	5. Eels (*Kooyang*)	Late summer
6. Tree exudates	All year	6. Salt bush	All year
7. Eggs (including turtles)	Early spring	7. Oyster (*Barrabool*)	Spring/summer
8. Eels (*Kooyang*)	Late summer	8. Mussel (*Moorabool*)	Spring/summer
9. Grass flowers	Spring/all year	9. Flathead fish	Spring/summer
10. Yam daisy (*Murnong*)	Spring/summer	10. Black fish	Spring/summer
Mallee	**Season**	**Bush and scrub**	**Season**
1. Emu (*Karwir*)	All year	1. Possums (*Wallart*)	All year
2. Witchety grubs (*Wooeekam*)	Spring/summer	2. Lizards (*Gaan*)	Summer
3. Tree exudates	All year	3. Snakes (*Kangalang*)	Summer
4. Lerp and manna (*Laap*)	Summer/winter	4. Witchety grubs (*Wooeekam*)	Spring/summer
5. Mallee fowls / Turkey	All year	5. Tree exudates	All year
6. Grass flowers	Spring/all year	6. Lerp and manna (*Laap*)	Summer/winter
		7. Gallahs/pigeons/birds	All year

The Wadawurrung bush tucker (food sources) list was provided during the interviews in 2014 by late Aboriginal Elder - Tandop David Tournier

fertilise the soil before the rains of autumn starts. During this time the Stringybark, Moonah and Banksia trees come into blossom, providing sweet nectar that attracts many insects and birds. The saltbush on the coastal dunes provides fruits for mixing into meals. The Eel Season starts with the migration of large quantities of the female Eels down the rivers to the sea, to take up their long journey up the east coast of Australia to the South Pacific. Eel traps are set, and the *Wadawurrung* clans set up camps along the rivers. The roots of water plants are part of a vegetable and fruit diet that includes the berries from late summer fruits. As birds start to flock to migrate to the warmer north, more frequent heavy rains start to signal the start of early winter.

- **Autumn/Early Winter (April to May)**
  The soil is still warm after the rains, and the mushrooms sprout up in circles at Freshwater Creek.[10] Large groups of women, young and the Elders sit inside the circles and harvest the mushrooms, while talking about their travels, confirming between them the Dreaming stories of the past and future, endorsing the importance of caring for Country. The billabongs start to fill with water after more frequent rains. Moths emerge; native bees swarm and move across country collecting nectar from the Flax-lily grass and Cranberry Heath. The clans start to move inland, to higher grounds as the rain of winter start to settle in.

[10] The location of the mushroom circles is in the region of Freshwater Creek. During the visit in 2014 to this special site by the Author and late Aboriginal Elder *Tandop* David Tournier, the importance of the formation of the circles and centres were discussed, highlighting the structure of order in social engagement and the connection to nature's forces. When the elderly and young women sit in a circle, there are no levels of dominance, everyone is at an equal level, and hierarchy is established by the respect for the Elders. The centre acts as a natural phenomenon of gravity and wholeness. This also reflects the *Wadawurrung* peoples beliefs that the position of the Aboriginal person in nature is that of oneness with *Country*.

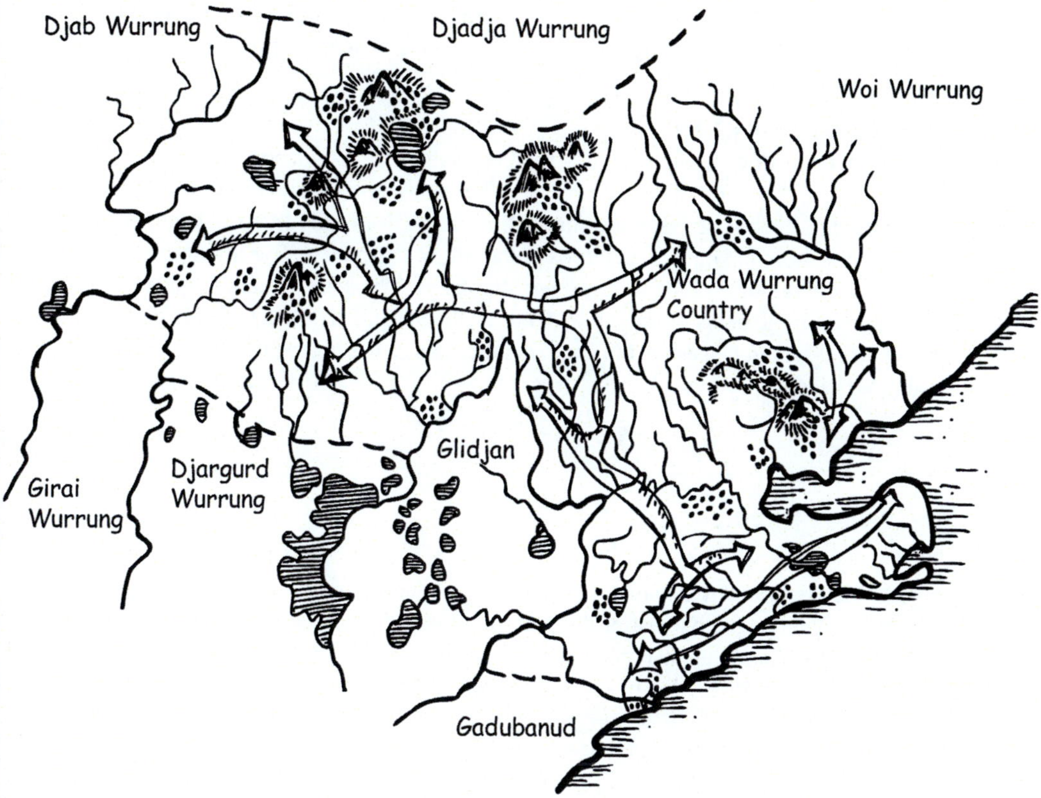

Fig. 5.5 Cultural landscape migration relationship of Wadawurrung Country (Drawing by Jesse Delmo, derived from Roös, 2017)

- **Deep Winter (June to mid-July)**
  Winter rains result in the flats near the rivers and creeks to be frequently flooded, low lands and salt marshes are mostly wet, unsuitable for camping and people move uplands to higher sheltered spots. Possums, koalas, wombats and Grey Kangaroos are also seeking shelter in the higher grounds, and people caught them close to the campsites. Herbs and the Bower Spinach that grows on top of mud roofs of the semi-permanent rock wall shelters are convenient for food supply while people stay inside the shelters and keep fires burning for warmth. Possum skins stitched together are wrapped around family members to protect against the intense cold spells that occur as the westerly winds rip through the country ahead of the cold fronts.

- **Early Spring (Mid July to August)**
  The Silver Wattle and the Coast Beard-heath start to flower, the first signs of a change in season, that winter is coming to an end and that warmer weather is on its way. People move towards the lower lands and plains, wetlands and lakes where they catch ducks, wildfowl and other birds. Eggs are in abundance and are collected from nests, including from turtles along the beach. While the men hunt, the woman collected large quantities of *Murnong*[11] in string bags to take back to the

[11] *Murnong* was a favored food of the Aboriginal people in Victoria. This is a small plant with a yellow flower that looks like a daisy, and has an underground tuber, Similar to a yam. It grows all year round, flowers in spring, and is particularly good to eat in spring (Clark & Kostamski, 2006, p. 20).

camps. Temporary bark shelters are constructed in camp clusters along the watercourses, dry wood is collected and fires are kept burning during the cold nights to keep warm.

- **Spring (September to October)**
  Murnong continues to flower, and provides root vegetables in abundance. Water lilies, orchards, sea berry, saltbush and bower spinach flowers and large amounts of greens are used in food. This is a time of abundance, trees, grasses and shrubs flower and bees and insects collect nectar. Snakes and lizards become active, young Kangaroos come out of the pouch. The floodplains are flooded and tadpoles appear in ponds, billabongs, rivers and lakes. Traps are set and tadpoles, snakes and lizards are roasted on the coals of the circle ovens in the ground. The different clans start to move towards the special sacred places for the annual gatherings of the tribes that will happen during the coming summer season.

5.7 Conclusion Statement

This chapter recorded the longer-term history of human and landscape interaction of the land through investigating the Indigenous knowledge of the past, and the local Aboriginal history of the region of the state of Victoria, in Australia. It is noticeable from this chapter that the Aboriginal peoples of Australia, as well as the *Wadawurrung* of Victoria, retain a wide knowledge of their seasons, seasonal activities and long-term cycles of climatic events. Further the intimate relationship to the environment result in an *integral interconnectedness* between humans and nature. This knowledge provides important opportunities to explore and discuss further with Aboriginal peoples their aspirations for working with nature to adapt with climate change, as well as sharing their knowledge with non-Aboriginals about *Country* and *Sea Country*, landscape management, the understanding of nature's patterns, as well as opportunities to assist in the design and planning of regenerative-adaptive futures. Additionally, Indigenous knowledge contribute to the appreciation of place (Roös, 2014a, p.671).

Therefore:

Deep cultural and spiritual connections with nature exists within the bioregions of the lands; the connection with *Country* is evident in the *Dreamtime Stories* of the *Wadawurrung* and other Aboriginal peoples across Australia. This is similar for all Indigenous people's knowledge across the world. Not necessarily obvious and evident to European occupants, but certainly these storylines and knowledge are engraved within the living structures of the land, constantly present in its evolutionary process. This *advanced Indigenous knowledge* need to be investigated, recorded and shared to assist in the further understanding of a deeper human-nature relationship, and the interconnectedness of humans and nature.

5.8 Fundamental Pattern 5

ADVANCED INDIGENOUS KNOWLEDGE [5].

When any human settlement is considered, first obtain from the local Indigenous custodians a deep knowledge of the land and the longer-term history of human and landscape interaction. Indigenous knowledge of seasons, seasonal activities, long-term cycles of natural and climatic events will provide insight to Nature's long-term patterns for adaptation planning (Fig. 5.6).

Fig. 5.6 Sketch of Fundamental Pattern – ADVANCED INDIGENOUS KNOWLEDGE [5] (Drawing by Jesse Delmo, synthesised by Roös)

Downward links:
Use this fundamental pattern to help consider how the Indigenous knowledge of the land can be applied to settlement planning in the future, which will respect the environment. This deep understanding of human-nature connection will help us to comprehend that humans are part of nature, accentuating our deep innate sense of the importance of Biophilia – LOVE FOR NATURE [6] – …

Acknowledgement of Country I acknowledge the Traditional Owners of the lands and waters that we live and work on across Australia and pay my respect to Elders past, present and future. I recognise that Aboriginal and Torres Strait Islander peoples have made and will continue to make extraordinary contributions to all aspects of Australian life including culture, economy and science.

I specifically wish to respectfully acknowledge and pay my respects to the *Wadawurrung* peoples and their Elders, past, present and future, the Traditional Custodians of the lands and waters upon which my research and writing of this book focuses and has been concluded. I duly acknowledge the rich cultural and intrinsic connections they have in their *Country*. I also recognise and acknowledge the contribution and interest of other Aboriginal peoples, Traditional Owners and organisations in the management of their lands, waters and natural resources in this region.

Disclaimer All descriptions, opinions and comments raised in this chapter by the author on Indigenous knowledge systems are the results from the author's research and interest in the subject matter. The author by no means claims expertise in Indigenous knowledge. It is the author's opinion that expertise in Indigenous knowledge are only invested and claimable by the Indigenous people themselves.

References

Alexander, C. (1964). *Notes on the synthesis of form*. Cambridge, MA: Harvard University Press.

Alexander, C., Ishikawa, S., Silverstein, M., Jacobson, M., Fiksdahl-King, I., & Angel, S. (1977). *A pattern language: Towns, buildings, construction*. New York, NY: Oxford University Press.

Berkes, F. (1998). *Sacred ecology: Traditional ecological knowledge and resource management*. Philadelphia and London: Taylor and Francis.

Bird, E. C. F. (1993). *The coast of Victoria: The shaping of scenery*. Melbourne, Australia: Melbourne University Press.

Brodie, V. (1994). *Kudlyo the black swan dreaming: Veronica Brodie and the continuity of Kaurna history at Glanville and Lefevre peninsula*. Adelaide: City of Port Adelaide.

Clark, I., & Kostanski, L. (2006). *An indigenous history of Stonnington. University of Ballarat, School of Business*. Stonnington: City of Stonnington.

Clark, I. D. (1990). Aboriginal languages and clans: An historical atlas of Western and Central Victoria, 1800-1900. *Geography, 37*.

COGG. (2014). *Kareenga Aboriginal Action Plan*. City of greater Geelong and aboriginal affairs Victoria, Community Development Unit. Geelong: City of Greater Geelong.

Coutts, P. (1978). *Aboriginal engineers of the Western District, Victoria*. Melbourne, Australia: Ministry for Conservation.

DEPI. (2014). *Department of Environment and Primary Industries - Victorian Sources Online*. Retrieved 11 July 2014 from salinity Indicator plants: http://www.vro.depi.vic.gov.au

Elkin, A. P. (1977). *Aboriginal men of high degree* (2nd ed.). St Lucia, Australia: University of Queensland Press.

Frankel, D. (1991). *Remains to be seen: Archaeological insights into Australian prehistory* (Vol. 1). Melbourne, Australia: Pearson Education Australia.

Gott, B. (2011). *Seasonal Calendars for the Melbourne Area*. School of Biological Sciences, Monash University. Retrieved 18 August 2014 from http://www.herringisland.org/seasons1.htm

Howitt, A. W. (1904). *The native tribes of south-East Australia*. London: Macmillan.

Johnson, I., & Jones, D. (2014). Incorporating indigenous Australian knowledge and perspectives into planning practice. In *UHPH 14, 12th Australasian urban history planning history conference 1* (pp. 341–360). Wellington: Victoria University of Wellington.

Jones, D. S. (2010). Kaurna cultural heritage. In C. B. Daniels (Ed.), *Adelaide: Water of a city* (pp. 132–145). Kent Town, Australia: Wakefield Press.

Kingsley, D. (2003, August 14). *The Lost Seasons*. Retrieved 8November 2013 from ABC science: http://www.abc.net.au/science/

Low Choy, D., & Jones, D. (2013). Words, Oral libraries and environmental responses: An Australian indigenous perspective to climate change adaptation. In *IFLA50: Shared wisdom in an age of change* (pp. 176–185). Wellington: International Federation of Landscape Architects and The New Zealand Institute for Landscape Architects.

Markwell, K. (2010). *Bonye Buru Booburrgan Ngmmunge: Bunya Mountains and aboriginal aspirations and caring for country plan*. Beenleigh, Queensland: Markwell and Associates.

Marshall, B., & MacCulloch, J. (2012). *Aboriginal cultural heritage conservation manual: GORCC crown land reserves between Torquay and Lorne*. Melbourne, Australia: Terra Culture.

Marshall, B. (1997). *Archaelogical Investigations of an Aboriginal Shel Midden at point Roadknight*. Anglesea foreshore committee of management. Anglesea: Anglesea Foreshore Committee of Management.

Memmott, P. (2007). *Gunyah goondie and Wurley: The aboriginal architecture of Australia*. St Lucia, Australia: University of Queensland Press.

Mulvaney, D. (1961). Archaeological excavations on the Aire River, Otway peninsula, Victori. *Proceedings of the Royal Society of Victoria, 75*, 1–15.

Ngarrindjeri Tendi. (2006). *Ngarrindjeri Nation Yarluwar-Ruwe Plan: Caring for Ngarrindjeri Sea Country and Culture*. The Ngarrindjeri Heritage Committee, Ngarrindjeri Native Title Management Committee, Meningie, SA: Ngarrindjeri Land and Progress Association.

Niewójt, L. (2013). Gadubanud society in the Otway ranges, Victoria: An environmental history. *Aboriginal History, 33*, 175–199.

NOO. (2004). South-east regional marine plan, Implementing Australia's Oceans Policy in the South-east Marine Region. National Oceans Office (NOO), Australian Government, Hobart. https://www.environment.gov.au/system/files/resources/83f69107-6e52-4876-b724-74eff9cdf347/files/sermp.pdf

Nunn, P. D., & Reid, N. I. (2016). Aboriginal memories of inundation of the Australian coast dating from more than 7000 years ago. *Australian Geographer, 47*(1), 11–47. https://doi.org/10.1080/00049182.2015.1077539

O'Connor, M., & Prober, S. M. (2010). *A calendar of Ngadju seasonal knowledge. A report to Ngadju Community and Working Group*. Ngadju aboriginal corporation, CSIRO sustainable Ecosustems. Floreat, WA: CSIRO.

Victoria, P.. (2014). You Yangs Regional Park. Retreived on 18 November 2014 from https://www.parks.vic.gov.au/places-to-see/parks/you-yangs-regional-park

Powell, B., & Tournier, D. (2019). Welcome to *Wadawurrung* country. In D. S. Jones & P. B. Roös (Eds.), *Geelong's changing landscape: Ecology, development and conservation*. Melbourne, Australia: CSIRO Publishing.

Riessman, C. (1993). *Narrative analysis*. London: Sage.

Richards, T. (1998). *A predictive model of aboriginal archeological site distribution in the Otway range, occasional report no 49*. Melbourne, Australia: Aboriginal Affairs Victoria.

Roberts, R. J. (1994). The human colonisation of Australia: Optical dates of 53,00 and 60,000 years bracket human arrival at deaf adder gorge, northern territory. *Quaternary Science Reviews, 13*, 575–583.

Roberts, A. (1975). *Dreamtime stories*. Sydney, NSW: Rigby.

Roös, P. (2014a). Landscapes in transition: Local perceptions of indigenous and European settlement changes along the Victorian coast. In *UHPH 14: Urban history planning history conference proceedings*. Wellington, New Zealand: UHPH/Victoria University of Wellington.

Roös, P. B. (2014b). Indigenous knowledge and climate change: Settlement patterns of the past to adaptation of the future, in the *International Journal of Climate Change: Impacts and Responses,* Common Ground Publishing, University of Illinois, Illinois, USA.

Roös, P.B. (2015). Indigenous knowledge and climate change: Settlement patterns of the past to adaptation of the future. *International Journal of Climate Change: Impacts & Responses, 7*(1), 13–31. https://doi-org.ezproxy-f.deakin.edu.au/10.18848/1835-7156/CGP/v07i01/37225

Roös, P. B. (2017). *Regenerative-Adaptive Design for Coastal Settlements: A Pattern Language Approach to Future Resilience*. PhD Thesis, Deakin University, Geelong, Victoria, Australia.

Smylie, J., Martin, C. M., Kaplan-Myrth, N., Steele, L., Tait, C. & Hogg, N. (2003). *Knowledge translation and indigenous knowledge*. Culture & Mental Health Research Unit, Department of Family Medicine. Ottawa: University of Otewa.

Smyth, D. (2004). *Kooyang Sea Country Plan*. Framlingham aboriginal trust and Winda Mara aboriginal corporation. Heywood: Winda Mara Aboriginal Corporation.

Thomson, D. F. (1949). *Economic structure and the ceremonial exchange cycle in Arnhem Land*. Melbourne, Australia: Macmillan.

Thomson, D. F. (1939). The seasonal factor in human culture illustrated from the life of contemporary nomadic group. *Prehistoric Society Proceedings, 5*, 209–221. London: Prehistoric Society.

Van der Ryn, S. (2005). *Design for life: The architecture of Sim Van der Ryn* (Vol. 1, 1st ed.). Utah, Layton: Gibbs Smith Publisher.

Victoria. Department of Aboriginal Affairs & Kerrup Jmara Elders Aboriginal Corporation (1993). *Lake Condah heritage management strategy and plan, 1993*. Aboriginal Affairs Victoria, [Melbourne]

Victorian Government. (2006). *Aboriginal Heritage Act 2006*. Office of the Chief Parliamentary Councel, Parliament of Victoria. Melbourne Victoria, Australia.

WAC. (2014). *Wathaurong Aboriginal Co-operative*. Retrieved 11 December 2018 from Wathaurong History and Profile: www.wathaurong.org.au

Wood, V. (2007). *Kaurna Cultural Heritage Survey. City of Port Adelaide and hemisphere design and GHD*. Adelaide: City of Port Adelaide Council.

Yawuru RNTBC. (2011). *Planning for the future: Yawuru cultural management plan* (Vol. 1). Broome, Western Australia, Australia: Pindan Printing Pty Ltd..

Yunkaporta, T. (2019). *Sand talk: How indigenous thinking can save the world*. Melbourne: Text Publishing.

Chapter 6
Interconnections Between People and Their Natural Environments

Abstract Progressing from Chap. 5 that investigated the longer history of humans and nature, through the narrative of Indigenous knowledge systems, this chapter explores the *phenomena of Biophilia*, the innate tendency of our human attraction to nature. It brings the narrative to a culmination within the contemporary context of our current twenty-first century, reflecting on the vast and complex interconnections and patterns between people and their natural environments, through the lens of *a nature language*. Further, the chapter highlights the issues of the declining health and well-being of urbanites, noting the need for human-nature connections by establishing future Biophilic Cities (Beatley, Biophilic cities. Washington, DC: Island Press, 2011). The principles of biophilic design practice are explored through the *15 biophilic design patterns*, identifying the key requirements of improving both physiological and psychological well-being resulting in a deeper embedded human-nature affiliation. The chapter concludes with the *fundamental pattern* LOVE FOR NATURE [6], stipulating an emphasis on the need to apply this deep affiliation with nature to our design practice.

Keywords Biophilic design · Biophilia · Human-nature relationship · Ecopsychology · Nature language · Natural attraction · Love for Nature · Pattern language

Upward links:

... the fundamental principle that will help us as humans to change the way that we depleting the Earth's resources, and rather moving into a state of regeneration, is to acknowledge and embrace our LOVE FOR NATURE [6]. Acknowledging that humans are part of nature and not above nature through the gaining of ADVANCED INDIGENOUS KNOWLEDGE [5], will result in a swift to a global society that embraces an INTEGRAL SUSTAINABLE WORLDVIEW [2].

6.1 Introduction

> The same stream of life that runs through my veins night and day runs through the world and dances in rhythmic measures... It is the same life that shoots in joy through the dust of the earth in numberless blades of grass and breaks into tumultuous waves of leaves and flowers... I feel my limbs are made glorious by the touch of this world of life... (Rabindranath Tagore, in Vaughan-Lee, 2016, p. 133)

The interrelationship between humans and nature gained large interest in the last century. Interest focused on the understanding the human-nature relationship, the various ways humans are connected to the natural world, and sustainability. Recently, practitioners in the design field have been investigat-

P. B. Roös, *Regenerative-Adaptive Design for Sustainable Development*, Sustainable Development Goals Series, https://doi.org/10.1007/978-3-030-53234-5_6

ing the health benefits associated with engaging with nature (Seymour, 2016), as they consider elements of biophilic design to potentially improve health and wellbeing in the built environment. Since the late nineteenth century a number of descriptive models have attempted to encapsulate the dimensions of human health and ecosystems, as well as their interrelationships (Seymour, 2016, p. 2). What is evident in the literature is that the interrelationship of humans with the natural environment does have an impact on human wellbeing both from a psychological and physiological perspective. According to Kellert and Wilson, humans do have an inherent biological-based connection to nature (Kellert & Wilson, 1993) (Fig. 6.1).

Where we have identified in Chap. 5 the long history of human and nature interaction, in this chapter I explore further the phenomena of *biophilia*, the innate tendency of our human attraction to nature, and the deep connection that exists between the two. Through the description of a *nature language*, this brings the narrative to a culmination within the contemporary context of the twenty-first century, reflecting on the vast and complex interconnections and patterns between people and their natural environments.

Further, the chapter explores the key requirements for improving both physiological and psychological well-being by identifying the need for human-nature affiliations and addressing a design paradigm that includes biophilic design and biophilic patterns, interaction patterns of a nature language, and the alignment between a physical design lens and an ecopsychology narrative.

6.2 Innate Affiliation with Nature

In the *Biophilia Hypothesis* (1986), Edward O. Wilson noted that humans needed daily contact with nature to be healthy and gain longevity. His reasoning was that humans have co-evolved with nature and are part of it, noting that biophilia is "the innately emotional affiliation of human beings to other living organisms. Innate means hereditary, and hence, part of ultimate human nature" (Kellert & Wilson, 1993, p. 31). Biophilia supports the proposition that our environment needs to include the essentials of nature for providing us with psychological and physiological health:

> Over thousands of generations the mind evolved within a ripening culture, creating itself out of symbols and tools, and genetic advantage accrued from planned modifications of the environment. The unique operations of the brain are the result of natural selection operating through the filter of culture. They have suspended us between the two antipodal ideas of nature and machine, forest and city, the natural and artificial, relentlessly seeking, in the words of the geographer Yi-Fu Tuan, an equilibrium not of this world (Wilson, 1984, p. 12).

This affiliation with nature continues to be critical in modern-day human health and wellbeing literature and practice (Kellert, 1997, 2012), and has been strongly identified as a crucial issue in the health sciences. A growing body of research reveals that exposure to nature continues to result in positive health benefits in a wide range of sectors; at work, home, recreation, community areas and even within the urban environments where people work and live (Browning, Ryan, & Clancy, 2014; Kellert, 2012). While humans may have an inherent affiliation with nature, the benefits from contact with it depend on repeated experience of biophilic effect elements (Roös, Jones, Downton, & Zeunert, 2018).

Alexander and his colleagues highly argued that this affiliation to nature is needed. In *A Pattern Language* (1977), Alexander et al. state:

> People need contact with trees and plants and water. In some way, which is hard to express, people are able to be more whole in the presence of nature, are able to go deeper into themselves, and are somehow able to draw sustaining energy from the life of plants and trees and water – Pattern 173 'Garden Wall' (Alexander et al., 1977, p. 806)

Fig. 6.1 Love for Nature. (Photo by Author)

This indicates that psychological benefits for us as humans are deeply rooted in the connection between humans and nature. The fundamental benefits of physiological and psychological wellness due to nature exposure were explored by Kellert (2005), with findings that demonstrate several health advantages and a better quality of life (Kellert, 2005; Kellert, Heerwagen, & Mador, 2008). These benefits can be achieved if we design and plan our built environments to include the considerations of Wilson's *biophilia*, and more specifically identifying the structures and patterns that occur in the form-making processes of the living systems of nature, as well as designing and planning with a *regenerative-adaptive process* that unfolds the living environments of place (Roös, Jones, Downton, & Zeunert, 2018; Salingaros, 2013).

6.3 Biophilic Design and Patterns

By nature, human beings are subconscious, constantly engaging with *patterns of biophilia* that exist in the surrounding environment (Downton et al., 2017b). Many patterns of biophilia are available to trigger and work with our senses. In this way, it is possible to create places that connect us with these patterns. While biophilia is the theory, biophilic design as advocated internationally by Kellert et al. (2008) and Beatley (2011) involves a process that incorporates reconnecting people with the natural environment. Beatley has evidenced the validity of this approach in *Biophilic Cities* (Beatley, 2011), pointing to numerous exemplars and precedents that can enable the successful implementation of this process. He has advocated putting the biophilia hypothesis into practice at an urban scale, proposing the essential elements of a *biophilic city*, and tabling examples and stories about cities that have successfully integrated biophilic elements (Beatley, 2011).

In health and wellbeing, Ryan, Browning, Clancy, Andrews, and Kallianpurkar (2014) have validated the relevance of biophilic design to humans whereby research in the neurosciences, endocrinology and other fields have scientifically validated the positive psychophysiological and cognitive benefits afforded by biophilia in design interventions. In the built environment sector, Söderlund and Newman (2015) have proposed a new set of design principles and practices where nature needs to play a bigger part called 'biophilic architecture', asserting that humans have an innate connection with nature that can assist in making buildings and cities more effective human abodes. Söderlund and Newman (2015) have also concluded that biophilic design is emerging as a social movement.

In *Creating Healthy Places* (2017), authors Downton, Jones, Zeunert, and Roös (2017a) have sought to apply biophilic design patterns as design and performance parameters in the new underground railway system in Melbourne. Subsequently, Downton et al. have argued that the research findings could potentially support the measurable, positive impacts of biophilic design on human health, strengthening the empirical evidence for the human-nature connection and highlighting its importance within both design research and design practice (Downton et al., 2017a). This argument is clearly validated by Kellert et al. (2008), who state that biophilic design has the potential to enrich nature and humanity, and propose that:

> …combining the biophilic desire to harmonise with nature together with the design of the built environment, results in some degree of deliberate refashioning of nature to satisfy human needs, but in ways that celebrate the integrity and utility of the natural world (Kellert et al., 2008, p. 14).

To implement biophilic design, Browning et al. (2014) have proposed '*14 Patterns of Biophilic Design*' – a framework for relating human biological science and nature to the design of the built environment, offering tools for understanding opportunities and avenues for design applications as a way to effectively enhance health and wellbeing for individuals and society. Building further upon these 14 patterns, Downton et al. (2017a) added an additional pattern, 'Virtual connection to nature', to deal with complex constraints in the built environment (Table 6.1). These virtual representations of nature could be implemented in design solutions where the provision of actual natural elements such as vegetation is impossible (Downton et al., 2017a, p. 25). Further, to apply biophilic design in its simplest form, Potteiger and Purinton (1998), Browning et al. (2014) and Kellert and Calabrese (2015) grouped biophilic design patterns in three categories of reference: *Nature in the Space*; *Natural Analogues*, and *Nature of the Space*.

Table 6.1 The 15 patterns of biophilic design

Biophilic design pattern	Context and application
NATURE IN THE SPACE	
1. **Visual Connection with Nature** A view to elements of nature, living systems and natural processes	Ensure visual access to real presentations of nature in preference to simulated nature and non-nature representations
2. **Non-Visual Connection with Nature** Auditory, haptic, olfactory, or gustatory stimuli that engender a deliberate and positive reference to nature, living systems or natural processes	Enhance opportunities for sensory connections (audible, smell, texture, temperature) to nature in preference to urban simulated or constructed representations
3. **Non-Rhythmic Sensory Stimuli** Stochastic and ephemeral connections with nature that may be analysed statistically but may not be predicted precisely	Instil patterns of nature's movements and seasonality using real or artistic representations where necessary
4. **Thermal & Airflow Variability** Subtle changes in air temperature, relative humidity, airflow across the skin, and surface temperatures that mimic natural environments	Consider sequential changes in thermal and airflow variability to refresh spaces and to enable comfortability
5. **Presence of Water** A condition that enhances the experience of a place through the seeing, hearing or touching of water	Use water as a static, dynamic and or variable design element to achieve multi-sensory experiences
6. **Dynamic & Diffuse Light** Leveraging varying intensities of light and shadow that change over time to create conditions that occur in nature	Use mixtures of dynamic, diffuse and changeable lighting arrangements and patterns (including illuminance and colour) to evoke movement, time, seasonality, while maximizing solar access
7. **Connection with Natural Systems** Awareness of natural processes, especially seasonal and temporal changes characteristic of a healthy ecosystem.	Use natural systems (weather, hydrology, geology, terrestrial and aquatic wildlife, diurnal and seasonal patterns) as design inspirations
NATURAL ANALOGUES	
8. **Biomorphic Forms & Patterns** Symbolic references to contoured, patterned, textured or numerical arrangements that persist in nature.	Ensure biomorphic patterns legibility and interest in the built environment
9. **Material Connection with Nature** Material and elements from nature that, through minimal processing, reflect the local ecology or geology to create a distinct sense of place.	Consider the richness of material colour, warmth, authenticity and tactility
10. **Complexity & Order** Rich sensory information that adheres to a spatial hierarchy similar to those encountered in nature.	Prioritise pattern compositional and order use enabling stimulation, interest and legibility, including artwork
NATURE OF THE SPACE	
11. **Prospect** An unimpeded view over a distance for surveillance and planning.	Provide a sense of arrival, and provide a sense of prospect, for key areas of habitation
12. **Refuge** A place for withdrawal, from environmental conditions or the main flow of activity, in which the individual is protected from behind and overhead.	Provide opportunities for retreat, contemplation, waiting, meeting, refuge, for key areas within the urban environment
13. **Mystery** The promise of more information achieved through partially obscured views or other sensory devices that entice the individual to travel deeper into the environment.	Provide a sense of journey in pedestrian environments that ensures sightlines, permeability, and variability in edges and planes
14. **Risk/Peril** An identifiable threat coupled with a reliable safeguard	Provide places that stimulates as sense of danger
15. **Virtual Connection with Nature** A view to a simulacrum of natural elements, living systems and natural processes	Provide virtual connections with nature viewed through mediated means or evoked by simulacrums of nature, living systems and natural processes. Examples include artificial skies, animatronics and portrayal of nature via virtual reality

Adapted from Downton et al. (2017a); derived from Browning et al. (2014)

It is interesting to note that Browning et al. (2014) propose the use of the term *patterns* rather than principles, values, or attributes. According to the authors, inspired by Alexander's *A Pattern Language* (Alexander et al., 1977), this term is used because:

> … to propose a clear and standardized terminology for biophilic design; to avoid confusion with multiple terms (metric, attribute, condition, characteristic, typology, etc.) that have been used to explain biophilia; and to maximize accessibility for designers and planners by upholding familiar terminology (Browning et al., 2014, p. 21)

The use of these 15 biophilic design patterns can potentially support and enhance the design of buildings in our cities, as noted by Tim Beatley:

> … Nature in cities, we increasingly recognize, is not something optional, but absolutely essential to leading a happy, healthy and meaningful life. And this extends to the design of every element in the city, including transit stations. (Beatley, in Downton et al., 2017a, p. 9).

However, when we consider the context of biophilia in the design of our built environments, the danger is that biophilic design becomes a buzz word, a superficial trend similar to the misused 'sustainability'. In 2017, as part of an online round table discussion of the *Nature of Cities*, I argued that "biophilic" and "biophilia" can become a trend where there is no *substance under the surface*. To avoid a greenwash approach, we need to embrace the concept of *the whole*. To be part of the whole is not simply to reflect on the need to link aesthetic preferences and attributes alone, but rather a *deep connection* to the geometric structures and patterns that occur in the form-making processes of the living systems of nature (Roös and Jones, 2017; Salingaros, 2015). The biophilic effect is more than what we can see on the surface, which is only a reflection of the complexity of *living structures underneath* (Roös, 2017). Because the effect of biophilia is more than what we can see, we need to grasp the occurrences of a deeper *nature language*.

6.4 A Nature Language

Nature talks to us, not in words but through a *nature language* that connects with our senses. This is evident when we engage with Nature, and distinct natural attraction senses and sensitivities are triggered. Nature is multi-sensory, informed by our conscious and unconscious innate connection to it. One thing we often overlook is that the natural world is a non-verbal community. In nature, the desire to breathe has no name. However, we have the sense that air exists and we feel our desire to breathe. This is what keeps us alive, and our 'inner child' or 'inner nature' knows it too (Cohen, 2013; Roös, Jones, & Wilson, 2018).

The phenomena of our experience of the natural senses has been researched intensively since the mid 1900s. During the years 1961–1978, researcher Guy Murchie made an exhaustive inquiry into the multi-sensory experiences that connect us to nature, which we can observe as humans (Murchie, 1978). He condensed together over 80 different biological sensitivities that pervade the natural world, into a group of 31 senses. From these, Cohen identified, documented and included 54 natural senses that connect us to nature in the practice of ecopsychology (Cohen, 2013, p. 169). This effect of *biophilia patterns* as a result of many multi-sensory phenomena, when experienced first-hand, is difficult to describe in words. A good example of this phenomenon is the *attraction to natural light* (Fig. 6.2). As human beings, subconsciously we know that light from the sun is the source of all life on earth. Without this light, the earth will be enclosed in darkness and our existence will be absolute (Roös, Jones, & Wilson, 2018). The 'sense of light' is listed as number 1 of the *54 Natural Senses* (Cohen, 2013). The 'Sense of light and sight, including polarised light' is core to the inherent natural attraction to the webstring of life, a genetically rooted web-of-life strand that helps to hold the human-nature existence together in this universe (Cohen, 2013, p. 170).

When we consider a *nature language*, it is important to acknowledge that we are going beyond the physical patterns in nature, such as geometric forms and systems. In a nature language the patterns are the phenomena of the *interaction with nature*. The interaction patterns that contain a nature language

Fig. 6.2 Attraction to natural light. The phenomena of natural light and its many characteristics results in multi-sensory experiences. (Photo by Author during an experiment in the Otway Ranges)

embody the structure of an engaged and meaningful human activity in, through part of, and with nature (Kahn Jr., Ruckert, & Hasbach, 2012, p. 57). An interesting observation when analysing the nature language proposed by Kahn Jr. et al. (2012), and indeed acknowledged by them, is that interaction patterns are not fundamentally distinct from each other, but they rather exist within a networked relationship. This relationship is similar to the proposition by Cohen (2013) that these patterns are an inherent part of the webstring of life, as described earlier.

It is evident that the human-nature interaction is beyond the physical patterns that we know, and as Greenway (2009) argued, we need a more robust language that can reflect our experience of nature. He stated that:

> I work towards and ecopsychology that will find within the language an accurate articulation of the human-nature relationship. This will of course be based on experience, but will be couched in a language… It must take up the deepest meanings of relationships in general and relationships in "mind" and "nature" in particular. It will be based on a variety of "modes of knowing" – neither ignoring or privileging science (Greenway, 2009, p. 50).

In the context of the *regenerative-adaptive pattern language*, it is this language of connecting "mind" with "nature" that is fundamental in seeking a *deep sustainability*, going beyond and acknowledging that we as humans are part of nature and not above it.

6.5 Conclusion Statement

The interconnections between humans and nature exists on two levels. One is where *physical interactions* enhance living structures and are part of the whole that interacts with our organic and inorganic surroundings on Earth to form a synergistic, self-regulating, complex system that helps to maintain and perpetuate the conditions for life on the planet. It is this deeper connection to the organised complex order of abiotic and biotic systems that results in us being nature, or being part of the whole (Roös, 2017). The second level of connection is the biophilic effect, which is more than what we can see on the surface; it is what is underneath, the *psychological phenomenon* where we are part of a *natural attraction*, a *sensory language* that is formulated in a *nature language*.

Therefore:

Our biophilic design agenda thus needs to include the characteristics of a complex order and the complexity of the traditional ornament make-up (Salingaros, 2015), inclusive of the rules of living structures (Alexander, 2001–2005). Rules for how ornament make-up contributes to a healing environment can be derived from understanding how our brain is wired to respond to our surroundings. These rules are part of *deep interaction patterns*. Both the physical perspective of biophilic design, as well as the ecopsychology perspective of *natural attraction* and sensory responses of a nature language embedded in biophilia, needs to be applied to design and planning practice. This *nature language* is fundamental to enhance and strengthen our *love for nature*.

6.6 Fundamental Pattern 6

Love For Nature [6]

The positive enhancement of connections between humans and their natural environments is fundamental to the health and wellbeing of a global society. Therefore, employ Biophilia to connect people to nature at physical, psychological and consciousness levels. Engage a deep holistic *nature language* that emphasises our love for nature (Fig. 6.3).

Downward links:

Where the understanding of Biophilia – our love for nature – and the practice of biophilic design helps strengthen the interconnection between people and their natural environments. It is important that at the large scale of environmental change we know how to design and plan in harmony with the forces of the landscape – Nature's Design [7]. This will help us practice sustainable development that entrenches Regenerative Ecosystems [8] as part of the process …

Fig. 6.3 Sketch of fundamental pattern – Love For Nature [6]. (Drawing by Jesse Delmo, synthesised by Roös)

References

Alexander, C. (2001–2005). *The nature of order – An essay on the art of building and the nature of the universe, book one: The phenomenon of life*. Berkeley, CA: The Center for Environmental Structure.

Alexander, C., Ishikawa, S., Silverstein, M., Jacobson, M., Fiksdahl-King, I., & Angel, S. (1977). *A pattern language: Towns, buildings, construction*. New York: Oxford University Press.

Beatley, T. (2011). *Biophilic cities*. Washington, DC: Island Press.

Browning, W. D., Ryan, C. O., & Clancy, J. O. (2014). *14 patterns of biophilic design*. New York: Terrapin Bright Green, LLC.

Cohen, M. J. (2013). *Nature as higher power: creating moments that let Earth teach*. Friday Harbor, WA: Institute of Applied Ecopsychology, Akamai University, Institute of Global Education.

Downton, P. F., Jones, D. S., Zeunert, J., & Roös, P. B. (2017a). *Creating healthy places: Railway stations, biophilic design and the metro tunnel project*. Geelong, VIC: Melbourne Metro Rail Authority, Deakin University.

Downton, P. F., Jones, D. S., Zeunert, J., & Roös, P. B. (2017b). Biophilic design applications: Theory and patterns into built environment education. In *DesTech 2016: Proceedings of the international conference on design and technology* (pp. 59–65). Dubai, United Arab Emirates: E-Knowledge. https://doi.org/10.18502/keg.v2i2.596

Greenway, R. (2009). Robert Greenway: The ecopsycholgy interview. *Ecopsychology, 1*, 47–52.

Kahn Jr., P. H., Ruckert, J. H., & Hasbach, P. H. (2012). A nature language. In P. H. Kahn Jr. & P. H. Hasbach (Eds.), *Ecopsychology – Science, totems, and the technological species*. Cambridge, MA/London: MIT Press.

Kellert, S. (1997). *Kinship to mastery: Biophilia in human evolution and development*. Washington, DC: Island Press.

Kellert, S. (2012). *Birthright: People and nature in the modern world*. New Haven, CT: Yale University Press.

Kellert, S., & Calabrese, E. (2015). *The practice of biophilic design*. www.biophilic-design.com. Accessed 1 June 2018.

Kellert, S., Heerwagen, J. H., & Mador, M. L. (2008). *Biophilic design: The theory, science & practice of bringing buildings to life*. Hoboken, NJ: Wiley.

Kellert, S., & Wilson, E. O. (1993). *The biophilia hypothesis*. Washington, DC: Island Press.

Kellert, S. R. (2005). *Building for life: Designing and understanding the human-nature connection*. Washington, DC: Island Press.

Murchie, G. (1978). *Seven mysteries of life*. Boston: Houghton Mifflin.

Potteiger, M., & Purinton, J. (1998). *Landscape narratives: Design practices for telling stories*. New York: Wiley.

Roös, P. B. (2017). *The nature of cities roundtable: What are we trying to accomplish with biophilic cities? What are ambitious goals and targets, and measures of success?* The Nature of Cities. Retrieved on 10 February 2020 from https://www.thenatureofcities.com/2017/03/27/ambitious-goals-targets-biophilic-cities-right-metrics-progress-toward-goals/

Roös, P. B., & Jones, D. S. (2017). *Weaving landscape fabrics of ecological cities: Patterns for regenerative-adaptive futures*. Presented at the EcoCity World Summit 2017, 12-14 July 2017, Melbourne, VIC. Retrieved on 10 February 2020 from https://www.ecocity2017.com/

Roös, P. B., Jones, D. S., Downton, P. B., & Zeunert, J. (2018). Biophilic railway stations: Re-imagine the nature of transit design. In *IFLA 2018: Biophilic city, smart nation, and future resilience: Proceedings of the 55th International Federation of Landscape Architects World Congress 2018* (pp. 800–813). Singapore, Singapore: International Federation of Landscape Architects.

Roös, P. B., Jones, D. S., & Wilson, A. S. (2018). *The biophilic effect: Hidden living patterns within the dance of light*. Melbourne, VIC: University of Melbourne, Faculty of Architecture, Building and Planning & UNESCO Observatory for Multi-Disciplinary Research in the Arts. Unpublished Manuscript submitted for publication.

Ryan, C. O., Browning, W. D., Clancy, J. O., Andrews, S. L., & Kallianpurkar, N. B. (2014). Biophilic design patterns: Emerging nature-based parameters for health and well-being in the built environment. *International Journal of Architectural Research, 8*(2), 62–76.

Salingaros, N. A. (2013). Unified architectural theory: Form, language, complexity - a companion to Christopher Alexander's "the phenomenon of life: The nature of order, book 1". Portland, OR; Kathmandu, Nepal: Sustasis Press.

Salingaros, N. A. (2015). *Biophilia & healing environments: Healthy principles for designing the built environment*. New York, LLC: Terrapin Bright Green.

Seymour, V. (2016). The human–nature relationship and its impact on health: A critical review. *Frontiers in Public Health, 4*, 260. https://doi.org/10.3389/fpubh.2016.00260

Söderlund, J., & Newman, P. (2015). Biophilic architecture: A review of the rationale and outcomes. *AIMS Environmental Science, 2*(4), 950–969.

Vaughan-Lee, L. (Ed.). (2016). *Spiritual ecology: The cry of the earth* (2nd ed.). Point Reyes, CA: The Golden Sufi Center.

Wilson, E. O. (1984). *Biophilia*. Cambridge, MA/London: Harvard University Press.

Chapter 7
Design and Planning with Nature

Abstract Planning and design has had a considerable impact upon the structure and evolution of settlements and urban cities worldwide and in Australia. Most cities and settlements have been planned by way of survey, plan and subdivision. Over time the planning of cities evolved, and in the 20th and twenty-first century planning practice evolved to cope with the urgent need to enable the rapid expansion of cities due to unprecedented population growth. The result was that planning movements focused on urban renewal and tried to grasp issues brought about by increasing populations due to city migration and internal growth. The result was that the consideration of environmental and ecological stewardship and the need to look after our natural environment have been lost. In this chapter, the concept of 'planning' is regarded as *one of the components* in a collective approach to the practice of *design and planning* of the built environment and its *interconnectedness with nature*, amongst the disciplines of architecture, urban design and landscape architecture. The chapter explores the existing knowledge systems of design and planning of our human settlements through the lens of contemporary environmental design practice, and progress to chart the seminal ecological planning and design theory of Ian McHarg, identifying the consideration of '*the whole*' in his work. The chapter concludes with how *Design with Nature* (1992 [1969]) can be used today in a contemporary practice to align with the *regenerative-adaptive pattern language*, and how to apply the *fundamental pattern* – NATURE'S DESIGN [7] – to the design and planning of our current and future human settlements.

Keywords Design with nature · Plan with nature · Ecological planning and design · Nature's design · Ecology · Regenerative · Planning and design · Deep roots · Sustainable

Upward links:
… so far, we have established that we need to consider THE WHOLE [1] when we design and plan our human settlements and that our interconnections with nature can be enhanced due to our LOVE FOR NATURE [6]. However, this innate attraction to nature can also be misused in a destructive manner, where humans see it as a resource to fuel the needs of settlements. This tendency of a human-centric worldview must be avoided at all costs, and an INTEGRAL SUSTAINABLE WORLDVIEW [2] needs to be adopted which embraces the practice of NATURE'S DESIGN [7].

P. B. Roös, *Regenerative-Adaptive Design for Sustainable Development*, Sustainable Development Goals Series, https://doi.org/10.1007/978-3-030-53234-5_7

7.1 Introduction

> Glaciation, lava flow, reef, mountain range, rainforest, mallee, desert, river and lake, mangrove swamp. Coastal wallum, wildlife habitat, ancient cremation site, rock engraving, mimi figure, wandjina, quinkan, dreaming track, scarred tree, fish trap, landing point, contact site, primitive cottage, barrack. Penal settlement, church, custom house, bank building, poppet head, miner's hovel, gold town, homestead, woolshed, shepherd's hut, log gaol, silo, machine shop, bottle kiln, pub. Parliament, opera house, government domain, housing estate, company town, civic centre, public square, parkway, regional park, redevelopment project, new town (Yenken, 1981, cited in Freestone, 2010, p. 2).

As an introduction in *Urban Nation: Australia's Planning Heritage* (2010), Freestone used Yenken's list of generic landscapes from the remnants of ancient glaciers to the Sydney Opera House as demonstrations of the rich history of Australia (Freestone, 2010; Yenken, 1981). In the context of generative processes of design and planning, this list reflects the *inherent patterns* that we need to acknowledge in design and planning methods for creating our settlements and cities. It encompasses the spectrum of Aboriginal natural and cultural places, values and phenomena that collectively help define the distinctive character of places in Australia (Freestone, 2010). Planning and design have had a considerable impact upon the structure and evolution of settlements and urban cities worldwide and in Australia. Much of Australia has been planned by way of survey, plan and subdivision. Planning has had a complex evolution; the term means different things to different people, and this meaning has changed in history (Freestone, 2010). Planning has evolved over time, from around the third millennium BC by Egyptian civilizations to recent modern urban planning movements. The ideology of planning in the 1800s to late 1900s evolved from a focus on social and political control (Gridiron town), to town and country planning (Garden city), to energy efficiency, land economy and networked cities (Compact city, Polycentric city), into the 2000s where the importance of liveable, walkable communities, as well as affordability, compactness, and solar access (New Urbanism, Ecologically sustainable development, Master planned community) have become key themes. Although there was some focus in the 1970s on environment-led design methods (Ecological planning), it seems that the need for looking after our natural environment was replaced by the urgent need to enable the rapid expansion of twentieth century cities due to unprecedented population growth. Planning movements focused on urban renewal and tried to grasp issues brought about by increasing populations due to city migration and internal growth.

In the narrative of this book, the concept of 'planning' is regarded as one of the ingredients in a collective approach to the practice of design and planning of the built environment and its *interconnectedness with nature* at the intersection of the disciplines of architecture, urban design and landscape architecture. Further, the concept of a 'sustainable human settlement' and how it functions (that is, a human settlement that will be sustainable and resilient alongside its natural environment, where the settlement does not deplete the natural resources) can be abstract (Fig. 7.1).

Arguments have been put forward that small European towns in the Middle Ages and prehistoric hamlets that evolved through vernacular growth were 'sustainable' (Ruano, 2002). These ancient settlements, however, were based on a natural resource use paradigm; they extracted resources from the environment and dumped waste back into it. They were considered 'sustainable' as their disruption to the natural environment was minor. The fact that they were small and did not over-consume their natural resources does not necessarily make them sustainable. The habit of consuming natural resources by these settlements escalated as they evolved into today's urban civilizations, with very little regard for the protection of the environment. Large cities (due to rapid urbanisation), the concentration of large urban populations, and the sprawl of cities into wider geographical areas have resulted in the rapid growth of mega-cities and the most significant transformations of Earth's human settlements. With the addition of issues of climate change, as designers and planners we face a challenge to plan for more sustainable, adaptable human settlements in the future.

Fig. 7.1 Tane Mahuta – the lord of the forest, one of the most ancient of trees. (Photo by Author)

7.2 Deep Roots

Throughout the narrative of this book the message brought to the forefront is that we aim to find a way to make our settlements more sustainable and adaptable, to make settlements that are regenerative and adaptive to changing environments, and to plan and design settlements that consider *the whole*. This calls for a holistic and integrated process that includes many complex factors. This integrated process needs to include and consider our *deep roots* to the natural environment, of which we as human beings are a part. Such a holistic integrated process, where the complex integrated environment of ecological systems, nature, climate and regenerative processes, built environment, people values and social structures are all the ingredients of a planned landscape, need multi-disciplinary actors. In this context, the local residents of a place (including disciplines of architecture, landscape architecture, planning and ecology) are critical in the placemaking of settlements that are resilient in a future landscape of a changing climate. Thus, theories in architecture, planning and urban design, ecological design and landscape architecture are further investigated, especially those that consider the *relationships between the built and natural environments* to inform a *regenerative-adaptive design and pattern language* hypothesis, and as a result contribute to the knowledge of the design and planning discourse. This chapter thus explores the history of planning and design of the built environment, more specifically theories that consider the inclusion or protection of the natural environment.

7.3 History of Planning and Design that Considers the Environment

The relationship between the built and natural environment can be traced back in Western civilisation's design and planning approaches as far back as 50 BC in Rome, through Vitruvius's recommendation on issues such as natural lighting, orientation and location. As described by Biermann et al. (2003), Vitruvius stated in his book *De architectura libri decem* (The Ten Books on Architecture) that a building must exhibit the three qualities of *firmitas* (firmness, robustness, structural strength), *utilitas* (commodity, usefulness), and *venustas* (delight, beautiful). This three-fold purpose of architecture, known as the *Vitruvian Triad*, is further supported by Vitruvius' belief that architecture[1] is an imitation of nature (Biermann et al., 2003). Vitruvius' approach however was that nature was a resource for satisfying human needs. This perspective did not change much over the next two millennia.

Planning and architecture went through a paradigm shift in the nineteenth century where extremely unhealthy conditions linked to industrialised cities gave rise to the consideration for healthier environments. A tendency developed to provide green spaces for people, such as Ebenezer Howard's 'Garden Cities', which were planned to be self-contained communities with surrounding 'greenbelts'. The planning of greenbelts included large areas of parks, containing separate areas for residences, industry and agriculture in a proportionate fashion (Hall, Hardy, Howard, & Ward, 2003). Howard's vision was that the Garden City was part of a linked system of healthy green cities, a whole system which he called the 'Social City' (see Fig. 7.2).

A similar example is the planning of the extension of Barcelona in 1859 by Ildefons Cerda. In this design, Cerda proposed that the extension of the city would become a liveable place with an approach including greenery of people's surroundings (Ruano, 2002; Soria y Puig, 1999). This led to new concepts of planning that included the preservation of nature, including the 'City Beautiful' movement during the late 19th and early 20th centuries, and Stein's concepts of 'New Towns of America' (Stein, 1951). Stein's work changed the stale urban environment into vibrant neighbourhoods, reflecting his belief of moulding urban construction into nature. His work had considerable impact on North American urban planning, and his concepts are still reused today (Parsons, 1998). Van der Rohe, Gropius, and Le Corbusier, the prominent figures of the Modernist Movement, established their reputations in the early twentieth century with designs engaging with sunlight and natural ventilation as essential factors, and their 'modernist planning' which considered the environment as the backdrop to urbanisation (Crouch, 1999). Fundamentally though, the approach in all these theories was that nature was there for human advantage; these proponents promulgated that the environment needs to be preserved for the benefit of the physical and mental health of residents.

It was not until after the Second World War that 'modernist planning' gained mass popularity and was implemented as a solution to the previous failure of architecture and urban planning to meet basic social needs (Henket, 2002). This thinking was further fuelled by the 'Green Revolution' that increased agriculture production. At the same time, new research was conducted into technologies that would harness energy sources such as nuclear, solar, wind, thermal, and other renewables, resulting in the belief that science can solve humanity's problems (Ruano, 2002). Nature was explored and studied so as to be more efficiently used to the benefit of humankind. During this time, architect Buckminster Fuller put forward the principles of the interdependency of humans and nature, advocating diversity and harmony between both (Fuller, 1938; Sieden, 2000).

[1] Architecture in the Roman times was a much broader subject than at present, and included engineering, construction, and urban planning. Roman architects can be described as engineers, architects, landscape architects, craftsman and artists combined. For example, the first of the ten books of *De architectura libri decem* deals with what is now contemporary known as Landscape Architecture (Thoenes in Biermann et al., 2003).

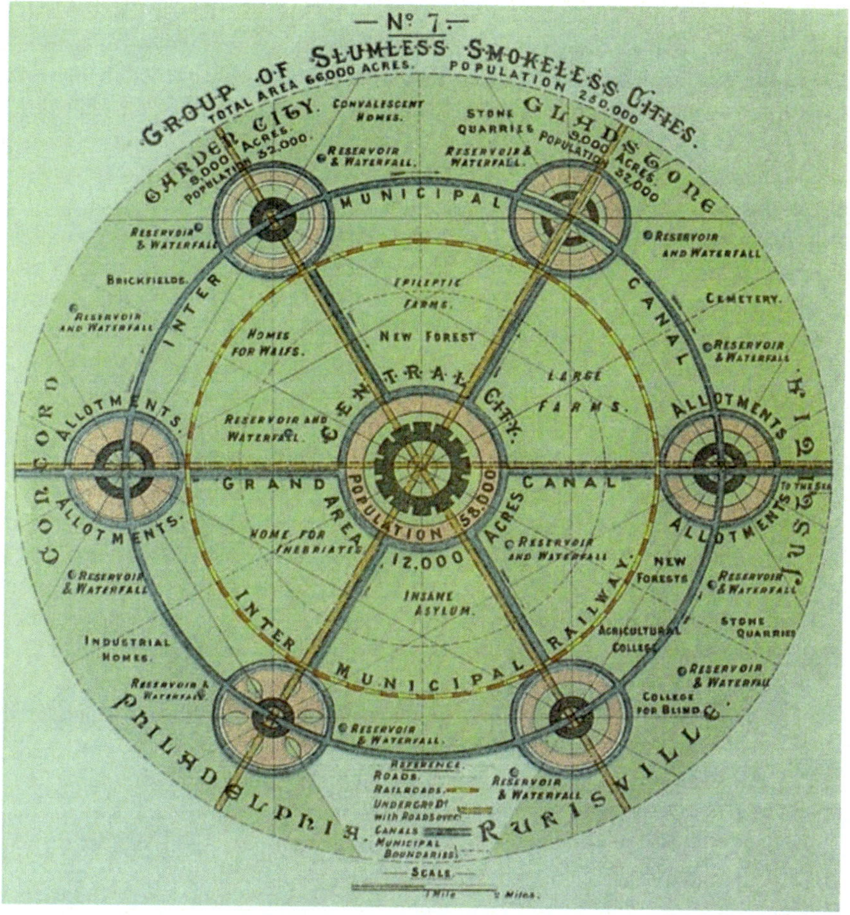

Fig. 7.2 Ebenezer Howard's vision of Social Cities – The Garden City. (Ebenezer Howard, 1850–1928. [Public domain])

7.4 Back to Nature

Lack of confidence in scientific and technological processes emerged in the late 1960s and early 1970s, and 'back to nature' became a strong belief embodied in the 'hippie era'. Inspiration was often drawn from Eastern cultures, where harmony with nature was considered essential for human wellbeing as well as to keep a cosmic balance. It can be strongly argued that it was a period when new theories of ecological design, landscape planning and urban design were initiated – theories that have evolved ever since. Work by McHarg established the foundations for ecological design in the context of landscape planning, through *Design with Nature* (1992 [1969]), while Alexander established a new perspective on the processes of architecture and planning in *The Timeless Way of Building* (Alexander, 1979), which eventually led to his famous, and also controversial, *A Pattern Language* (Alexander et al., 1977). Subsequently, *A New Theory of Urban Design* (1987) by Alexander, Neis, Anninou, and King (1987) expressed renewed interest in urbanism and articulated an understanding of generative processes as being a fundamental part of urban design.

During this period other theorists emerged using ecology as a focus, including Soleri who coined the term 'Arcology' (Architecture and Ecology) and who built *Arcosanti* in the desert, a city designed to maximise human population and ecology interactions (Soleri, 1969). In the "Gaia Hypothesis"

(Lovelock & Margulis, 1974), Lovelock and Margulis put forward the theory that the Earth is a self-regulated organism, a complex ecological system with everything within the earth and the biosphere interconnected and evolving. This 'interconnectedness' has become instrumental in the works of landscape architect Lyle, who promoted the duality of 'man and nature' in *Design for Human Ecosystems* (Lyle, 1985); he also stressed the importance of regenerating our natural environment in *Regenerative Design for Sustainable Development* (Lyle, 1994).

Despite concerns about ecological degradation, and efforts to better understand and improve the integration of our built and natural environments, economic and technical progress in the industrial sector has been the overriding human goal in the late 1980s and early 1990s, including providing for material comfort at the expense of nature. Notwithstanding the emergence of 'sustainability', embodied in the Brundtland Commission's inquiry in 1987, sustainability has become a mantra for the 1990s, offering one perspective as to a balance and permanence in the world (Van der Ryn & Cowan, 1996). This perspective is also despite extensive evidence of the exploitation of fossil fuels to provide energy, to fuel cars and to support the ever-expanding industrial sector that continues to poison the Earth. The demands of human needs and lifestyle pumped carbon emissions into the atmosphere like never before in the history of humankind. The proponents of technological sustainability assert that fundamental changes in the way we do things are not necessary; minimising the harm to nature is enough. Sim Van der Ryn challenges the efforts of the Brundtland Commission about the definition and effectiveness of '*Sustainable Development*', given its assumption of the principles of technological sustainability, and instead promotes '*Ecological Design*' as the correct vehicle for improving the built environment. In *Ecological Design* (1996), Van der Ryn and Cowan support David W. Orr's definition of *ecological sustainability*, in contrast to *technological sustainability*, defining *Ecological Design* as "… any form of design that minimises environmentally destructive impacts by integrating itself with living processes" (Van der Ryn & Cowan, 1996, p. 18).

Van der Ryn and Cowan further argue that, in many ways, *the environmental crisis is a design crisis*, and that impacts upon the environment are as a consequence of how things are made, buildings are constructed, and landscapes are used. Supporting this design paradigm, Van der Ryn and Cowan defined design as the intentional shaping of matter, energy, and process to meet a perceived need or desire. They summarise the *failure of design and planning* discourse as follows:

> … Our present forms of agriculture, architecture, engineering and industry are derived from design epistemologies incompatible with nature's own. It is clear that we have not given design a rich enough context. We have used design cleverly in the service of narrowly defined human interests but have neglected its relationship with our fellow creatures. Such myopic design cannot fail to degrade the living world, and, by extension, our own health (Van der Ryn & Cowan, 1996, p. 9)

7.5 Design and Planning with Nature

> Matter, of this is the cosmos, sun, earth and life made. Sun, shine that we may live. Earth our home, oceans our ancient home. Atmosphere, protect and sustain us. Clouds, rain, rivers and streams, replenish us from the sea. Plants, live and breathe what we may breathe, eat and live. Animals, kin. Decomposers, reconstitute the wastes of life and death so that life may endure. Man, seek the path of benign planetary enzyme, aspire to be the world's physician. Heal the earth and thyself (McHarg, 1992 [1969], p. vi).

McHarg wrote the above prayer, after 20 years of thought, as an introduction to *Design with Nature* (1992 [1969]). Planner Mumford commenced the introductory chapter to McHarg's book stating that 'there is still only a small shelf of books that deals with man's relation to his environment as a whole'. Mumford expressed that this book was a notable edition to a handful of important texts, because it drew upon science, reflected upon 'ecology' and confirmed that human life is intertwined in the forces of nature, so that nature needs to be treated as an ally and friend (McHarg, 1992 [1969], p. vii).

Trained as a town planner and a landscape architect, McHarg was described as an inspired ecologist. His work, of particular interest to the narrative of the *regenerative-adaptive pattern language*, offers a view that humans and the environment can be seen as a '*whole*'. Further, his identification of *patterns of nature* in the process of mapping, relates closely to the investigation of 'deep sustainability' (described in more detail later in this book), where we further investigate the pattern language and generative codes.

In *The Essential Ian McHarg: Writings on Design and Nature* (2006), Steiner summarised the influence of McHarg's theory on contemporary design and planning as 'Nothing is as good as a good theory'. The dictum of *Design with Nature* not only changed design and planning disciplines, but also enormously influenced fields as diverse as engineering, geography, forestry, environmental ethics, ecology and even soil science (McHarg & Steiner, 2006). *Design with Nature* acknowledges that humans are as ecological as any other species. Like other species humans are part of the whole web of life. The theory not only deals with 'man's relation to his environment' as a whole, but also with how an ecology-based design can shape and change the future for both.

7.6 Nature's Design

Design with Nature (1992 [1969]) was written by McHarg about landscape planning, essentially providing step-by-step instructions on how to analyse a landscape region and devise its appropriate uses. McHarg's theory, underpinned by an 'ecological approach', provides the basis for his landscape architecture and regional planning. The method sought to *reveal nature as a process containing intrinsic forms*. The method involved a *set of layers*, similar to layers that appear in nature, starting with the *geology of a place* and concluding with the *habitats of plants, animals and humans*. McHarg put forward a proposition that this ecological approach method offered opportunities and constraints having regard to that place, and that plants, animals and humans in that place are only comprehensible in terms of their physical and biological evolution. Historical geology explains the current form, physiography identifies the configuration of the place, and hydrology provides the patterns of rivers and flows, which support abundance or decline of plants; collectively these represents the climatic factors, and in turn provides specific habitats for animal species. All this information provides a *cross-ecological inventory* that can be interpreted as analysing existing and proposed human land use and management (Swaffield, 2002, p. 39; McHarg, 1967). The next step is to map all the values and attributes of the landscape, and identify the best-suited human uses. McHarg stresses that this is not a plan to enforce demand, but rather a process to understand nature. It reveals causality and that 'the place is because':

> …. This process permits us to interpret natural processes as resources, and to prescribe prospective land uses with compatible communities with nature (Swaffield, 2002, p. 41).

McHarg also states that the ecological method is useful for *creating form*. In the context of his method, form is explained as 'give' and 'made' forms. The former is a natural landscape, and the latter is the accumulation of changes that happen in the built environment overtime, resulting in a city or settlement. Each city is different, and by applying the method the layers of formation that shape a place can be evident. The successive stages of urbanisation are thus documented with the adaptations of the environment, and these outcomes can contribute to the *genius loci*. From a synthesis of this information a landscape plan can be prepared, which includes the elements of identity, natural process, and the palette for formal expression in the planning process (McHarg, 1967).

7.7 Ecological Planning and Design

The best way to understand McHarg's ecological method is to reflect upon a formative project worthy of study, the *Potomac River Basin Study* (1965–66), recorded in *Design with Nature* (1992 [1969]). This study used most of McHarg's methods and was the first project that combined the physiographic region and the river basin as the primary organizing context for ecological planning and designing a framework that links past, present, and the anticipated future in multiple landscape scales, from the region to the garden (Spirn, 2000, p. 106). The *Potomac River Basin Study* provides the opportunity to address all aspects of McHarg's methods, as the Basin is a single hydraulic unit that transcends a number of physiographic regions, and as such it could provide opportunities for different and specific land uses. For McHarg this was the best opportunity to test the ecological planning method – which, McHarg noted, employed a 'rationale method', whereby evidence is derived from the collection of environmental factors of the site, and the study then identifies merely those areas where certain land uses could occur with the least costs and greatest savings (McHarg, 1992 [1969], p. 105). This process of McHarg's ecological method can be summarised as:

1. Identify the area of study and set the boundaries
2. Compile and map data of the site in the following order:
   (a) Climate
   (b) Historical geology
   (c) Surficial geology
   (d) Physiography
   (e) Hydrology
   (f) Soils
   (g) Plant ecology
   (h) Wildlife habitats
   (i) Land use
3. Analyse data within a value system process
4. Report on the analysis with the best proposed land use based on the value system

Following the step-by-step process above, the mapping of the 'ecological inventory' of the *Potomac River Basin* recorded maps with longitudinal sections which included the diverse physiographic regions of the area. It also summarised patterns of topography, hydrology, geology, soils, vegetation, and current land use. These maps and cross sections made it possible to analyse a comparison of patterns from region to region within the river basin/catchment.

McHarg was very specific about this order, and his students called it '*the layer cake*' (see Fig. 7.3) (Spirn, 2000, p. 108). Analysis of sites always follows the same order (climate, geology, hydrology, soils, vegetation, wildlife), and McHarg described the reasons for this ordering as follows:

> Written on the place and upon its inhabitants lies mute all physical, biological and cultural history to be understood by those who can read it. This is the prerequisite for intelligent intervention adaptation. So, let us begin at the beginning. The place, any place, can only be understood through its physical evolution. Both climate and geology can be invoked to interpret physiography, the current configuration of the place. If one knows historical geology, climate, and physiography, then the water regimen becomes comprehensible: the pattern of rivers and aquifers, their physical properties and relative abundance, oscillation between flood and drought. Knowing the foregoing and the prior history of plant evolution, we can now comprehend the nature and pattern of soils. By identifying physiographic, climatic zones and soils, we can perceive order and predictability in the distribution of constituent plant communities. Animals are fundamentally related so that given the preceding information, with the addition of the stage of succession of the plant communities and their age, it is possible both to understand and to predict the species, abundance or scarcity of wild animal populations (Spirn, 2000; McHarg, 1992 [1969], pp. 105-7).

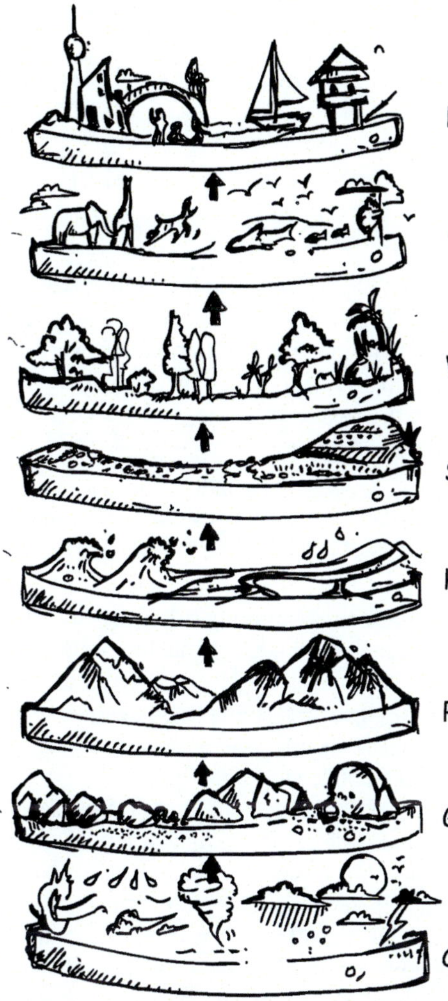

Fig. 7.3 The Layer cake method of Ian McHarg. (Drawing by Jesse Delmo, interpreted and synthesised by Roös, based on McHarg, 1992 [1969])

The ecological inventory summarized the patterns, with the outcomes represented as the land uses deemed suitable for particular locations, with alternatives indicated in the suitability maps. The development of McHarg's ecological approach through 'the ecological inventory' method, as previously indicated in the *Potomac River Basin Study* (1965–66), resulted in his scientific theory called *creative fitting,* which both explained and validated *Design with Nature* (1992 [1969]), and lent scientific integrity to his approach. McHarg's method was classified as *ecological scientific*, not only because it used ecological data, but also because the outcomes of the method *matched the processes of adaptation and evolution* (McHarg & Steiner, 2006, pp. 124–125). This theory helped determine where proposed uses of buildings, roads and infrastructure could fit on the land. *Creative fitting* locates the fittest environment for both human and natural uses integrated in various land use proposals, so that the land uses fit in with a changed landscape. Because of the 'fit for purpose' principles connecting with the landscape, it could be reasoned that creative fitting fulfilled the basic principles of ecological planning. Creative fitting was explained by McHarg as a theory that 'has absolutely no status whatsoever, except insofar as all the parts have been derived from excellent scientists' (McHarg, 2007, p. 21); it was a theory that conjoined the scientific theories of Charles Darwin's *The Origin of Species*

Table 7.1 Health as criterion for McHarg's model of creative fitting (derived from Swaffield, 2002)

Retrogression		Evolution	
Ill-health	Simplicity Uniformity Independence Instability Low number of species High entropy	Health	Complexity Diversity Interdependence (symbiosis) Stability (steady state) High number of species Low entropy

(Darwin, 2003 [1861]) and the lesser known physiologist and biochemist Lawrence Henderson's *The Fitness of the Environment* (Henderson, 1913).

McHarg defined *fit* as a blend of these two scientific theories, where every organism, system, and constitution are required to find the fittest environment to adapt to (McHarg & Steiner, 2006). Creative fitting theory takes *ecological planning* to the next level: *ecological design*. McHarg further strengthened his theory in his definition of *Ecological Design*:

> Ecological design follows planning and introduces the subject of form. There should be an intrinsically suitable location, processes with appropriate materials, and forms. Design requires an informed designer with a visual imagination, as well as graphic and creative skills. It selects for creative fitting revealed in intrinsic and expressive form (McHarg, 1967, p. 123).

Providing an example of the principles of creative fitting, McHarg noted the process of life and death. The form of life is through the evolution of a single egg into the complexity of an organism, and its death is the retrogression of an organism into a few decomposed elements. As summarised in Table 7.1, this model suggests that any retrogressing system is moving towards simplicity, uniformity, independence, instability, high entropy and ill health, where in contrast any system that evolves is moving towards a state of health (Swaffield, 2002; McHarg, 1967).

Throughout his works, McHarg argued that ecology should inform the schemes of designers and planners, as it would help to understand interactions between natural phenomena and landscape patterns. Steiner has explained that this approach is based upon collecting data in a chronological order. For example, regional climate helps shape the geology of a place, which in turn affects other abiotic processes, such as hydrology, that influence specific soils and microclimates (McHarg & Steiner, 2006). These abiotic processes come together in combinations that provide niches for plant and animal communities. Plant and animal communities can then be mapped to understand how they inform potential places for human settlement development that consider natural aspects. This process is reflective of the fundamentals of the *pattern language*, working across *levels of scale*.

7.8 Process of Place

A good example of considering the *processes of a place* and its natural aspects is McHarg's explanation of the *New Jersey Shoreline* in Chapter 2 of *Design with Nature* (1992 [1969]), which is very relevant to the coastal settlements and case study described in this book. McHarg addressed the issues of *land development on coastal dunes*, which indeed is still a problem in the twenty-first century (we still find that town planners approve proposed developments on primary dune and estuarine areas). The first important aspects to acknowledge are that dunes are small sand hills, formed by waves and winds, and are un-stabilized, extremely vulnerable to the forces of nature. The natural elements that help certain areas of the dunes to be more stable include different vegetation types with deep root systems that weave a dense mat network. The natural characteristics of the coastal shore, and areas viable for development or not, are indicated in Fig. 7.4.

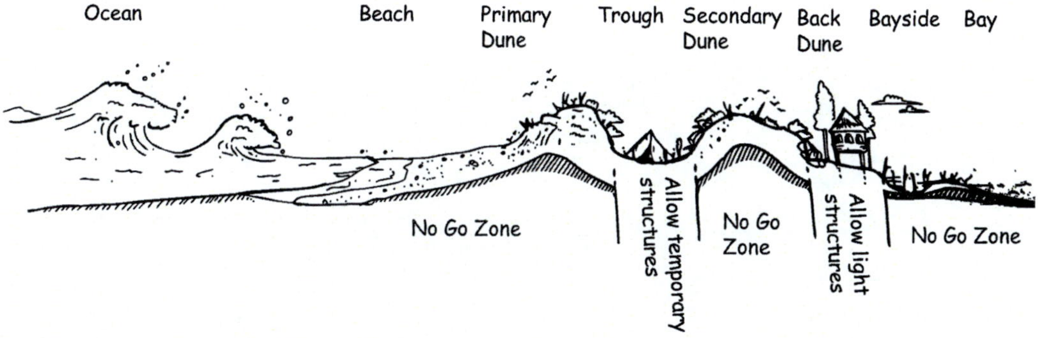

Fig. 7.4 Long section of dune systems, indicating areas of sensitivity. (Drawing by Jesse Delmo, synthesised by Roös, based on McHarg, 1992 [1969])

Below is a short summary of McHarg's analysis of the dunes and its probability of human use and development:

> The processes that determine dune creation are under the control of the waves and the wind. Waves usually approach beaches from an angle, and the sand carried by the receding wave is transported down drift of its origin. This is described as littoral drift and is a major factor in determines the formation of the beach and dunes. The winds close to the shore contribute salt spray, and thus influence the type of vegetation that grows on the dunes. As the dunes begin to form, vegetation colonise from seeds blown onto the dunes, and is further enhanced due to the arresting grains of sand and the prism of freshwater that is captured within a dune formation. Carried from inland, beach heather consolidate on the original dune, and in the trough vegetation grows to assist with protection of the secondary dune, which then results in woody plant material that grows on the lower third back dune. In the final stage the beach remains free of vegetation, but the primary dune is a thick dune with dune grass while in the trough nestle low beach heather, the face of the inland secondary dune, on the back dune there is stabilised woodlands, grass and beach heather on the perimeter and the landscape graduates into a swampy salt marsh and reed grass as part of the estuary (McHarg, 1992 [1969], pp. 7-18).

We can now draw conclusions from this simple analysis by McHarg, considering that since this is a continuous, changing and evolving environment, the tolerance or intolerance of human interference could guide settlement possibilities:

1. The *first zone* is *the beach*, and is tolerant of many interferences and impacts. Recreational uses will be adequate, such as fishing, swimming etc., but building structures are at risk due to the dynamic forces of tides, winds, storms and humidity.
2. The *second zone*, the *primary dune*, is absolutely intolerant of any interference, and cannot even be disturbed by trampling. This is an absolutely "no-go" zone, and no interference or any development of structures can be allowed.
3. The *next zone* is the *trough*, and this area is obviously protected from the winds, storms and blowing sand. It is possible that development could occur here, but the issue is groundwater, as this is the area where the groundwater accumulates. Interference on the hydrology of this area could result in the loss of plants and other natural species, which in turn will result in erosion and negative impacts on the whole system. Temporary shelter or structures would be a better option.
4. The *secondary dune*, or *inland dune*, is the second line of defence, and is as vulnerable as the primary dune system. This should not be developed at all, but the *back-dune system* with *stable woodlands* offers a better opportunity for establishing human settlement.
5. The *final zone* is the *bay and estuary area*. The bay and estuary are the most productive ecological systems; it is here where the infantile stage of fish occurs, shellfish dwell, and they are the breeding grounds for many birds. Salt marshes provide an important role in the protection and filtering system of nutrients, as well as playing an active role as a buffer between the sea and the main land. *This area should not be developed.*

Fig. 7.5 Breamlee, a settlement along the Victorian Coast, representative of settlement planning that allows development on vulnerable and sensitive dune systems. (Photo by Donna Squire, 2018)

McHarg's analysis and recommendations still stand today, and ecologists will support this argument. Unfortunately, if we go through this list, we quickly *realise that our human settlements of the past as well as in our current time, ignore these simple rules* and we develop built environments exactly in the most vulnerable areas (Fig. 7.5). This results in damage to the ecological system and nature, as well as causing major problems to the built environment, including flooding, structure degradation, hydrology issues, soil movement and very high maintenance issues due to the salt spray and salinity at foundations level. This knowledge of the coastal area is not new. Knowledge of this dynamic and vulnerable coastal environment was understood, respected, and used accordingly thousands of years ago by Indigenous people across the globe and by the Aboriginal peoples of Australia. This surely indicates to us how important it is to '*Design with Nature*'. McHarg concludes in his own words with the relevance of ecological planning and design:

> … Design with Nature is a personal statement to the power and importance of sun, moon, and stars, the changing seasons, seedtime and harvest, clouds, rains and rivers, phenomenal universe, participating in that timeless yearning that is evolution, vivid expression of the past, essential partners in survival and with us now involved in the creation of the future (McHarg, 1992 [1969], p. vi).

7.9 Conclusion Statement

We have identified in this chapter the important message that McHarg articulated throughout his career, which is that the *holistic connection between human and nature is very important*. This was supported by Lewis Mumford in the Introduction to *Design with Nature* (1992 [1969]), who stated that 'there is still only a small shelf of books that deals with man's relation to his environment as a whole'. Mumford contended that this theory is a notable edition to landscape planning and architecture, as it draws upon science, reflects on 'ecology' and confirm that man's life is bound up in the forces of nature (McHarg, 1992 [1969], p. vii). Further, as we have investigated McHarg's theory in more detail, we identified the principles of the *ecological method*, concluding that it has been successfully applied to various landscape planning projects. McHarg's multi-layered model of land uses (the layer-cake), and natural features of the landscape have certainly evolved into GIS mapping and relevant technology in use today for landscape planning and environmental assessments. While critics may have a point about the elaborate inventory process of McHarg's method, they lose sight of the most important considerations of the *ecological method*: the *systematic inclusiveness* and the *relationship of different structures* of the *natural environment and landscape*. We can certainly take note of these lessons.

Another criticism of McHarg's '*Design with Nature*' theory is that it falls short of the broader, holistic interconnected and 'cyclical ecology' and 'regenerative' context. It focuses mostly on the context of *patterns of land use* and morphology applicable to the *establishment of human settlements* in a specific area or region. This is understandable due to the time and era in which the theory was developed, but indeed McHarg succeeds in instructing us to *reintroduce nature*, or ecology for that matter, into the planning and design of the human built environment. Although McHarg reflects on the importance of ecological design, the method still encapsulates only the *principles and application* at a level of *planning practice*. However, we have identified fundamental principles of the *ecological method* that can be used in the construction of a *regenerative-adaptive pattern language*. These include the principles of *creative fitting*, the *vulnerability of coastal systems*, and *the layer cake,* which in essence represent an ecosystem process in levels of scale from the larger biosphere right down to the elements of soils, geology and hydrology.

The next step beyond this ecological method is *regenerative design and planning*, especially considering the context of cyclical ecology and regenerative processes, where regeneration is paramount. The following chapter looks at the theory of *Regenerative Design*, pioneered by landscape architect John Tillman Lyle, and introduces how the concepts of *patterns* and *wholeness* need to be part of the design and planning discourse. To align these theories with the *regenerative-adaptive pattern language* we further investigate *A Pattern Language* (1977) for the integration of a *whole systems* and *living systems* process.

Therefore:

The design and planning of human settlements needs to include the consideration of all *ecological systems* of a specific place, and the holistic connection between *human and nature* needs to be at the forefront of planning and design practice. The *ecological method* of McHarg through the application of *the layer cake,* which includes the principles of *creative fitting*, should be used. Human settlements should not be developed on sensitive coastline dunes and systems; rather, an approach should be applied that will use – NATURE'S DESIGN [7] –, forming an integral part of the *regenerative-adaptive pattern language*.

7.10 Fundamental Pattern 7

NATURE'S DESIGN [7]

Everything in nature is interconnected through a vast *webstring network* that works together to promote life. Using the ecological method of levels of scale, and the creative fitting process, Nature reveal itself as a process containing intrinsic forms, which includes a set of layers from the geology of a place to the plant, animal and human habitats. Therefore, locate places where human systems can be ordered in ways analogous to, and integrated with, the order of nature's systems (Fig. 7.6).

Fig. 7.6 Sketch of Fundamental Pattern – NATURE'S DESIGN [7]. (Drawing by Jesse Delmo, synthesised by Roös)

Downward links:

NATURE'S DESIGN [7] helps us to link and understand the context of levels of scale and creative fitting for ecological processes. However, within the *webstring* of life, an eco-systematic order is always present, and the phenomenon of order as part of a continued evolutionary process requires a regenerative-adaptive response to the forces of change within REGENERATIVE ECOSYSTEMS [8] ...

References

Alexander, C. (1979). *The timeless way of building* (Vol. 1). New York, NY: Oxford University Press.

Alexander, C., Ishikawa, S., Silverstein, M., Jacobson, M., Fiksdahl-King, I., & Angel, S. (1977). *A pattern language: Towns, buildings, construction*. New York, NY: Oxford University Press.

Alexander, C., Neis, H., Anninou, A., & King, I. (1987). *A new theory of Urban Design (Center for Environmental Structure)*. New York, NY/Oxford: Oxford University Press.

Biermann, V., Borngasser Klein, B., Evers, B., Friegang, C., Gronert, A., Jobst, C., et al. (2003). *Architectural theory – From the renaissance to the present, 89 essays on 117 treatises*. Kohn: Taschen.

Crouch, C. (1999). *Modernism in art, design and architecture*. New York, NY: St. Martins Press.

Darwin, C. (2003 [1861). *The origin of species by means of natural selection*. Broadway: D. Appleton and Company.

Freestone, R. (2010). *Urban nation: Australia's planning heritage*. Collingwood, Melbourne: CSIRO Publishing, Department of the Environment, Water, Heritage and Arts, Australian Government.

Fuller, R. B. (1938). *Nine Chains to the Moon. (first edition)*. Philadelphia, PA: Lippencott.

Hall, P., Hardy, D., Howard, E., & Ward, C. (2003). *To-morrow: A Peaceful Path to Real Reform (Original edition with commentary by Peter Hall, Dennis Hardy and Collin Ward)*. London: Routledge.

Henderson, L. (1913). *The fitness of the environment: An inquiry into the biological significance of the properties of matter*. New York, NY: Macmillan.

Henket, H. (2002). Modernity, modernism and the modern movement. In H.-J. Henket & H. Heynen (Eds.), *Back from Utopia: The challenge of the modern movement* (p. 411). Rotterdam: 010 Publishers.

Lovelock, J., & Margulis, L. (1974). Atmospheric homeostatis by and for the biosphere: The Gaia hypothesis. *Tellus. Series A, 26*(1–2), 2–10.

Lyle, J.T. (1991 [1985]). *Design for human ecosystems – landscape, land use, and natural resources*. California/ Washington, D.C.: Island Press.

Lyle, J. T. (1994). *Regenerative design for sustainable development*. New York, NY: Wiley.

McHarg, I. (1967). An ecological method for landscape architecture. *Landscape Architecture, 57*, 105–107.
McHarg, I. (1992 [1969]). *Design with nature.* Washington, D.C.: Wiley.
McHarg, I. (2007). *Ian McHarg: Conversations with students: Dwelling in nature.* (L. C. Margulis, Ed.). New York, NY: Princeton Architectural.
McHarg, I., & Steiner, F. (2006). *The essential Ian McHarg: Writings on design and nature.* Washington, D.C.: Island Press.
Parsons, K. C. (Ed.). (1998). *The writings of Clarence S. Stein: Architect of the planned community.* London: John Hopkins University Press.
Ruano, M. (2002). *Eco urbanism – sustainable human settlements: 60 case studies* (Second ed.). Barcelona: Editorial Gustavo Gili.
Sieden, L. S. (2000). *Buckminster Fuller's universe: His life and work.* New York, NY: Perseus Books Group.
Soleri, P. (1969). *Arcology: The City in the image of man.* Cambridge: MIT Press.
Soria y Puig, A. (Ed.). (1999). *Cerda: The five bases of general theory of urbanization.* Stockholm: Electa.
Spirn, A. (2000). *Ian McHarg, landscape architecture, and environmentalism: Ideas and methods in context.* Washington, D.C.: Dumbarton Oaks Publications.
Stein, C. (1951). *Toward New Towns of America.* (T. U. Michigan, Ed.). University Press.
Swaffield, S. (2002). *Theory in landscape architecture: A reader.* Philadelphia, Pennsylvania: University of Pennsylvania Press.
Van der Ryn, S., & Cowan, S. (1996). *Ecological design.* Washington, D.C.: Island Press.
Yenken, D. (1981). *The illustrated register of the National Estate – Introduction.* Melbourne: Macmillan.

Chapter 8
Regenerative Design, Ecology as Teacher

Abstract Regenerative design requires a change of our current processes of design and planning, moving from the piecemeal, technological and mechanical approach to a model that better reflects the understanding how the universe actually works as *a whole*. In this chapter we explore the importance of this shift in thinking, and provide information on how to design and develop systems that actually consider the *whole system*.

When considering the whole system, learnings from ecology need to be explored and these connections of the larger framework (scale linking) of *the whole* are dealt with in *levels of scale* and at a certain local *place-based* scale (Lyle, 1991 [1985]). To work within the larger context, a set of hierarchical scales in a larger organised system gets applied. As part of design thinking, the context of *scale* and *order* are fundamental in the process of 'ecosystem design', which considers *hierarchies of scale* and represents in the midst of complexity, the nature of *order*. This chapter discuss the *phenomenon of order* as suggested by Christopher Alexander (2001–2005a, 2001–2005b) and the three modes of *eco-systematic order* described by John T. Lyle (1991 [1985]), encapsulating modes of structure in *structural order*, function of materials and entities in *functional order*, and order in *locational patterns*. The chapter concludes with the REGENERATIVE ECOSYSTEMS [8] pattern, representing the reasoning that an ecosystem thinking needs to be fundamentally embedded in the discourse of regenerative design.

Keywords Regenerative ecosystems · Levels of scale · Eco-systematic order · Ecosystem design · Regenerative design · Generative processes · Sustainable systems · Sustainability

Upward links:
… along with NATURE'S DESIGN [7], which fundamentally informed us that everything is interconnected and that we need to consider the eco-systematic order that is always present, are the workings of THE WHOLE [1], which are constantly in a state of regeneration. For more than 4 billion years of adaptation and regeneration, life on Earth has calibrated itself through the process of REGENERATIVE ECOSYSTEMS [8]. The regenerative ecosystems process can help inform us about how to move towards the implementation of a regenerative-adaptive design paradigm that supports an INTEGRAL SUSTAINABLE WORLDVIEW [2].

P. B. Roös, *Regenerative-Adaptive Design for Sustainable Development*, Sustainable Development Goals Series, https://doi.org/10.1007/978-3-030-53234-5_8

8.1 Introduction

> In shaping the places where we live, we shape the patterns of our own behaviour. Over the past century or so, we have built into the landscape behaviour patterns that derive from attitudes about the nature of the earth and the human relationship with it that go back at least to the Renaissance. Expanded and driven by fossil fuels and exploding population, they are now not only outmoded but dangerous. For our culture to survive, for the human environment to become sustainable, we will have to change some of those patterns, which means, changing not only our behaviour but our environment. It is not just a matter of fine tuning, not even a matter of overhaul. What is needed is redesign (Lyle, 1991a [1985], p. ix).

Regenerative design theories emerged from earlier concepts of sustainable development. Sustainable development attempted to integrate environmental responsibility, social equity, and economic viability (Lyle, 1994). Sustainable design strives to minimise harm and be 'ecologically neutral'. In a sustainable system, lost ecological systems are not returned to existence or being regenerated. In a regenerative system, those lost systems can ultimately 'regenerate' themselves back into existence. Lyle's theory focuses on the ecological component of regenerative design, and concerns the community at a larger landscape scale. He contends that with respect to a given population, the landscape must be designed for supporting on going supplies of energy and materials for habitat, daily living and economic activity (Lyle, 1994). Accomplishing this regenerative state requires replacing the present linear system of material flow (Figs. 8.1 and 8.2), with cyclical flows at sources. A regenerative community or settlement provides for continuous replacement of the energy and materials used in its operation through its own functional processes (Lyle, 1991a [1985]; 1984). Many of the mechanisms and processes of regenerative design are inherently performed by nature (Cole, 2012). These processes are adaptive and constantly replacing and adapting resources to create living structures.

Regenerative design relates to people, and seeks to develop approaches that support 'the co-evolution of human and natural systems' so that both natural and social capital are supported (Cole,

Fig. 8.1 Saltmarshes and Wetlands are representative of interconnected complex ecological systems. (Photo by Author)

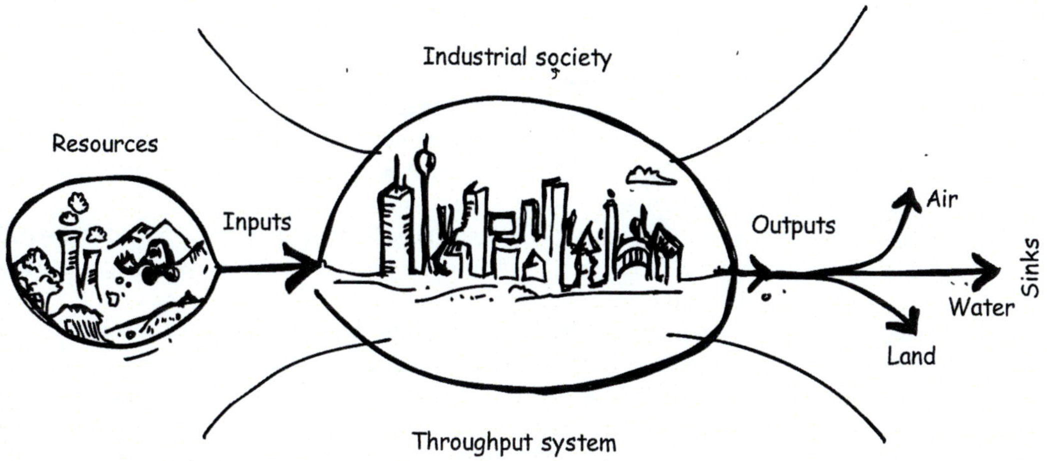

Fig. 8.2 Diagram indicating linear flows destructive to nature. (Drawing by Jesse Delmo, synthesised by Roös, based on Lyle, 1994)

2012). To achieve this balance, a consideration of cultural, social, economic and environmental aspirations is required. To do this, the 'uniqueness of place' needs to be understood and valued, and ideas about any changes in a locality need to be worked through with different people, representing different interests, over a period of time (Cole, 2012, p. 40).

The fundamentals in understanding 'the uniqueness of place' (where there is the unique, multi-layered network of living systems within a specific geographic region that results from the complex interactions of the natural ecology and human culture), are described in Lyle's (1991) article "Can floating seeds make deep forms?". In this work, Lyle asks landscape architects to think of creating form as floating seeds that disperse, lodge, and sometimes take root, resulting in shallow or deep forms. This reflects fundamentally the three basic modes of ecological order combining to create form. What a person visually experiences in the landscape is the direct expression of the latter's structural, functional and locational order, as they exist in that particular moment in time (Swaffield, 2002, p. 188; Lyle, 1991b). Lyle further adds human settlements into the mix of ecological order, and terms the full integrated system as 'human ecosystems'. An important aspect of 'human ecosystems' is to take into account that when human settlements intervene in the existing ecological systems, processes should be put in place to regenerate the resources used from nature. While McHarg's 'ecological method' helps to identify and map the ecological systems and land use patterns of a place, as noted in *Design with Nature* (McHarg, 1992 [1969]), Lyle's approach to regenerate ecosystems for achieving future sustainability is a strong expression of a more adaptive design approach (Swaffield, 2002, p. 171), as demonstrated in *Design for Human Ecosystems* (Lyle, 1991a [1985]). Various regenerative design methodologies developed later from Lyle's work, but essentially they are just different versions of the fundamental concept that regenerative design denotes the integration of human aspirations with the evolution of natural systems – essentially, the co-evolution of both (Mang & Reed, 2012, p. 26). This chapter thus investigates the principles and methods of regenerative design, but more specifically reviewing its connection to 'place', as well as the context of adaptation.

8.2 Design for Human Ecosystems

When introducing the concept of 'design for human ecosystems', Lyle reflected on a case study of the San Elijo Lagoon located in the coastal plains north of San Diego. The Lagoon is a narrow tangle of marches, mudflats, and shallow channels that push out of the Pacific Ocean into the coastal plain. Lyle used a ‘chain of effects’ process to identify the human impacts on the Lagoon system, and subsequently proposed design interventions to achieve a state of equilibrium where the Lagoon would be sustainable in regenerating resources similar than its natural state.

He proposed a ‘use frame’ for the development and management of the Lagoon that grew out of his analysis and studies (Lyle, 1991a [1985], p. 13). He included a ‘control by design’ schedule to minimise the damage to the natural environment, and established initiatives for regeneration. The pattern of uses that was proposed was very complex. Lyle acknowledges that to be able to achieve a successful regenerative system for the Lagoon, one would need to understand its initial natural state, as well as consider the total landscape integration of the ecology systems beyond the footprint of the Lagoon. This requires considering the *wholeness of the system*, even beyond its boundaries.

To solve this issue would be a mammoth task. Lyle proposed three organisational concepts in the shaping of ecosystems. The first is *scale*; the relative size of the landscape in question with its connections to larger and smaller ecosystems. The second is the *design process*; the pattern of thought that will deal with this landscape scale. The third is the underlying *order* that binds the systems together and makes the total human ecosystem work (Lyle, 1991a [1985], pp. 17–21). I found an interesting observation that in the works of Lyle and Alexander, *scale* and *order* feature as an important aspect in design thinking. Lyle (1991a [1985]) addresses this topic from the generative perspective, mostly aligned to the natural environment, and Alexander (2001–2005) from the form-making perspective, mostly aligned to the built environment.

Scale recognizes that the ecosystem is part of a larger and smaller subsystem, and that in turn it includes a number of smaller subsystems. In Lyle’s method of ‘ecosystem design’, this is a fundamental principle in understanding the larger framework that needs to be considered for designing a human ecosystem. Van der Ryn and Cowan (1996) refers to this as ‘scale linking’, where nature’s processes are inherently linked and intimately dependent upon the flow of energy and materials across scales. He explains this concept as follows:

> Consider a drop of rain. Hidden within it is an implicit history of places – water gathered from ancient fjords, alpine lakes, urban reservoirs, Antarctic ice, all running together in a single cycle, ever changing yet unitary. The flow of water in the biosphere links all of this together. Other natural cycles bind us to the living world as they carry nutrients and trace minerals between earth, air and water (Van der Ryn & Cowan, 1996, p. 33).

Lyle explains that despite all these connections of the larger framework (scale linking), a single human ecosystem has definite boundaries, and thus it can only be dealt with at a certain local scale (Lyle, 1991a [1985], p. 25). To work with this in the larger context, he proposes a set of hierarchical scales in a larger organised system. The hierarchy of scales ranges in levels from ‘construction’, to ‘site’, and through broader and broader hierarchical levels, to the ‘whole earth’, as indicated in Fig. 8.3.

The *design process*es of ‘ecosystem design’ consider hierarchies of scale and reflects in the midst of complexity the nature of *order*. Similarly, Christopher Alexander (2001–2005) argues that everything that exists in nature and the built environment includes the phenomenon of order (Alexander, 2001–2005a; 2001–2005b). Lyle defines order as visual in terms of landscape design. The visual order of nature reflects the geometry of the underlying eco-systematic order. Three modes of eco-systematic order are considered: the modes of structure in *structural order*, the function of materials and entities in *functional order*, and order in *locational patterns* (Lyle, 1991a [1985], p. 18).

Structural order describes the composition of biotic and abiotic elements, including rocks, soil, plants and animals. In consideration of the structure of an ecosystem, the interconnectivity between living and non-living elements is taken into account. Natural systems are continually reorganizing themselves through a specific structural process that is linked to place (location). Eugene Odum hypothesized that the input of energy in simple systems allows them to maintain stability (Lyle, 1994; Odum, 1975). Regenerative systems tend to have a need to reduce energy input and do this by means

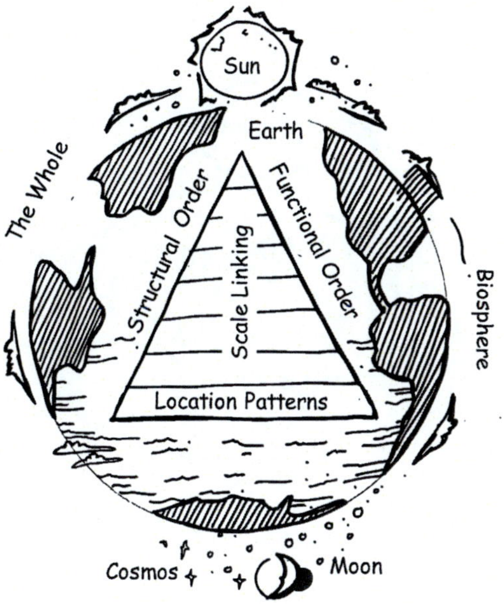

Fig. 8.3 Hierarchy of scale as described by Lyle. (Drawing by Jesse Delmo, synthesised by Roös – based on Lyle, 1991a [1985])

of a diverse structure, with interactions among elements and species within the system that benefit the whole; in this way, resilience is built into the structure (Lyle, 1994, p. 23).

Functional order derives from the flow of energy and materials that distribute the basic needs for life within the ecological system. The flow constitutes the dynamics of the system and creates the change that occurs according to a certain set of rules. An example of this is that landscapes receive energy from the sun, and this incoming energy then undergoes a series of transformations, one of which is photosynthesis. The energy harvested by photosynthesis makes its way through the web of food within the ecosystem, according to the second law of thermodynamics whereby more energy is passed on through transformation to other forms of energy. These forms of energy can be defined as regenerative forms of energy, constantly in a state of evolution-regeneration processes. They help maintain a balance in the ecosystem and maintain their functional integrity, contributing to wholeness (Lyle, 1994, p. 23).

Locational patterns are part of 'deep form', created before life forms existed. The geological composition of earth guided living species through the existing patterns, resulting in different landscape structures. An example of different *locational patterns* is that of a desert, whose landscape structure is very different to that of a rainforest. The number and type of species that the specific ecosystem can support are determined by the place and environment, which in turn is influenced by local conditions such as topography, the hydrological system, soil and climate (Lyle, 1994, p. 24). The whole system establishes a *structural order*, *functions* in a certain way, and has clear visible *patterns of location* that can be identified. It is within these patterns of the system that we can establish the elements that contribute to regeneration.

The above modes of order inform Lyle's argument that landscape is a process. Landscape is the physical context of a complex ecosystem and its visible manifestation (Lyle, 1994, p. 25). According to Lyle, the visible manifestation represents the regenerative capacities of an ecological system in six basic phases of functioning: conversion, distribution, filtration, assimilation, storage, and where human development occurs, human thought (Fig. 8.4).

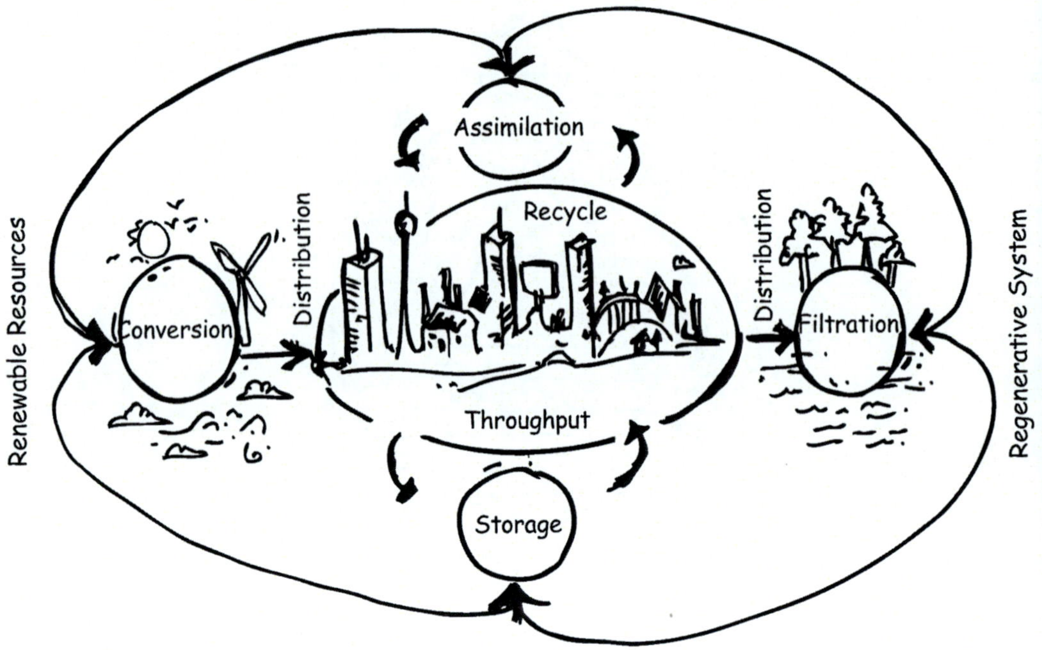

Fig. 8.4 The six phases of regenerative ecosystem functioning. (Drawing by Jesse Delmo, synthesised by Roös – based on Lyle, 1994)

Lyle argues that the regenerative capacities of the landscape are the key to the substance of life itself and thus to sustainability. The phases of functioning of the regenerative capacities of an ecological system can be summarised as follows (Lyle, 1994, pp. 26–27):

1. *Conversion*: During the process of photosynthesis, solar rays become life. The radiant energy of the sun goes through a series of conversions, from plant to biomass, to animal biomass, to heat in various forms and eventually is dissipated back into space.
2. *Distribution*: Energy materials reach the disbursed parts of an ecology system through distribution. In nature there are different ways of distribution: winds, water and gravity. Moving animals and insects move materials and energy with them.
3. *Filtration:* Plants and soils act as filters as air and water move across the landscape carrying materials and pollutants. These filtration functions of nature restore the relative purity of the air and water in preparation for the next phases of on-going ecological roles.
4. *Assimilation*: Everything produced in the landscape returns to the landscape. Most of the materials are dead biomass that provides food for vast populations of decomposing organisms. These are infinitely diverse, multitudinous processes of regeneration. Decomposition and re-assimilation are the basic processes of revitalising the earth.
5. *Storage*: During the journey of distribution and further ongoing cycles, materials are held inactive at some points in the process, awaiting eventual reuse. Water can be stored for centuries in underground aquifers, while on the other hand nutrients in biomass in rainforests are stored only for hours.
6. *Thought:* The human process of conscious thought, which has become the guiding force, joins nature's processes. These are influenced by human activity, more than 60% of the global landscape. Given the far-fetching impact of human activities on the atmosphere, the climate and water, there are actually no independent landscapes left in the world that do not get impacted by human thought.

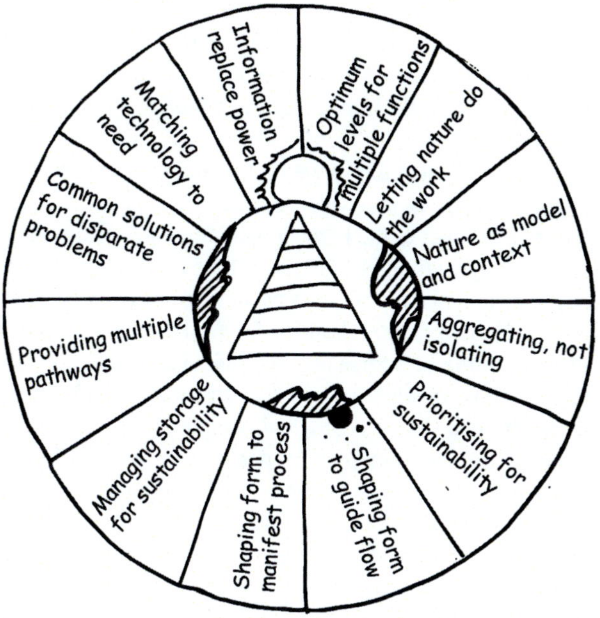

Fig. 8.5 Regenerative Design Strategies from Lyle. (Drawing by Jesse Delmo, synthesised by Roös, based on Lyle, 1994)

Following the phases of functioning, the 12 regenerative design strategies were put in place, as demonstrated in Fig. 8.5. These strategies have been applied to various case studies, such as the Centre for Regenerative Studies, the Rocky Mountain Institute, Etiwanda-Day Canyon Community Plan, and the Tijuana River Valley, amongst others.

It is this framework of 6 regenerative capacities, with the consequent 12 strategies, that forms the fundamental principles of regenerative design paradigms, and that have inspired other contemporary ones. It is the combination of the framework and the principles that moved Lyle's 'ecological design' theory to the next level of 'regenerative design'. This can be considered as a theory of systems thinking, and is a methodological directive that offers a somewhat unique identity to regenerative studies (Brown, 2008). Brown notes that the title of Lyle's book *Regenerative Design for Sustainable Development* suggests that studies of regenerative design are about a means (the method of systems thinking) to an end (achieving sustainability) (Brown, 2008, p. 5).

8.3 The Consideration of the Whole

Regenerative design requires a change of our mental model, from the piecemeal, technological and mechanical approach to a model that better reflects the understanding how the universe actually works as a whole. This shift helps to design and develop systems that actually consider the whole system, resulting in an integrated worldview. *Whole Systems and Living Systems Thinking* take the concept of sustainability into a thoughtful relationship with 'life support' systems (Reed, 2007, p. 675). Jaworski (1996) recapitulates the importance of responding to the integration of the whole system:

> Our mental model of the way the world works must shift from images of a clockwork, machine-like universe that is fixed and determined, to the model of a universe that is open, dynamic, inter-connected, and full of living qualities (Jaworski, 1996, quoted in Reed, 2007, p.675).

Concepts of regeneration such as *Whole Systems and Living Systems Thinking*, the *Regenerative Development*, *Permaculture*, and *Regenesis* begin with the notion that each place is dynamic with its own entity. It further acknowledges that the whole system needs to be included in an integrated regenerative design system where humans *participate as nature* (Mang & Reed, 2012), contributing to the co-evolution of the whole system.

When we consider the design and planning practice to achieve regenerative design outcomes as part of sustainable development, the designer or planner needs to design an ecosystem that integrates natural and human living systems as part of the regenerative process, considering the ongoing process of generation and adaptation. In addition to eco-literacy, the participatory nature of the regenerative process also requires psychological and cultural literacy, and the ability to tap into the creativity of the community (Mang & Reed, 2012; Mang, 2009; Landry, 2006). Regenerative design requires a participatory approach and an understanding of human and ecological systems.

Regenerative design brings with it challenges, and one aspect to consider is the changing environment. Kibert (2008) supports the strategy of ecological design as defined by Van der Ryn, whereby 'any form of design that minimizes environmentally destructive impacts by integrating itself with living processes', but states that it needs to include 'the effective adaptation to and integration with nature's processes' (Kibert, 2008, p. 34). Alexander et al. (1987) suggest that design and planning processes need 4 main characteristics to deal with this complexity, one of which is adaptation. Such characteristics are responsive to local conditions, they adapt to changing conditions, employ decentralized approaches, and are developed through the contribution and collaboration of many processes of bottom-up self-organization that follow patterns and certain generative rules (Alexander et al., 1987).

8.4 Conclusion Statement

In its native form, the concept of *Regenerative Design* is based on a process-oriented systems theory, including processes that restore, renew or revitalize their own sources of energy and materials, creating sustainable systems that integrate the needs of society with the integrity of nature. Regenerative Design matches ecosystems in that biotic and abiotic materials is not just metabolized but metamorphosed into viable ones. The central message from the regenerative paradigm is to shift the focus away from the concept of sustainability and sustainable development, whereby we should minimize the activities that are degenerative to the environment, towards also maximising human activities that restore and regenerate ecological systems. Cole articulates that regenerative design 'emphasizes a co-evolutionary, partnered relationship between humans and the natural environment, that builds, rather than diminishes, social and natural capitals' (Cole, 2012, p. 39). This fundamental re-conceptualization of sustainability means that the interdependence of the system's parts needs to create a whole that results in an outcome that is more positive than negative in terms of both human and nature systems (Reed, 2007). From this framework emerges the desire to have human systems improve both human and ecological system functioning. But to achieve this goal, systems must be adaptive, in that they must be designed to have the capacity to change and learn from experience, to self-correct through feedback (Gauthier, 2013, p. 2). This indicates that systems need to have the ability to be regenerative-adaptive in themselves.

However, as described in this chapter, we identified that the regenerative design paradigm mostly focuses on the input-output cycle, considering systems that restore, renew or revitalize their own sources of energy and materials, creating sustainable systems. What is lacking is a critical consideration of adaptation. Although many argue that systems must together contribute to the co-evolution of one holistic system (Mang & Reed, 2012), which by default include some levels of adaptation, the ability and adaptive capacity as well as the adaptive processes of these systems are not addressed. In the natural processes of the environment, adaptation is part of a systematic series of patterns that comprise a web of reciprocal relationships that are directed to some end.

To achieve regenerative outcomes that include the adaptive capacity, the need is there to identify the patterns of relationships between systems as clues to how systems and processes are sustained and self-organized, and how emerging adaptive outcomes are produced by these systems. The patterns have the potential to reveal the directionality, emergence and nature of the processes within the overall regenerative cycle (Gauthier, 2013; Michael & Meacham, 1998; Marvick & Murphy, 1998). Patterns should also reveal adaptability. These patterns collectively will result in a language that can be continuing, adaptive and resilient. Reed (2007) argues that the regenerative methodology should begin by 'attempting to understand how the systems of life work in each unique place' (Reed, 2007, p. 1). What this indicates is that it is about the understanding of the patterns that are unique to a place. In order to maximize the potential of regenerative development processes, patterns form the clues to the rationale behind the place-based process (Capra, 1996).

Therefore:

To understand the *regenerative process* of a *whole system* that is part of The Whole [1] is to understand the *patterns* and the generation of the system in the first place. Patterns provide a *place-based* understanding of the *integrated human and ecological systems*, which can only be appreciated through exploring the nature of generation processes and their relationship to patterns and pattern languages. It is thus important to apply to the practice of *regenerative design* the stipulations of *generative processes* from both human and ecological systems as part of the sustainable development discourse.

8.5 Fundamental Pattern 8

Regenerative Ecosystems [8]

Regenerative design engages the input-output cycle, considering processes that restore, renew or revitalize their own sources of energy and materials, thereby creating sustainable sys-

Fig. 8.6 Sketch of Fundamental Pattern – Regenerative Ecosystems [8]. (Drawing by Jesse Delmo, synthesised by Roös)

tems. Beyond this baseline approach, design human-nature ecosystems as co-spatial processes in which all underlying living processes contribute to the co-evolution of one holistic regenerative and adaptive structure (Fig. 8.6).

Downward links:
The pattern REGENERATIVE ECOSYSTEMS [8] demands a human-nature ecosystem that recognises all living systems as a manifestation of their underlying processes and flows of energy, information, materials and organisms, which ultimately contribute to THE WHOLE [1] through co-evolution of a regenerative-adaptive process. This requires us to participate to the TRANSFORMATIONS OF WHOLENESS [9] as part of nature, embedded in the unfolding of complex structures …

References

Alexander, C. (2001a). *The nature of order – An essay on the art of building and the nature of the universe, book one: The phenomenon of life*. Berkeley, California: The Center for Environmental Structure.

Alexander, C. (2001b). *The nature of order - An essay on the art of building and the nature of the universe, book two: The process of creating life*. Berkeley, California: The Center for Environmental Structure.

Alexander, C., Neis, H., Anninou, A., & King, I. (1987). *A new theory of urban design (Center for Environmental Structure)*. New York, Oxford: Oxford University Press.

Brown, K. D. (2008). *The future of regenerative studies as an area of inquiry. Lyle Center for Regenerative Studies*. Pomona, CA: Lyle Center, Col Poly Pomona University.

Capra, F. (1996). *The web of life: A new scientific understanding of living systems*. New York: Anchor Books.

Cole, R. J. (2012). Transitioning from green to regenerative design. *Building Research & Information, 40*(1), 39–53.

Gauthier, M. (2013). Regenerative development processes: Beyond systems thinking. SBSP Topics.

Jaworski, J. (1996). *Synchronicity: The Inner Path of Leadership, Berrett-Koehler*. San Francisco, CA.

Kibert, C. (2008). *Sustainable construction: Green building design and delivery* (Vol. 2). Hoboken, NJ: Wiley.

Landry, C. (2006).ccc London: Earthscan.

Lyle, J. T. (1991a [1985]). *Design for human ecosystems – Landscape, land use, and natural resources*. California, Washington, D.C.: Island Press.

Lyle, J. T. (1991b, Spring). Can floating seeds make deep forms? *Landscape Journal, 10*(1), 37–47.

Lyle, J. T. (1994). *Regenerative design for sustainable development*. New York: Wiley.

Mang, N. (2009). *Toward a regenerative psychology of urban planning*. San Francisco, CA: Saybrook Graduate School and Research Center.

Mang, P., & Reed, B. (2012). Designing from place: A regenerative framework and methodology. *Building Research and Information, 40*(1), 23–28.

Marvick, V., & Murphy, T. (1998). Patterning as process. *Permaculture Activist, 39*, 24–27.

McHarg, I. (1992 [1969]). Design with nature. Washington, D.C.: Wiley.

Michael, P., & Meacham, B. (1998). Pattern – key to the universe. *Permaculture Activist, 39*, 16–18.

Odum, E. (1975). Diversity as a function energy flow). In W. H. van Dobben & R. H. Lowe-McConnell (Eds.), *Unifying concepts of ecology: Report of the plenary sessions of the first international congress of ecology, The Hague, the Netherlands, September 8–14, 1974* (pp. 11–14). Berlin: SpringerLink.

Reed, B. (2007). Shifting from sustainability to regeneration. *Building Research & Information, 35*(6), 674–680.

Swaffield, S. (2002). *Theory in landscape architecture: A reader*. Philadelphia, Pennsylvania: University of Pennsylvania Press.

Van der Ryn, S., & Cowan, S. (1996). *Ecological design*. Washington, D.C.: Island Press.

Chapter 9
Living Structures: The Fundamental Properties of Wholeness

Abstract In ecological design and planning, as well as in regenerative design, the consideration of *the whole* is paramount, as indicated in the literature that were reviewed and in the previous chapters. If we are to shift from our current mechanical 'set of parts' thinking, to a holistic, integral based systems thinking, we must understand the nature of the change required. The foundation for the evolution of understanding, and thereby of the way we participate as part of nature, is through understanding *the patterns of the whole*. Wholeness thinking recognizes that the *entirety is interconnected*, and moves us beyond *mechanics* into a world activated by *complex interrelationships* such as natural systems, human social systems, and the conscious forces behind their actions (Seamon, Christopher Alexander and a phenomenology of wholeness. In EDRA – Environmental Design Research Association, Architecture Department. Manhattan, KS: Kansas State University, 2007). It is thus appropriate to introduce the concept of *wholeness* as a fundamental entity of nature, and in this chapter, we investigate the phenomena of the *unfolding of the whole* which creates living structures, and group these emergent characteristics as the *15 fundamental properties of wholeness*. The chapter further discuss the *generative processes* of design, including *patterns* and *pattern languages*, *morphogenetic sequences*, and the *generative code*. The chapter concludes with the *fundamental pattern* TRANSFORMATIONS OF WHOLENESS [9], providing guidance on an all-encompassing philosophy.

Keywords Pattern language · Wholeness · Transformations of wholeness · Fundamental patterns · Morphogenetic sequences · Properties of wholeness · Fifteen fundamental properties · Generative code · Generative process · The whole

Upward links:
… where the pattern REGENERATIVE ECOSYSTEMS [8] recognises all living systems as a manifestation of their underlying processes collectively in a human-nature ecosystems structure, TRANSFORMATIONS OF WHOLENESS [9] requires an essential mental shift towards the acceptance of the *fundamental properties of wholeness* evident in nature as well as in built structures.

9.1 Introduction

If we hope to bring our towns and buildings back to life, we must begin to recreate our languages, in such a way that all of us can use them: with the patterns in them so intense, so full of life again, that what we make within these languages will, almost of its own accord, begin to sing. … To work our way toward a shared and living

P. B. Roös, *Regenerative-Adaptive Design for Sustainable Development*, Sustainable Development Goals Series, https://doi.org/10.1007/978-3-030-53234-5_9

language once again, we must first learn how to discover patterns which are deep, and capable of generating life (Alexander et al., 1977, pp. 245–246).

In my quest to find an *Architecture for Life*, I often asked the question: what are the fundamental requirements for design and building to create objects that are inherently living structures? Nature is full of living structures that create wholeness. In this chapter I further investigate in detail Alexander's (2001–2005a, 2001–2005b) explanation of his hypothesis that *morphogenesis* is the *fundamental process of wholeness in nature*. Alexander argues that the process in nature does not use a 'plan' in the development of organisms, but rather it acts to transform an existing whole into a new whole through specific pattern languages (Alexander, 2001–2005b) (Fig. 9.1).

In this chapter we identify how nature preserves the structure of the earlier whole, but it often amplifies, articulates and deepens it in some important way. The discussions start at the *introduction of wholeness* and look at a scientific explanation of *the whole*, the structures of fractals and patterns in formations of nature and biological processes, and align these with the *notion of pattern languages*. DeKay (2011) explains that living systems are *integrated wholes* that possess properties which cannot be reduced to the properties of their smaller parts, due to the fact that the properties of wholeness are embedded emergent characteristics (DeKay, 2011, p. 302). Alexander describe these emergent phenomena as the *unfolding of the whole*, creating living structures, and groups these emergent characteristics as the *15 fundamental properties* – whether the unfolding happens in microseconds or over millions of years (Alexander, 2001–2005b, p. 32).

Fig. 9.1 The perfect structure of the Copper Amethyst African Daisy (*Osteospermum*). In nature symmetries and fractals complete the picture. (Photo by Author)

9.2 Introduction to Wholeness

An introduction to *wholeness* is encapsulated in the work of physicist Henri Bortoft, who argues that *the whole* cannot be explained through some analytical, sequential approach that puts parts together as in a mechanical system (Bortroft, 1996; Stefanovic, 1991). Instead, the whole can only be understood by analysing its parts through careful, intuitive and encounter thinking, feeling and perception. This means that there is a way to experience how the whole is present throughout its parts so that it can be found. Bortroft explains as follows:

> As one finds ways to better understand the parts, so the whole to which they belong becomes better defined; in turn, this progressive clarity of the whole sheds additional light on the parts, which become yet more understandable and say more about the whole; and
>
> If a part is to be an arena in which the whole can be present, it cannot be any old thing. Parts are not bits and pieces, because a part is only a part if it is such that it can bear the whole. There is a useful ambivalence here: "to bear," in the sense of "to pass through" and "to carry"; and "to bear" in the sense of "to suffer," where this is taken in the sense of "to undergo." By itself the part is nothing, not even a part, but the whole cannot be whole without the part. The part becomes significant itself through becoming a bearer of the whole (Bortroft, 1971 in Seamon, 2007, p. 3).

Bortroft's explanation of parts and the whole has a direct relationship to the works of Alexander, because throughout his life the latter tried to find ways to understand wholeness by attempting to identify the parts, or the patterns that connect them all together. This is best known through his works including: *A Pattern Language: Towns, Buildings, Construction* (1977) and *The Timeless Way of Building* (1979), and *A New Theory of Urban Design* (1987). During the 'pattern language period', Alexander and his colleagues tried to find the right parts by gathering examples of buildings and places that created a sense of order, identifying and explicating underlying physical qualities, with patterns that were constellations of the environment and human experience relationships (Grabow, 1983, p. 109). Alexander's *The Nature of Order* (2001–2005a, 2001–2005b) is a further attempt to provide more clarity on the importance of pattern language and wholeness. Alexander also provides a mathematical explanation of wholeness, of which a further description is provided in Appendix 1. This effort to incorporate life-evoking geometry and design methods into a process of place-making that sustains environmental wellbeing was fundamental in my approach to develop a *contemporary regenerative-adaptive design and pattern language theory*.

In the mammoth four-volume publication *The Nature of Order: An Essay on the Art of Building and the Nature of the Universe* (2001–2005a, 2001–2005b), Alexander described the formula that leads to a liveable and sustainable built environment as a *set of patterns*, a system of explicit steps for *creating living structures* within the fabric of a place. These '*generative codes*' are specific to the environment, and can be found in nature through the process of *morphogenesis*. Alexander's theory defines the word 'generative' as a process where there is always *a sequence, an order, and instructions that follow the rule of centres,* which appear within the larger whole as distinct and noticeable parts (Alexander, 2001–2005a, 2001–2005b). He argues that generative codes are capable of driving the organic unfolding of a place in such a way that the people who live and work in that place have an opportunity to be sustainable and flourish personally, economically and ecologically. The theory further argues that by creating a living structure, a place is created that is 'alive'. Living structures like this are referred to as places that have 'soul', and have an inherent pattern language that is part of a generative process. Places with soul inevitably arise whenever living structures are used, and should provide for long-lasting sustainability and resilience. Similar to places with place character and 'sense of place', people living there experience deep connections with nature, and refer to these as healthy environments.

9.3 Generative Process and Design

The *generative process* offers an unconventional way of conceptualising the mechanisms and design of our built environment. Research in generative systems is closely linked to the general concept of synthesis, mostly apparent in the systems of nature. Nature has devised specific mechanisms for generalised synthesis, using the structure of DNA, protein synthesis and biochemistry (McCormack, Dorin, & Innocent, 2004). The diversity and adaptability of life on earth through the *generative process* demonstrates how to potentially overcome problems in the built environment. It may be helpful to describe the roots of the word 'generative', which have relevant definitions and meanings relating to the concept of generative design and pattern languages. Further, 'regenerative' has a very close relationship to the processes that we will investigate in this chapter. This relationship is evident in the basic definition of 'generative', to bring something into existence: 'having the power or function of generating, originating, producing, or reproducing', as defined in the Merriam-Webster Dictionary (http://www.merriam-webster.com/dictionary/generative). Neis, Brown, Gurr, and Schmidt (2012) describe a definition of the generative process in building relative to the origination of many into one: '…many inputs, actors and activities that are relevant to the design and production of a building or a urban neighbourhood' (Neis et al., 2012, p. 4). These definitions lead us to look at the properties of *generative systems*, *generative process* and *generative design*.

Generative systems offer a philosophy that views the world in terms of dynamic processes and their outcomes. In Kuhn's terminology (Kuhn, 1962) the properties of generative systems can be summarised as:

- The ability to generate complexity, creating many orders of magnitude greater than their specification. This is known as database amplification, where aggregates create a dynamic hierarchy;
- The ability to self-maintain and self-repair. Generative systems may adapt to maintain stable configurations within a changing environment;
- The complex and interconnected relationship of organism and environment. Organisms not only evolve to adapt to their environment, their presence can affect and change the environment self. Inter-species dependencies form a complex web of relationships with connected feedback loops; and
- The ability to generate novel structures, behaviours and relationships. Novel structures in this instance refer to the quality of being new, original and different from anything else (McCormack et al., 2004).

Generative design offers modes of experience based on the incorporation of generative system dynamics into the production of objects and experience. In this context, generative design tries to understand and conceptualise the world and its complex structures, and attempts to approach problems through the generative process, formulating a number of parameters, principles and rules that interact with each other and create form, shape and place (Neis et al., 2012, p. 4).

Generative process, as defined in the context of urban planning and architectural design, needs to explore the challenges of human settlements, neighbourhoods, connectivity, liveable streets, green spaces, identity, and whole communities. The *generative process* explores the processes within the built environment at different levels of scale and in different modes that include environmental, physical, social, cultural and economic themes considering the urban design and planning needs. Neis et al. (2012) distinguish between three different generative processes within the overall pattern language approach (Neis et al., 2012, pp. 4–5):

1. *Patterns and pattern languages* are considered as the first kind of generative system or process. The applications of patterns started in the mid-sixties and extended into many fields of science, are still widely in use, and;

2. *Morphogenetic sequences* and the system of rules are considered the second kind of generative process. This process works within a dynamic system of organized rules and is applied in a sequential manner. In the context of architecture, planning and urban design this has been described in the book *A New Theory of Urban Design* (Alexander et al., 1987); and
3. The third kind of generative process is reflected in the '*generative code*' (urban), which includes systems to design, develop, and build a coherent, living neighbourhood that is 'alive', sustainable, active and beautiful, and which unfolds through the people that live there.

An explanation of generative processes, as above, raises a few questions. What is the relationship of patterns in nature with patterns in the built environment, in the context of Alexander's pattern language? Do the morphogenetic sequences and the generative code reflect the living properties of nature? We will explore in further detail the three generative processes in relation to the context of the *regenerative-adaptive pattern language* proposed in this book.

9.4 Pattern Language for Design

Undeniably, the identification of the nature of design problems and the proposed solution through a mathematical method in Alexander's *Notes on the Synthesis of Form* (1964) set the foundation for *patterns* and *pattern language*. Logical structures form the order of the method, and the design solution include a process of conceptual order and patterns. The rationale behind the method is the use of *logical structures*, similar to what was used in the creation of form by *unself-conscious cultures.*[1] Alexander makes a comparison between 'unself-conscious' and 'self-conscious' cultures to explain the concept of logic and patterns, especially in the method of making things and buildings. In unself-conscious cultures there is very little thought about aesthetics and architecture as such (Alexander, 1964, p. 33), whereas in our self-conscious cultures, engineering, architecture and art have been prominent in the design of things and buildings, classified within styles and periods. Unself-conscious cultures repeat patterns of tradition, where form is informed by symbols (Van der Ryn, 2005), and there is a right way or a wrong way to make and build things (Alexander, 1964). In unself-conscious cultures, form-making is learnt informally, through imitation and correction. Primitive forms have good structure from a process of gradual adaptation, thus the system passes through history adapting forms to the context of the environment.

These complex patterns and forms appear continuously across generations in Africa, in representations of phenomenal complex fractals in settlement architecture, cross-cultural engagements, and logarithmic scaling in designs (Eglash, 1999 [2005]). Adaptation through scale linking is clearly demonstrated in circular fractals in the Ba-ila settlement in southern Zambia, as demonstrated by Eglash (1999 [2005], p. 27), as well as the settlements of traditional Zulu villages and many others (Fig. 9.2).

Based on these principles of *primitive form-making*, Alexander introduced a method that could potentially solve problems of the design and development of the buildings of self-conscious cultures (as opposed to unself-conscious cultures), introducing the idea of a pattern, and consequently formulating the basis of the 'pattern language' (Alexander, 1964). Although Alexander argued that the structure of a pattern is very simple, the results of this kind of fusing relationships can be *highly complex*, similar to what can be found in traditional environments as demonstrated in the Ba-ila, traditional Zulu villages, and many other African settlements. This notion of simplicity generating

[1] Unself-conscious cultures noted by Alexander relates to traditional, indigenous, or primitive cultures that were the first peoples of the land (Alexander, 1964, p. 33)

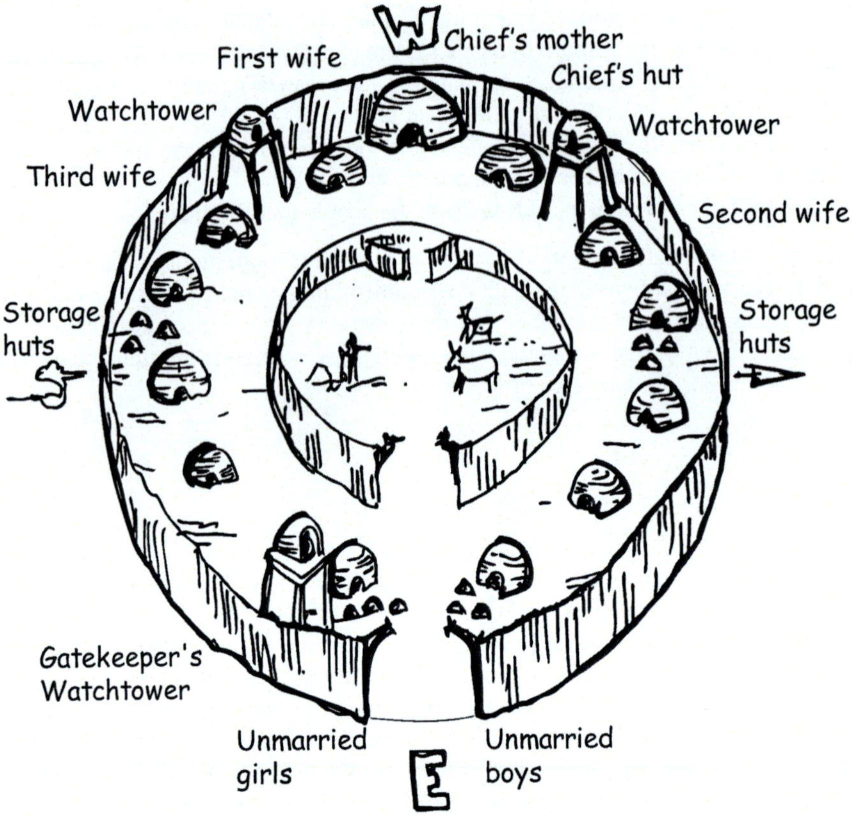

Fig. 9.2 Traditional Zulu village settlement representing circular fractals. (Drawing by Jesse Delmo, synthesised by Roös)

complexity is also evident in complexity science,[2] where simple processes can generate the complex structures of the natural world through adaptive iteration (Mehaffy, 2009a). *Improvisional patterns* therefore have built-in dynamics of repetitive principles and solutions combined with extreme variability, and the complex fusing relationships that can be found in fractal geometry (Stark, 2012). The *Mandelbrot Set*, indicated in Fig. 9.3, is a good example of *patterns at self-similarity at all scales.* What Mandelbrot discovered is that many natural systems and entities have an *underlying geometric order* that results from these self-similarity patterns, which can be either spatial or temporal.

The cycles that connect phenomena in nature at very different scales are named by Van der Ryn as *scale linking*. Van der Ryn and Cowan (1996) further provide a comparison of nature's geometric forms and cycles with a description of the *Koch curve*. Taking a single line segment and dividing it into three segments of equal length, as shown in Fig. 9.4, the repetition constructs a curve. Then an equilateral triangle is drawn from the middle segment from segment *n1*, erasing the base, which result in segment *n2*. This process can then be applied to each line segment as shown in segment *n3*. Continuing this process of pattern shaping to the segments *n4* and *n5*, the *Koch curve* has infinite length because each time step *n3* is undertaken for each line segment, the number of line segments are generated 4 times. At each stage the new line segments are one-third the length of those in the preceding stage, resulting in symmetry of self-similarity. The symmetry in the curve is built up by sub-forms that echo the whole. The self-similarity replications are named fractals, similar to those in the

[2] Complexity Science is the scientific study of complex systems, systems with many parts that interact to produce global behaviour that cannot easily be explained in terms of interactions between the individual constituent elements.

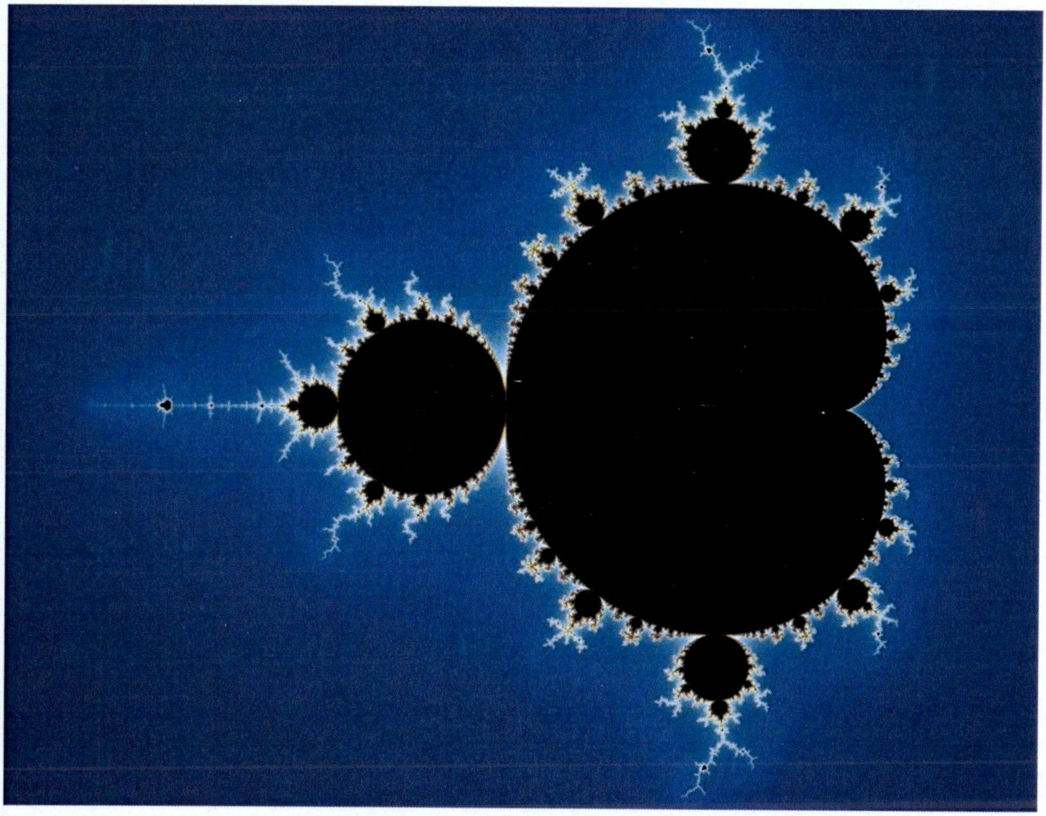

Fig. 9.3 The Mandelbrot Set. (From Wikimedia Commons, Created by Wolfgang Beyer with the program Ultra Fractal 3. https://commons.wikimedia.org/wiki/File:Mandel_zoom_00_mandelbrot_set.jpg. [CC BY-SA])

Mandelbrot set. The self-similarity in nature's geometry is a direct consequence of identical processes shaping form across many scales to form the whole (Van der Ryn & Cowan, 1996, p. 38).

The fusion of these patterns results in a product that is called the *Koch Snowflake*. As depicted in Fig. 9.5, the Koch Snowflake has similar properties to snowflakes in nature, with the astounding property that while it can have a boundary of infinite length, the area of the snowflake can never exceed the area of the circle that connects the 3 points of the original equilateral triangle. A snowflake in nature exhibits usually a sixfold radial symmetry (Fig. 9.6).

Consider the complex inherent *scale linking* process that nature had to undertake to form the symmetrical patterns of one snowflake. Complex shapes are created from patterns as the flake moves through differing temperature and humidity regimes; the moisture contains vast microscopic molecules gathered from the atmosphere, and results in individual snowflakes that are each unique in structure and form. These examples indicate that *nature's orderly structure* within patterns and geometry could guide us in understanding a pattern language for developing and designing our built environments.

Further, we can argue that outcomes of the fusing relationships of patterns that happens at different *levels of scale* in the built and natural environment shape the essence of a place. *The character of a place is given to it by the events, or patterns that happened there.* The events create patterns of space, and can be summarised as:

> These patterns of events are always interlocked with certain geometric patterns in the space. Indeed… each building and each town is ultimately made out of these patterns in the space, and out of nothing else: they are the atoms and the molecules from which a building or town is made (Alexander, 1979, p. 78).

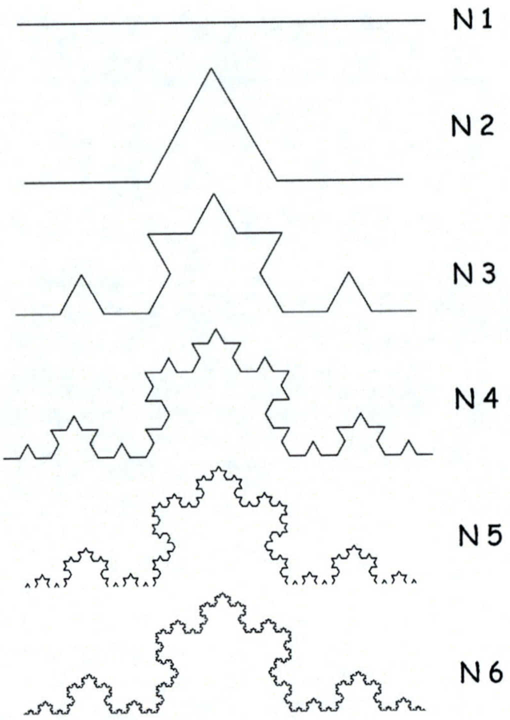

Fig. 9.4 The Koch Curve in the repetitive process of Scale Linking. (By Author, adapted from User Romero Schmidtke on es.wikipedia, https://commons.wikimedia.org/wiki/File:Koch_snowflake0192.png. [CC BY-SA])

Fig. 9.5 The Koch Snowflake and Nature's Snowflake. (Koch Snowflake adapted from https://commons.wikimedia.org/w/index.php?curid=466496, [CC BY-SA 3.0]) (Snowflake5 from Bentley, https://commons.wikimedia.org/wiki/File:Snowflake5.png, detail of image: SnowflakesWilsonBentley.jpg [Public Domain])

Fig. 9.6 Snowflake forms indicating the six-fold radial symmetry. (Adapted from Wilson Bentley, https://commons.wikimedia.org/wiki/File:SnowflakesWilsonBentley.jpg. [Public Domain])

Each place has a unique character, and the visible (and non-visible) patterns of that place form a pattern language. When people of a community shape buildings and a town, these patterns grow through the *process of unfolding*. Alexander calls this 'genetic code' a 'pattern language' (Alexander, 1979). According to Borchers (2008), the *notion of patterns* in a pattern language, used in the fields of architecture, software design or sustainable innovation, will follow the following formal principles:

> Each pattern is a set $p = \{n, f_i \ldots f_i, s, e_i \ldots e_i\}$ of a name n, forces $f_i \ldots f_i$, the solution s, and examples $e_i \ldots e_i$. It describes a commonly encountered design problem and suggests a solution that has been proven useful in this situation. A pattern language is a directed acyclic graph with each mode represents a pattern. There is a directed edge from pattern $p1$ to $p2$ if $p1$ recruits $p2$ to complete its solution. Edges pointing away from a pattern are its consequences, showing what lower levels of pattern need to be applied next. Edges pointing to a pattern are its

context, the situations in which it can be applied. This relationship establishes a hierarchy within the pattern language (Borchers, 2008; in Neis et al., 2012, p. 92).

Within this hierarchy of relationships in the pattern language (which follow formal principles as indicated above), there are also overlaps, cross-linkages and ambiguities. Alexander identified this problem, and in *A Pattern Language* (1977) he developed 253 environmental patterns to solve it. Through many years of study and testing, synthesised in the four-volume *The Nature of Order* (2001–2005a, 2001–2005b), he attempted to solve these issues.

A Pattern Language and its further development *The Nature of Order* included two major topics: the *identification of life-forming geometrical properties* that might provide a link between the pattern language and its physical manifestation, and the *method or process* of construction, identifying a way of building whereby each step of design and construction follows from the preceding steps to the next steps in the building process. These patterns may be described as combined fragments of geometric configurations that obey a kind of grammar, much as a natural language does. Just as a natural language uses a fairly limited set of elements recombined into endless possible configurations, similarly the pattern language is intended to be recombined in a much more flexible, networked configuration within the next level of the language – the so-called *Morphogenetic sequences* (Mehaffy, 2009b).

9.5 Morphogenetic Sequences

If the process of *morphogenesis* is applied to the design and planning of the built environment 'deep sustainability' can be achieved. This hypothesis was presented by Alexander in his Schumacher Lecture in Bristol (2004) entitled *Sustainability and Morphogenesis: The Birth of a Living World.* This lecture explores the phenomenon of *wholeness of a place*, and proposes that the morphological growth of a place gives it character and allows the built environment to adapt to changes. Based on this natural process, the place achieves sustainability (Alexander, 2004). Alexander referenced most of his lecture content from *The Nature of Order: An Essay on the Art of Building and the Nature of the Universe* (2001–2005a, 2001–2005b). To explain its application in the built environment, Alexander described the context of morphogenesis in the natural world as follows:

> Things in the biological world, almost by definition, are created continuously by morphogenesis, that is by a process which is all the time growing and adapting, whether it be in a growing embryo or in a forest or a field, and which gives form, progressively, while growth and change and adaptation are happening. In *real morphogenesis* the form of what is coming, or what is about to be, is always drawn from the form of what was in the moment just before. That is, things are always going like that. If a tree is growing for 500 years, it is continuously unfolding from its previous state, and then what we see and recognize is first of all in itself a process. But even if you just look at it in its static state, it is at that moment the end product of transformations that have been going on, and on, and on. And these are the things, which give it shape, form, and substance…Traditional society also managed to do something very much like that – that is to say, morphogenesis…Whatever it was, it was shaped, modified, shaped again, and adjusted and so on, and so on, and so forth (Alexander, 2004, p. 6).

Morphogenesis is the *fundamental process* of *wholeness in nature*. Alexander argues that process in nature does not use a 'plan' in the development of organisms, but rather it acts to transform an existing whole into a new one. In doing so, nature preserves the structure of the earlier whole, but it often amplifies, articulates and deepens it in some important way (Mehaffy, 2009b). He used the example of the formation of the seed of the Shepherds Purse (*Capsella bursa-pastoris*) to identify this process and demonstrate the unfolding of the whole (Fig. 9.7). Other examples include the unfolding of embryos, the development of the foot of a mouse, a bird's beak, river patterns carved in the landscape, frog embryo growth, crystal growth, and the sequence in the Belousov-Zhabotinski reaction (Alexander, 2001–2005b, pp. 18–44).

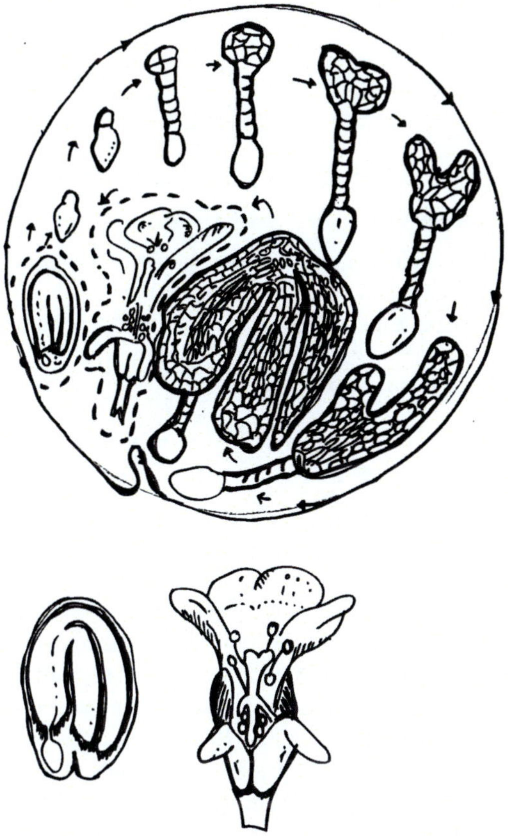

Fig. 9.7 Development of angiosperm seed – Shepherds Purse (*Capsella bursa-pastoris*). (Drawing by Jesse Delmo, based on Alexander, 2001–2005b, p. 34)

In biology the step-by-step process is well analysed and understood in the context of the natural processes, chemical composition, DNA structure and so forth, but the context of *configurations and form* is less known (Alexander, 2004, p. 20). To deal with the similar step-by-step process in the context of the built environment and the natural environment in form-making, Alexander developed and used *15 fundamental properties* (Alexander, 2001–2005b, p. 32).

9.6 Fifteen Fundamental Properties

To explain the creation of geometry and form, *15 fundamental properties* in nature can be identified, and are always present. The *15 fundamental properties*[3] occur in nature as well as in buildings, places and situations that sustain *the wholeness* and life (Alexander, 2001–2005a, pp. 244–296):

[3]The *15 fundamental properties* are described and referred to in various publications of Alexander, starting in *The Nature of Order, Book 1, The Phenomenon of Life* (2001-Alexander, 2005a) as the *15 fundamental properties* (p. 143–242), and the *15 properties in nature* (pp. 244–296). In *The Nature of Order, Book 2, The Process of Creating Life* (2001-Alexander, 2005b) as the *structure-preserving transformations* (pp. 65–84). In *The Battle for the Life and Beauty of the Earth* (Alexander et al., 2012), the properties are referred to as the *15 geometric properties of wholeness* (pp. 421–430), and the *15 transformations* (pp. 431–436).

1. Levels of scale	8. Deep interlock & ambiguity
2. Strong centres	9. Contrast
3. Boundaries	10. Gradients
4. Altering repetition	11. Roughness
5. Positive space	12. Echoes
6. Good shape	13. Void
7.Local symmetries	14. Simplicity & inner calm
	15. Not separateness

Important to the creation of wholeness is the identification of strong '*centres*'. Centres are key to the generative process, and part of the 15 properties. But the integration of all *15 fundamental properties [of wholeness]*[4] needs to happen to support regeneration and adaption, and to form the whole, summarised as follows:

9.6.1 Levels of Scale

Levels of scale are formed naturally from the way the whole system brings itself into order. This is evident in all natural systems, for example a tree consists of decreasing geometric forms, from the largest to the smallest; trunk, limbs, branches, twigs, leaves, flowers, seeds (Fig. 9.8). Similarly, the cell consists of cell wall, organelles, nucleus, and chromosomes. In the landscape, a mountain range consists of high mountains, individual peaks, foot hills, surrounding smaller hills, valleys, plains, rivers. Within these *levels of scale* there is an obvious functional coherence, with recognizable hierarchies in the organization of these systems (Alexander, 2001–2005a, p. 246). *Levels of scale* are the most obvious in the context of the biophysical world, identified in the levels of biological organisation, and the hierarchy of complex biological structures that define life form (Solomon, Berg, & Martin, 2010).

Lovelock describes in *The Vanishing Face of Gaia* (2009) that the Earth with all these *levels of organising* is a self-regulating complex system involving the biosphere, the atmosphere, the hydrosphere and the pedosphere, tightly connected in an evolving system. His Gaia theory argues that the Earth system is a whole, an interconnected system that consists of a physical, biophysical and chemical environment optimal for contemporary life (Lovelock, 2009). As a self-organising system, the Earth continues with creating conditions for life in an ecological *levels of scale* structure.

9.6.2 Strong Centres

Throughout the physical world, *strong centres* are always visible. Natural processes have centres of action, and the reaction from centres radiate outwards. In physics, spatially symmetrical fields always carry electric, gravitational, magnetic, nuclear and gravitational patterns outwards. Alexander uses the famous photograph of the milk drop by Harold Edgerton, describing the radial symmetry and notion of a *strong centre*, indicated in Fig. 9.9. This ordinary quick process results in a remarkable structure preserving transformation, with deep interconnections of the parts to form a whole. This occurs not just in one single step, but continuously through a sequential transformation order of form and time (Mehaffy, 2009a, p. 107). The geometry of this process results in an intensified crown shape with similar droplets in a smaller scale, just to return back into the original form. From the point of view of the droplet's configurations, the process produces the qualities of a *strong centre* (Alexander, 2004, p. 20).

[4]The author adopted the 15 fundamental properties for application to the *regenerative-adaptive pattern language*, and rephrased them as '*15 fundamental properties of wholeness*' due to the results of this research indicating the inherent wholeness and connection of the properties in the context of regeneration and adaptation.

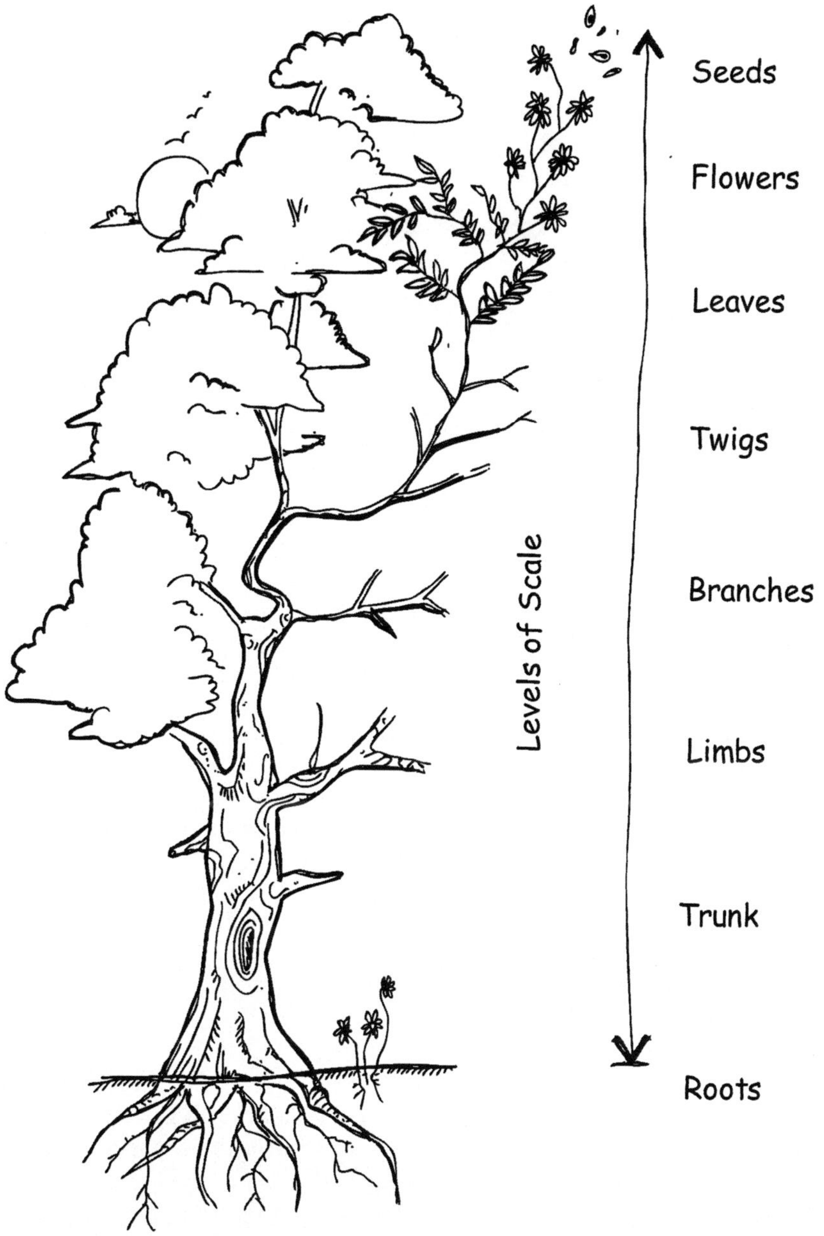

Fig. 9.8 Structure of biological organisation evident in the levels of scale as noted in the structures of a tree. (Drawing by Jesse Delmo, synthesised by Roös)

9.6.3 *Boundaries*

In nature *strong boundaries* exists to achieve functional separations and transitions between different systems. Between different phenomena there is the area of interaction from one zone to the other. A good example is the wall of a plant cell, a thick structure where the flow in and out of the cell is controlled. Yet, the plant cell wall creates a definite *boundary* between itself and other cells in a multicellular structure. Another example is the landscape of a river (Fig. 9.10), where the banks of the river

Fig. 9.9 Strong centres: Radial symmetry in a milk drop. (Drawing by Jesse Delmo, synthesised by Roös)

form a natural boundary, and the natural habitat around the river has its own ecosystems, with its own definite *ecological boundaries*, laws and habitats (Alexander, 2001–2005a, p. 255).

9.6.4 Altering Repetition

Formations of spiral galaxies, the repartition of sand waves driven by the wind, and the forming and breaking of waves in the ocean all reflect principles of *fractal geometry,* reminding us that nature is constantly linking scales in *altering repetitions*. The famous drawing of Hokusai, *The Great Wave of Kanagawa,* illustrates the principles of altering repetition, with the formation of self-similar waves at many levels of scale. The drawing is not based on imagination, but on a constant repetitive reality that is evident in waves (Alexander, 2001–2005a, p. 258; Van der Ryn & Cowan, 1996, p. 43). When examining and looking closely at waves, one finds that the on-going wave is extended, maintained, developed and repeated with a continued *altering repetition*, but never violated (Fig. 9.11).

9.6.5 Positive Space

In the variety of natural systems, the nature of *positive space* is necessary to preserve the *wholeness of the system*. In the majority of natural developed wholes, the wholes and the spaces between them form an unbroken continuous arrangement of *positive space*. When crystals grow, they form coherent polyhedral shapes as they butt into each other. An outward thrusting crystal, resulting in positive

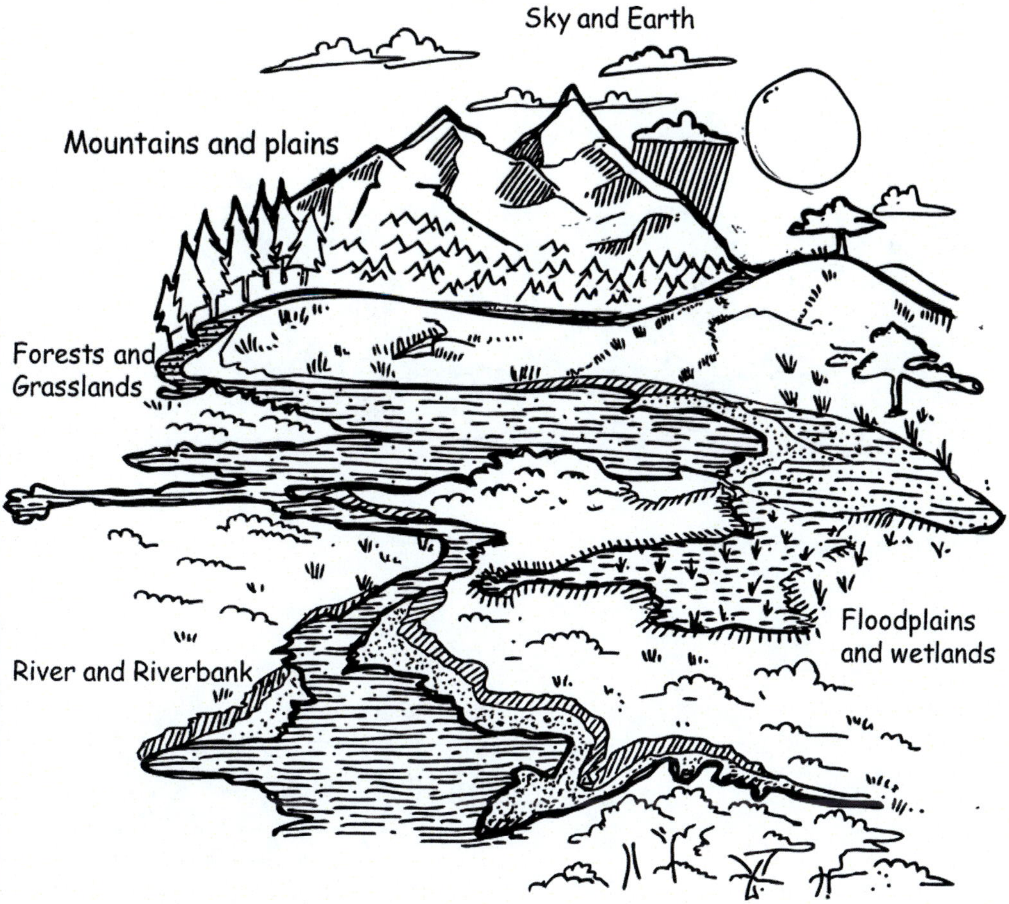

Fig. 9.10 Natural boundaries of a river system and ecological boundaries evident in the landscape. (Drawing by Jesse Delmo, synthesised by Roös)

space, occupies each bit of space in the group of crystals. The *positiveness of the space* results from the compactness of the crystals in an outward manifestation of the internal coherence of the physical system. In the crazing of porcelain, when the surface cools it forms a coherent shape of cracks (Fig. 9.12). The cracks follow maximum stress lines and form in such a way to relieve the stress. This results in good, compact shape with *positive space* in between the cracks (Alexander, 2001–2005a, p. 262).

9.6.6 Good Shape

Systems in nature form patterns that result in *good shapes*. We can see this in the beautiful shapes of plant structures, leaves, crystals, animals, birds, shells, the bone in a skull, and the curl of the breaking wave. The beauty of these shapes in nature is a result of harmony, of geometric forms that are often curved (Fig. 9.13). *Meanders* reflect topological mixing and are visible in the bends and curves in rivers, the shapes on shells, and the vortex streets in lakes and flood plains, all reflective of good shapes. *Cracks* in nature provide similarity in degrees and joints when a surface is relieved from stress, as we can see on the surface of old pottery, the drying mud of dams and lakes, cooling lava

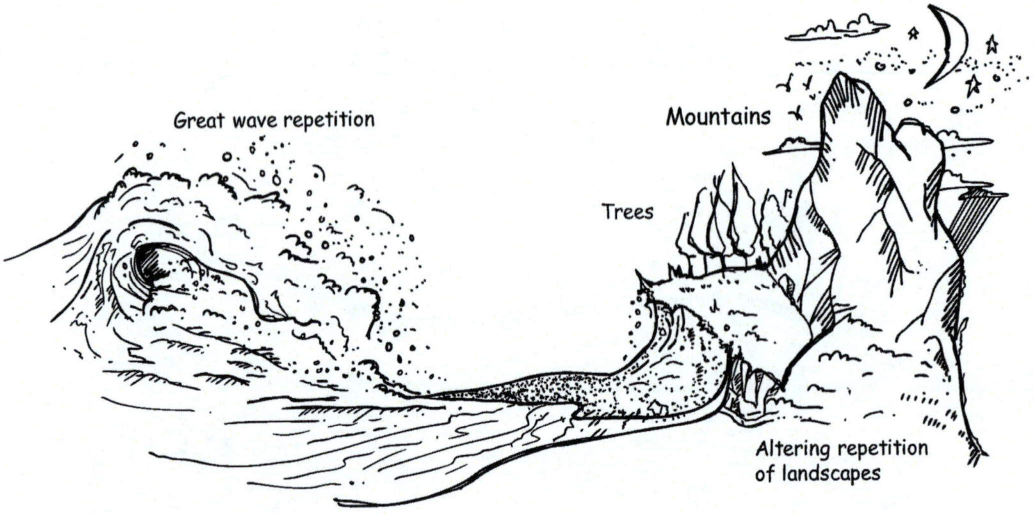

Fig. 9.11 Altering repetition of a great wave, spirals and curves. Similar altering repetition of shapes evident in the landscape of mountains and trees. (Drawing by Jesse Delmo, synthesised by Roös)

Fig. 9.12 Positive space in the cracks of porcelain, and formation of crystals. (Drawing by Jesse Delmo, synthesised by Roös)

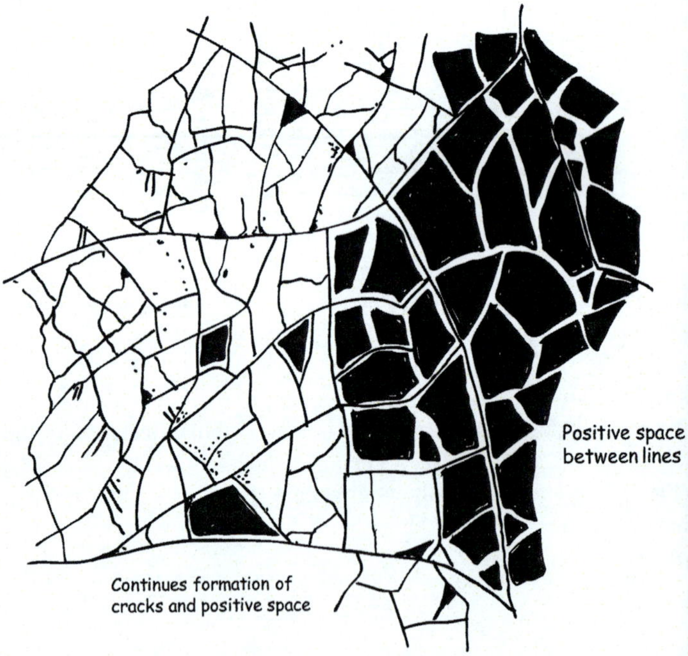

basalt, and the vertical branching cracks of a palm tree trunk (Stevens, 1974, p. 208). *Pattern formations* that are spontaneously created include the mosaic of a honeycomb, the branching of leaves, the transition of feathers of the Guinea fowl, and the lattice framework of the male cone of the queen sago cycad (Ball, 2009, pp. 168–180). Alexander argues that these good shapes all have the same property: the presence of a centre surrounded by a combination of minor centres. Good shapes arise because each part exists as a geometric centre, which gets intensified by the other centres (Alexander, 2001–2005a, p. 264).

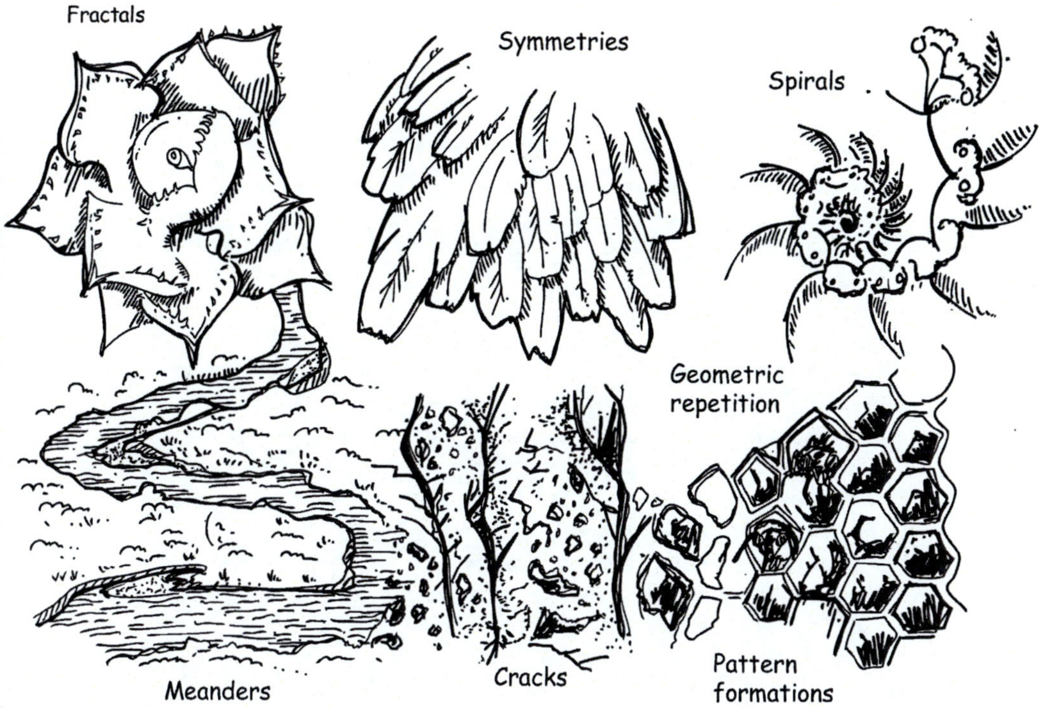

Fig. 9.13 Good shapes formed in nature with geometric centres and pattern formations. (Drawing by Jesse Delmo, synthesised by Roös)

9.6.7 Local Symmetries

Symmetries are evident and persistent throughout nature. We can see this in the shapes of trees where symmetry is evident around the tree trunk; in the human body we find major bilateral symmetry, as well as symmetry in its parts such as the finger around its length, the eye, the finger nail, the knee and the woman's breast. The different types of patterns in nature result in a widespread beautiful harmony of repetitive symmetries, as depicted in Fig. 9.14. *Symmetry* is apparent in all living things; in the stripes of a tiger we see bilateral symmetry, the beautiful spots of a leopard reflects repetitive symmetry, five-fold symmetry in the starfish, six-fold symmetry in the snowflake, and rotational symmetry in flowers (Stewart, 2001). *Fractals* create shapes of infinitely self-similar symmetry forms, such as spirals in the Romanesco broccoli, the Angelica flower head, and the constant repetitive forms in clouds, river networks, fault lines, mountain ranges, and ocean waves (Addison, 1997). *Spirals* are reflective of complex mathematical repetition, such as the logarithmic spiral in the phyllotaxis of the spiral aloe, the nautilus shell and the seed head of the sunflower, with symmetry evident in each shape (Ball, 2009, pp. 29–32).

9.6.8 Deep Interlock and Ambiguity

This property in the forming of shapes in nature comes about where *neighbouring systems connect* with each other, and forms interpenetrate to link together. A good example exists in the surface of the cerebellum. To be able to maximise the number of connections with the surrounding tissue, the cerebellum is crinkled deeply to provide increased surface area. Similarly, fractals show that lines tend to

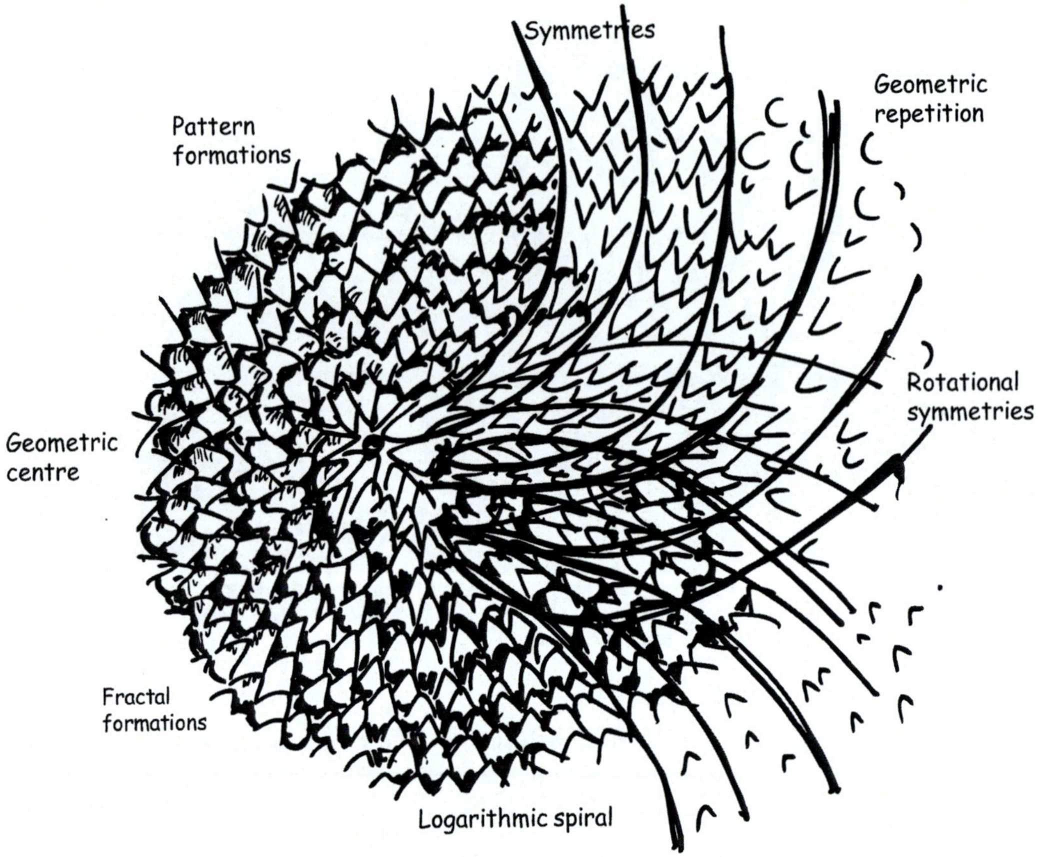

Fig. 9.14 Fractals create shapes of infinitely self-similar symmetry forms in nature, as represented in the fractals of the Angelica flowerhead. (Drawing by Jesse Delmo, synthesised by Roös)

fill the portions of space, and that surfaces grow with accretions. In the example of lichens growing on the tree trunk, the surface forms interpenetrate links to maximise interconnections (Fig. 9.15). Defining '*deep interlock and ambiguity*', Alexander states that living structures contain some form of interlock where centres are 'hooked' into their surroundings (Alexander, 2001–2005a, pp. 195, 270).

9.6.9 Contrast

In the organisation of natural systems, *contrast* is the most obvious phenomena from the *interaction of opposites*. This is evident in the *contrast* of male and female in almost every kind of organism; it appears in the cycles of day and night (Fig. 9.16), land and water, dark and light, and the positive and negative currents in electric charges (Alexander, 2001–2005a, p. 274). In geometry *contrasts* result in distinct subunits, which help us to distinguish between adjoining units, and provide a *figure-ground symmetry of opposites* (Salingaros & Mehaffy, 2006).

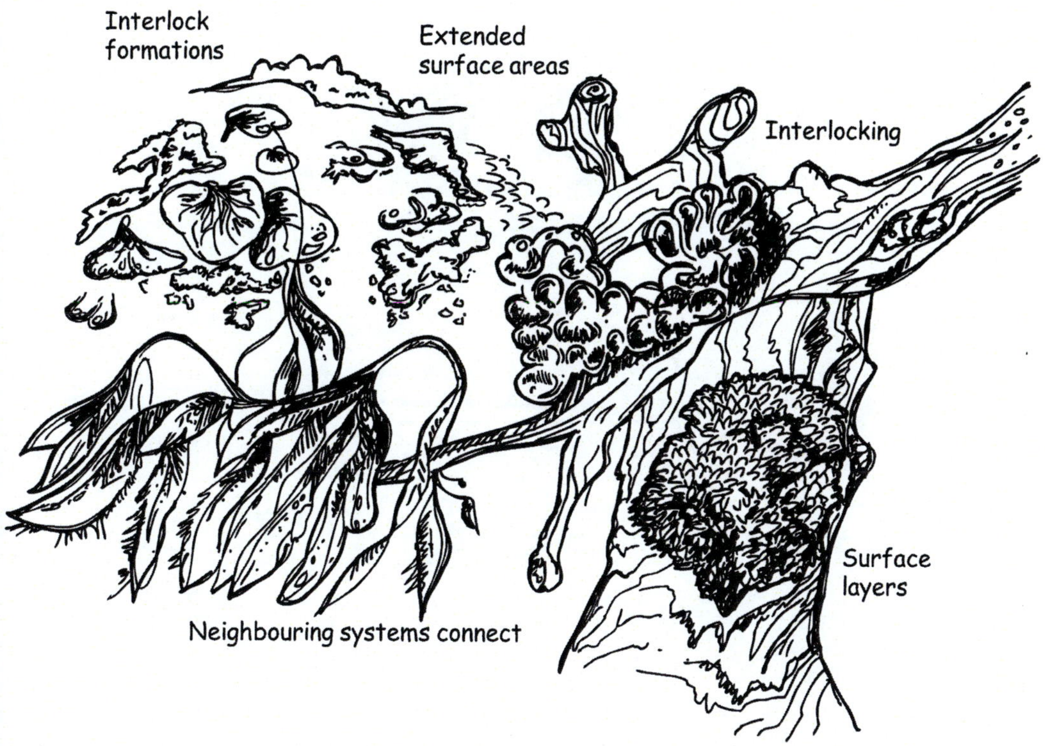

Fig. 9.15 Deep interlock in lichens on a tree trunk forming interpenetrate links to maximise interconnections of the surface layers, with similar interconnections forming across elements in nature. (Drawing by Jesse Delmo, synthesised by Roös)

9.6.10 *Gradients*

Every time there is a transition from one space to another in nature, *gradients* are established. Change happens gradually. For example, in an electric field the field strength changes the further it moves away from the source, forming a gradient of intensity. In the flow of a river, *gradients of turbulence* and velocity occur near the riverbank. The size of twigs in a tree varies in *gradient* from the centre of the tree to the fine edge of the leaves. *Gradients* are visible in the element size growth of a nautilus shell, formed over a period of growth (Alexander, 2001–2005a, p. 275), as depicted in Fig. 9.17.

9.6.11 *Roughness*

The smoothness of a crystal is interrupted by irregularities known as dislocations, with small increments of error that make it possible for the crystal to continue with the grid structure. *Roughness* allows the crystals to continue their growth. Roughness appears in nature as a result of the interplay between well-defined order and the constraints of three-dimensional space. An excellent example of the *roughness in creation* of form is the varying groove sizes of the paper nautilus shell of the *Argonauta Argo*, depicted in Fig. 9.18. As the octopus grows, the shell forms grooves in *varying sizes due to the nature of the space*, resulting in a 'roughness'. This *force of irregularity* happens to create an end result of greater regularity and order. This *roughness* is also evident in the forming of stripes

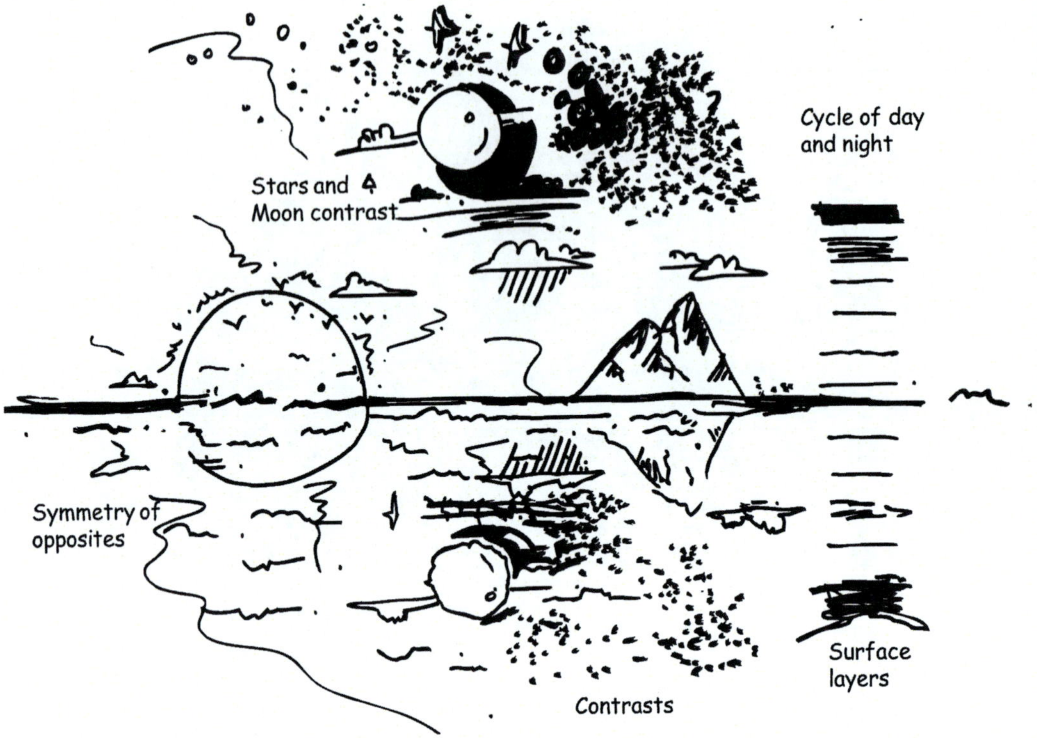

Fig. 9.16 Contrast in nature clearly evident between night and day, land and water. (Drawing by Jesse Delmo, synthesised by Roös)

on a Zebra, the virtual archetypes in atoms, the slight variations in the cells of a honeycomb, and even in the formations of waves in the sea.

9.6.12 Echoes

During the processes of growing and forming in nature, *geometric forms* are left behind in the static structure of systems. The structure indicates the repetition of certain angles, proportions and shapes over and over again, which determines the morphological character of the system and its parts. Alexander named the repetitive forms *echoes*; distinctive characteristics that repeat themselves over and over '*echoing*' what was there before. Similar growth processes in the different components of the Lily flower shows that characteristic curves and the same delicate proportions are repetitive (Fig. 9.19). The lips of the petal, the stems, and the stamens all have the same proportions, reflecting *echoes in the system*.

It can be argued that the *morphological echoes* of growth and their traces over time result in deep character, especially evident in the aging of a human face. The aging face has a certain rugged character which appears in the nose, eyebrows, cheeks and chin. The process of skin tightening, sagging, and weathering overtime repeats similar combinations of angles over and over again (Alexander, 2001–2005a, p. 281). This is what gives a unique character to an old person's face – in essence, a landscape painted over years, resulting in *echoes of time*, as indicated in Fig. 9.20.

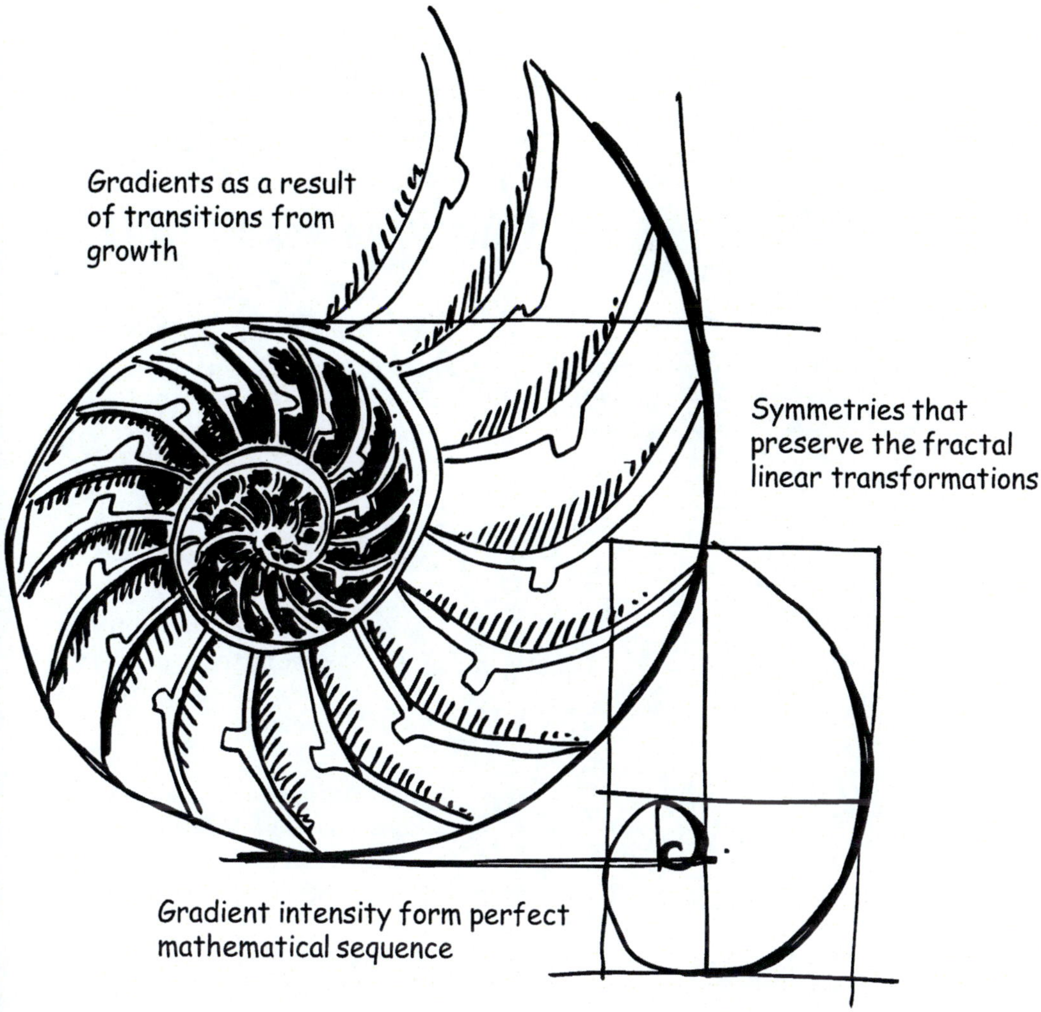

Fig. 9.17 Gradients of element-size formations by growth in Nautilus, typical of gradients in nature. (Drawing by Jesse Delmo, synthesised by Roös)

9.6.13 The Void

In nature, a *minor system* almost always occurs in relation to the 'quietness' of the larger and stable systems it belongs to. In fractal geometry it is not possible to fill in the whole fractal with detail, thus the 'centres' of the complex boundary result in a focused middle, *the void* (Salingaros & Mehaffy, 2006). In nature a good example of *void* is the 'eye' of a hurricane, cyclone or tornado (Fig. 9.21). The *minor system* within the *larger system* is calm, almost as if there is 'nothingness' to it. This has a homogenous zone in the middle, bound to more intense zones and structures around it (Alexander, 2001–2005a, pp. 284–286).

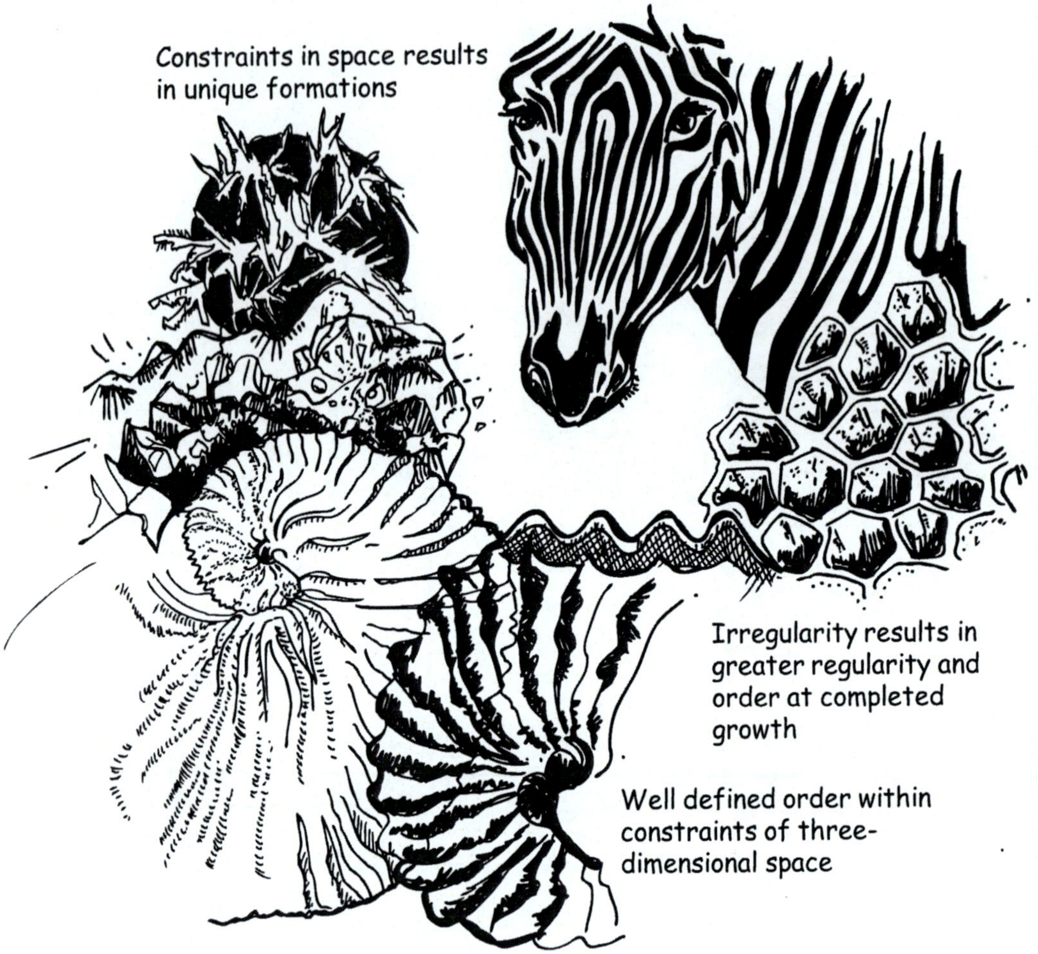

Fig. 9.18 Roughness of the Paper Nautilus, and variants in nature formations. (Drawing by Jesse Delmo, synthesised by Roös)

9.6.14 Simplicity and Inner Calm

Simplicity and inner calm can be identified when each configuration in nature is at its *simplest consistent with its own relative conditions*. Nature follows minimum energy principles to create form. This principle of the least action in nature is the ancient formulation of *simplicity*. A good example is the least-weight structure of a leaf at a cantilever state to support the uniformly distributed load (Fig. 9.22). The result is that the natural form of the leaf closely approaches the ideal, least weight and *simplest form to create balance* in the overall system (Alexander, 2001–2005a, p. 287). It is as if an inner calmness exists in the overall structure.

9.6.15 Not-Separateness

There is no isolation in the systems of nature, and it is evident that the *interconnectedness of all things* is visible in the analysis of things in science, quantum physics, and biological research. *Not-separateness* corresponds to the fact that each part of every system is always part of larger systems,

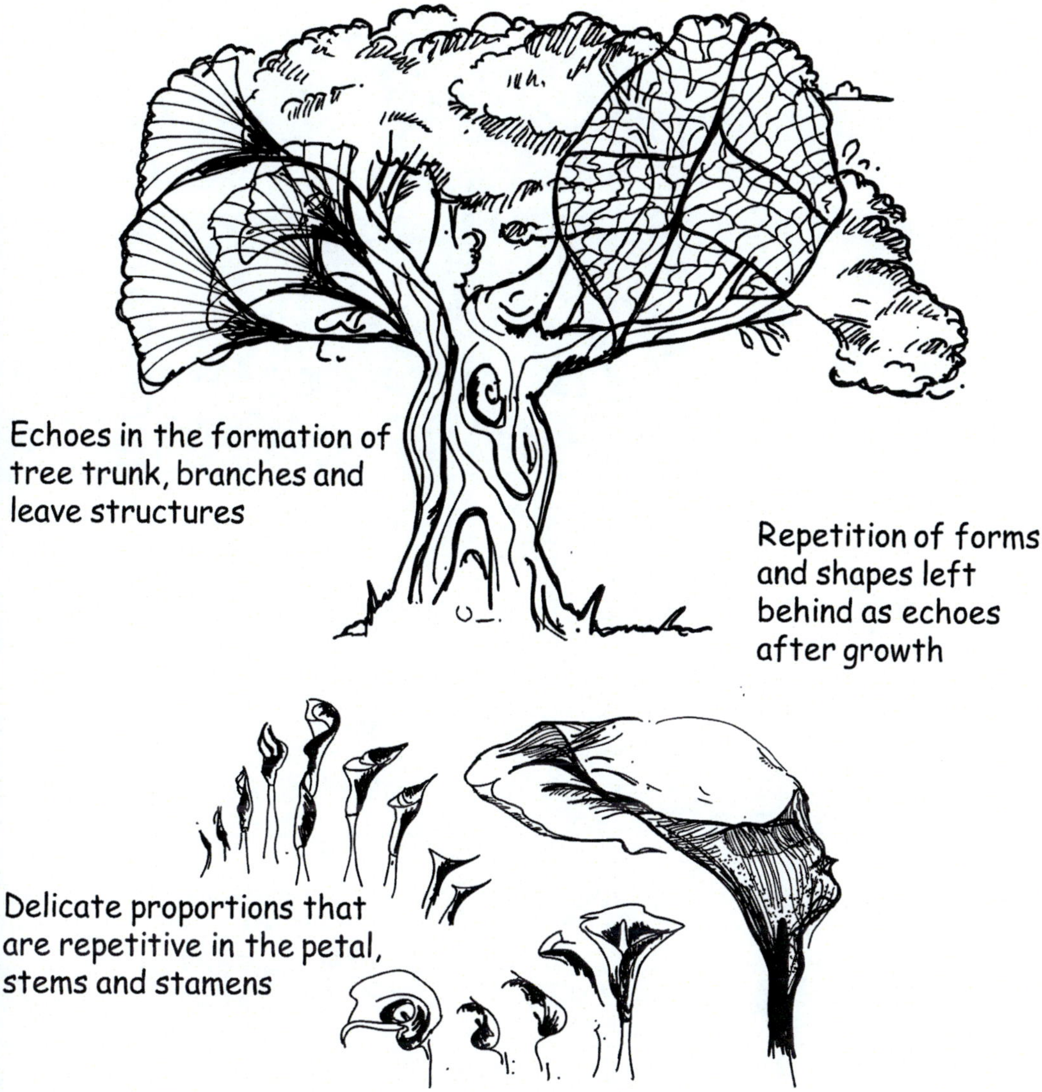

Fig. 9.19 The echoes in the forming system of a Lily, and elements of trees. (Drawing by Jesse Delmo, synthesised by Roös)

and they are connected in their behaviour (Alexander, 2001–2005a, p. 288). Lovelock and Margulis have put forward in the *Gaia Hypothesis* (1974) that organisms interact with their inorganic surroundings on Earth to form a self-regulating, complex system that contributes to maintaining the conditions for life on the planet (Lovelock & Margulis, 1974). This self-regulating complex system of the Earth is *interconnected within many systems* on the planet and in the biosphere and reflects the *characteristics of not-separateness*, as depicted in Fig. 9.23.

The examples of the *15 fundamental properties* [*of wholeness*] reflected in this section clearly indicate that these properties appear repeatedly in nature; they occur in different scales, and they occur in a specific order. These properties arise naturally as a result of the *natural transformation* in the process of morphogenesis, unfolding in a specific order (Mehaffy, 2009b; Seamon, 2007; Alexander, 2001–2005a).

Fig. 9.20 The echoes in the aging and weather-beaten face of an old person. (From Zach Dischner, CC BY 2.0)

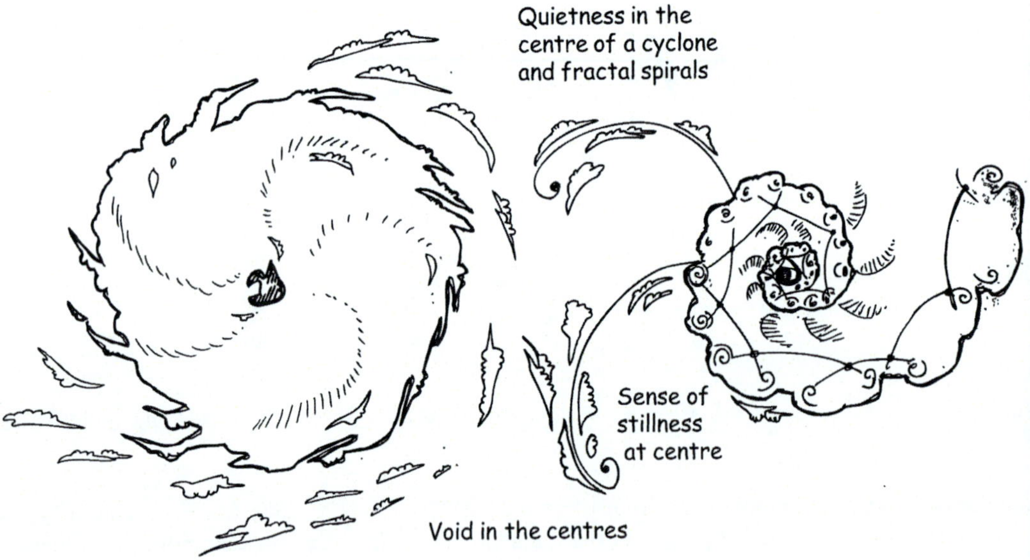

Fig. 9.21 The void visible in the centre of a hurricane, in forming of fractals, and elements of nature. (Drawing by Jesse Delmo, synthesised by Roös)

The *notion of the transformations* thus exhibits features of organisation, with a unity of form even without the overall composition. In the example of the milk drop, the structure has arisen as a direct result of the steps in the transformation process, and it can be understood that the structure *preserving transformation* is a reflection of a *generative process*.

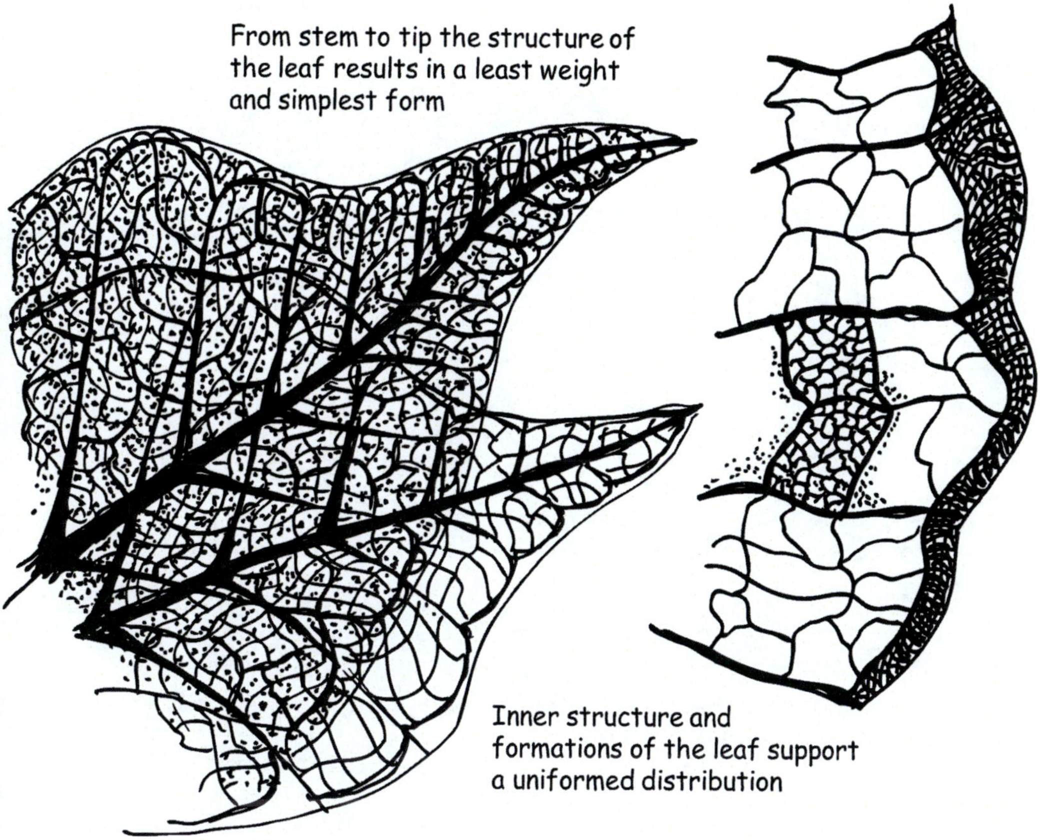

Fig. 9.22 Simplicity and inner calm structure of a leaf. (Drawing by Jesse Delmo, synthesised by Roös)

9.7 Generative Process and the Generative Code

Every object, matter, or form is the outcome of a *generative process*. A frog grows from a fertilised egg into a tadpole and then into an adult; matter creation results from the conversion of massless particles to massive particles known as pair production; a city develops from a small settlement to a metropolis, and a piece of art is created from rock to sculpture according to the artist's design. The patterns and *morphological processes* described earlier in this chapter, and contended by Alexander, McHarg, Lyle, Van der Ryn and Cowan, are in essence different observations of a *generative approach*. These generative approaches argue that an object, form or phenomenon that we 'see' in nature or in the built form is the result of an underlying process. This is a recognizable one, with patterns that occur over and over again, which produces complex objects. The promise of generative processes is productive and supported in biology by Thompson (1961), and in physics and computer science by Wolfram (2002). Generative processes in biological formations, and similarly in physics, are underpinned by a specific *code* (Thompson, 1961; Wolfram, 2002). The basic foundation of Wolfram's 'new kind of science' is the simple abstract rule of an elementary programming language, a *generative code* that is very similar to the principles of *A Pattern Language*. However, *generative codes* are not a new concept; they are used as the codes of a biological system that helps the generative process to unfold a plant from its seed into the finest detail, listed in the genetic information of its DNA (Alexander et al., 2005, p. 3).

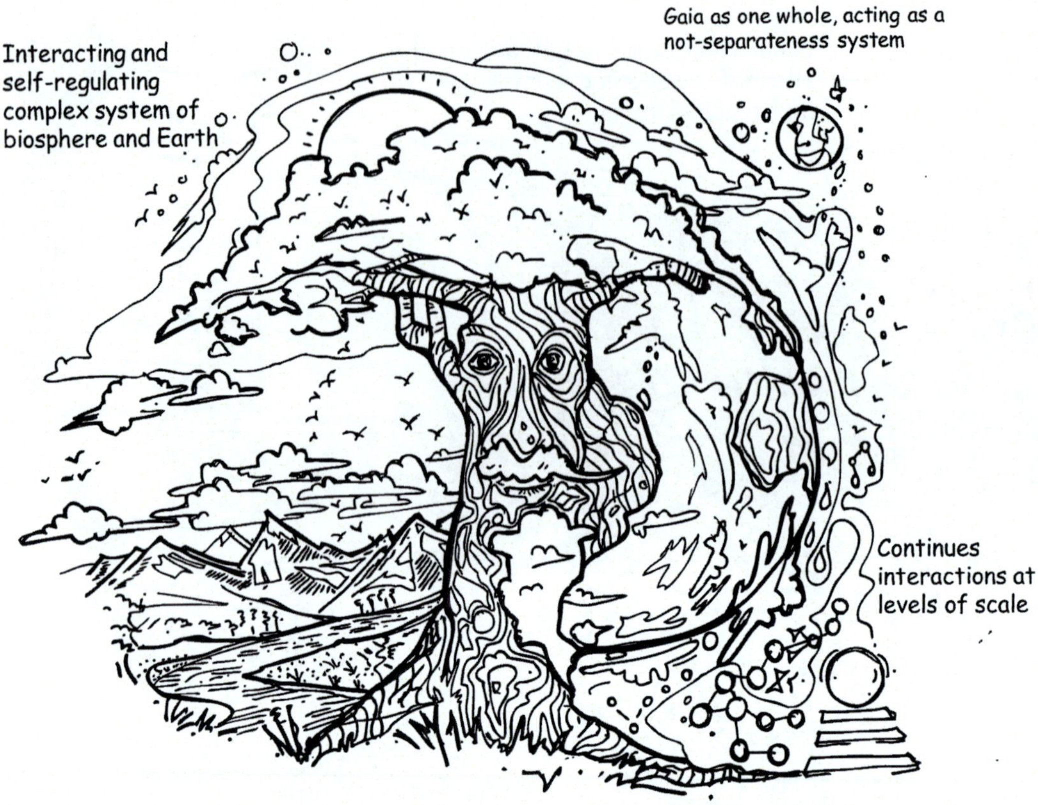

Fig. 9.23 The earth and the biosphere as a not-separateness system – Mother *Gaia*. (Drawing by Jesse Delmo, synthesised by Roös)

A *Generative code* resulted from the need to apply these principles of a *Pattern Language* and *Morphogenetic sequences* to the built environment. Alexander realized that his pattern language was guilty of a typical 'template' limitation (Alexander et al., 2005). If language is used to come up with a design planned in advance without considering a careful generative process for adapting the form, then the form will not have a 'living' quality that is so obvious in the forms of nature (Mehaffy, 2009b, p. 6). The *15 fundamental properties* arose naturally as a result of the natural transformations in nature, and living structures are a result of the process of morphogenesis. To overcome the static 'template' limitation in the design and planning of the urban environment, a generative code would follow the steps of 'unfolding' similar to those in nature. Thus the *15 fundamental properties* need to be considered.

Neis explains that the *Generative code* (an urban code) is a new process that is supposed to include all the information needed to actually design, develop and build a neighbourhood in a way that *creates living neighbourhoods*. This includes not only the order in which the planning and building process must proceed, but also needs a coherent form that:

> … describes all the human interaction, practical and legal and procedural details, to get an actual living neighbourhood to appear on the ground as a result of interactions among the people who live and work there (Neis et al., 2012, p. 5).

When *generative codes* are used in the process of designing and building a new neighbourhood or precinct, the following characteristics are visible (Alexander et al., 2005, p. 4):

1. A more beautiful and coherent geometric form that is natural to the land;
2. More probable successful integration and adaptation to plants, trees, animals, and land form – resulting in communities and built areas, which like traditional towns and villages, seem part of nature;
3. Successful fine-tuning and deep adaptation;
4. More successful integration with the living process in the daily life of the inhabitants;
5. Better fit with individual local needs of any given building, garden, space, or enclosure;
6. Far greater likelihood that genuine community will emerge in the new place;
7. More uniqueness of each place, each street, each building, and each project;
8. More profound linkage to sustainability and environmental objectives; and
9. An easier path to the desired end state, as described above.

The *Generative Code* as an overall process, which includes the principles of *A Pattern Language*, has been tested on various projects. One project worthy of mention is the Eishin Campus in Japan, a combined college and high school campus in a suburban location in Tokyo. It exhibits elements of the generative code's sequences that have unfolded over many years due to integrated design and construction with active community participation. This project, which started in 1991, slowly unfolded over several years resulting in many lessons learned, and was subsequently recorded by Alexander, Neis and Maggie Moore Alexander in *The Battle for the Life and Beauty of the Earth – A Struggle Between Two World Systems* (2012). Two important themes are recognised in the book: first, the crucial importance of *local adaptation*, and second, the search for coherence in the physical, ecological, emotional and spiritual realm, thereby healing communities, resulting in *wholeness*, or creating *the whole* (Alexander et al., 2012, pp. 19, 87). It is very evident throughout the literature review that the theme of *wholeness* and *the whole* continuously emerges as an important ingredient that is reflected in the principles of ecological design and planning (McHarg, 1992 [1969]), regenerative design (Lyle, 1994), and generative design [code] (Alexander, 2005). Further investigation of this phenomenon of the whole and wholeness is explored in the next section.

9.8 Wholeness and the Whole

In science research the tendency is to investigate specific parts, and in general the philosophical position accepted is that a complex system is nothing but the *sum of its parts* (Goldstein, 1999). In contrast, Einstein's *general theory of relativity* ties together space, time and matter in an intrinsic relationship that *cannot exist without the whole*. *Chaos theory* identifies that systems are exquisitely sensitive to their circumstances and that therefore they must be considered in the context of their whole (Gleick, 1988). *Biological systems* are a sum of biological cells with great inter-related complexity (Noble, 2006). *Complexity theory* identifies that the behaviour of complex systems constitutes an interrelated pattern connection that not only exchanges between constituents, but also exhibits a kind of holistic pattern-forming capability with active information forming the whole (Kauffman, 1993; Polkinghorne, 1998).

Here again we can see that the concept of 'the whole' appears over and over again in the literature. In relation to the design and planning of the built environment and the understanding of the whole, Alexander and Neis ask the question: How do we make intellectual sense of the word wholeness? They note that 'wholeness' suggests the understanding of things in their entirety and that it speaks of the oneness of all things, the connection of wholes (Alexander et al., 2012, p. 87). It is thus relevant to reflect on the explanations of wholeness by Alexander and Neis in *The Battle for the Life and Beauty of the Earth* (2012, pp. 87–96), as follows:

1. Wholeness can be understood as a structure. When we try to find the wholeness of a place, we can identify the specific structure of that place, and thus be able to reflect on its wholeness;
2. Wholeness, when we explain the feeling or the intuition of the whole, always extends beyond the object in question;
3. When we identify wholeness of a place, an object, it always eludes comprehension. Wholeness thus comprehends the intangible; the object is unique;
4. The presence of unity is constant. What we refer to as wholeness, is always a quality of being one, interlaced and unified. It is somehow at peace;
5. Each whole contains and is composed of many other wholes. There are specific geometric qualities and properties that identify relationships between these wholes; and
6. Wholeness encompasses the idea of healing. When something is whole, we consider that the object has been healed.

Sustainability activists in essence try to 'heal' many issues of the world, but in reality, only address technical matters that are far from the context of *the whole*. Alexander refers to a second meaning of the word sustainability. He observes that 'Sustainability' tries to stitch together a world that is falling apart, tries to minimise damage to nature, or tries to put in place technical solutions to sustain the resources humans depend upon:

> The second meaning of the word sustainability, which is so far removed it is almost another topic, refers to the wholeness of the land, the extent to which we see our land (rural, urban, or wilderness) as sacred, and the extent to which we treat our interaction with the land as a sacrament. The extent to which we recognize the beauty of what we make in the land is of paramount importance. It is not just an add-on or a luxury. Rather, it goes to the core of what sustainability really is (Alexander, 2004, p. 5).

Finally, Alexander and Neis have proposed a procedural definition of wholeness as:

> … a general specification for a process that is capable of increasing wholeness in a domain, and particularly in the physical domain where something, some place, some part of the world is being conceived, or built, or repaired.

and they recommend that:

> If each of us is to engage in the very personal journey toward wholeness, and thereby heal our communities … we must keep wholeness at the forefront of our thinking and our activities. (Alexander et al., 2012, p. 87).

This collective argument aligns with the view of Australian Aboriginal knowledge on *the whole*. As noted in Chap. 6, Aboriginal consciousness is formulated in the context of wholeness. Such is reflected in the words of Aboriginal Elder, Aunty Veronica Brodie: *'life is part of one vast network of relationships'*. Aboriginal people believe that the position of an Aboriginal person in nature is that of oneness; it is *Law*[5] that you need to look after nature, and thereby be able to heal nature. Humans are part of nature, not above nature, but part of a sequence of natural phenomena. These principles are *centres of nature*, the collective phenomena of an environment that create *wholeness*, resulting in *the whole*. This supports Alexander's and Van der Ryn's argument that 'unselfconscious cultures' (Indigenous cultures) understood the whole, and were part of the whole in such a way that in building structures and settlements they *unconsciously created form in oneness with nature*.

9.9 Conclusion Statement

This Chapter started by investigating *wholeness* in the context of physics, where Bortroft argued that the whole cannot be explained through some analytical, sequential approach that puts parts together as in a mechanical system (Bortroft, 1996; Stefanovic, 1991). Wholeness is present due to the exis-

[5]Aboriginal Law is the law of the land, in essence aligned with nature and the 'lore' relates to the Dreamtime, learned from childhood the rules on how to interact with the land, kinship and community (*Working with Indigenous Australians*: http://www.workingwithindigenousaustralians.info/content/Culture_4_The_Law_and_the_Lore.html – Accessed 20 February 2019)

tence of all of its parts, and a part itself cannot exist if it is not part of *the whole*. We identified in the discourse of *Pattern language, Morphogenetic sequences* and *the Generative code* that there are *15* specific *fundamental properties* present, both in good structures and in Nature, supporting an argument that this formula leads to a good liveable and sustainable built environment; this includes a set of patterns, a system of explicit steps for creating living structures within the fabric of a place.

The theory of wholeness defines a *generative process* where there is always a sequence, an order, instructions that follow the rule of centres that appear within the larger whole as distinct and noticeable parts. This generative process (the generative code) is capable of driving the *organic unfolding of a place* in such a way that people who live and work in that place have a healthy connection to *the whole*, and this makes the place alive, exhibiting properties of living structures (Neis et al., 2012). Through the *15 fundamental properties* [*of wholeness*], we identified that throughout nature it is evident that patterns link all the parts together. This is evident in the geometric order of all phenomena, as demonstrated in the Mandelbrot Set, the Koch Curve, and complex symmetrical structures in nature such as snowflakes, flowers, the honeycomb, and the nautilus shell amongst others (Stark, 2012; Mehaffy, 2009a; Nelson, 2005).

The review of this Chapter hence results in the formulation of the following questions: How can these principles of the *pattern language, morphogenetic sequences,* and the *generative code* be applied in the context of *adaptation, regeneration and a changing environment*? How can generative processes in the context of regenerative and adaptive design help to solve contemporary problems of the built environment and its connected natural ones? The message is made clear in Alexander's work, whereby the *patterns* of a healing kind and *morphogenetic growth* are already all around us. We can find them in the structures of nature and in the processes that produce them. We can find patterns in the collective intelligence of traditional structures and Aboriginal knowledge, ready to be revived and regenerated, as part of the living structure of our collective future environment (Mehaffy, 2009b). The *regenerative-adaptive pattern language* offers a holistic approach, informed by nature, for settlement planning and design (Roös & Jones, 2017). We explore in the next chapter (Chap. 11) the details of the development of the *regenerative-adaptive pattern language.*

Therefore:

When we practice ecological design and planning, as well as in regenerative design for sustainable development, the consideration of *the whole* is essential. Wholeness thinking recognizes that the *entirety is interconnected*, and the *unfolding of the whole* creates living structures. The *15 fundamental properties of wholeness* must be used in association with the *generative processes* of design, which are *patterns* and *pattern languages*, *morphogenetic sequences*, and the *generative code*.

9.10 Fundamental Pattern 9

TRANSFORMATIONS OF WHOLENESS [9]

The unfolding of Nature's wholeness at later stages is embedded in earlier stages' process patterns and structure. Wholeness is a product of ordered generative instructional sequences that follow the 'rule of centres' in which noticeable parts appear distinctly within the larger whole. Therefore, practice regeneration as a process of wholeness transformations (Fig. 9.24).

Downward links:

The incremental process of TRANSFORMATIONS OF WHOLENESS [9] recognises that the *entirety of all is interconnected*, and the unfolding of THE WHOLE [1] through the *15 fundamental properties of wholeness* creates living structures. These transformations within REGENERATIVE ECOSYSTEMS [8] require us to consider the dynamic forces of nature and culture, and to deal with this complexity of significant change through the application of the NOTION OF REGENERATIVE-ADAPTIVE PATTERNS [10] …

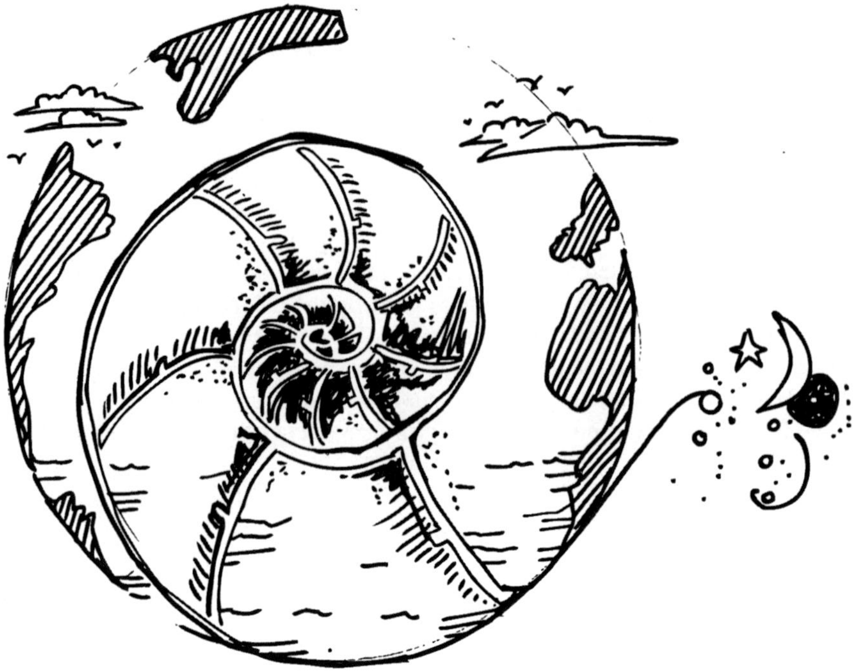

Fig. 9.24 Sketch of fundamental pattern – TRANSFORMATIONS OF WHOLENESS [9]. (Drawing by Jesse Delmo, synthesised by Roös)

References

Addison, P. (1997). *Fractals and chaos: An illustrated course*. Edinburgh, UK: IOP Publishing – Napier University.

Alexander, C. (1964). *Notes on the synthesis of form*. Cambridge, MA: Harverd University Press.

Alexander, C. (1979). *The timeless way of building* (Vol. 1). New York: Oxford University Press.

Alexander, C., Neis, H., Anninou, A., & King, I. (1987). *A new theory of urban design* (Center for Environmental Structure). New York, Oxford: Oxford University Press.

Alexander, C. (2004). *Sustainability and morphogenesis: The birth of a living world*. Berkeley, CA: Centre for Environmental Structure.

Alexander, C., Schmidt, M., Alexander, M., Hanson, B., & Mehaffy, M. (2005). *Generative codes: The path to building welcoming, beautiful, sustainable neighbourhoods*. Berkley, CA: Centre for Environmental Structure.

Alexander, C. (2001–2005a). *The nature of order – An essay on the art of building and the nature of the universe, book one: The phenomenon of life*. Berkeley, CA: The Center for Environmental Structure

Alexander, C. (2001–2005b). *The nature of order – An essay on the art of building and the nature of the universe, book two: The process of creating life*. Berkeley, CA: The Center for Environmental Structure.

Alexander, C., Ishikawa, S., Silverstein, M., Jacobson, M., Fiksdahl-King, I., & Angel, S. (1977). *A pattern language: Towns, buildings, construction*. New York: Oxford University Press.

Alexander, C., Neis, H. J., & Moore Alexander, M. (2012). *The battle for the life and beauty of the earth: A struggle between two world-systems*. New York: Oxford University Press.

Ball, P. (2009). *Shapes: Natures patterns: A tapestry in three parts*. London: Oxford University Press.

Borchers, J. (2008). A pattern approach to interaction design. In P. G. Satinder (Ed.), *Cognition, communication and interaction: Transdisciplinary perspectives on interactive technology* (pp. 114–129). London: Springer.

Bortroft, H. (1996). *The wholeness of nature*. Hudson, NY: Lindesfarne Press.

Bortroft, H, (1971). *The Whole: Counterfeit and Authentic*. Systematics, *9*(2), 43–73.

DeKay, M. (2011). *Integral sustainable design: Transformative perspectives*. London/Washington, DC: Earthscan.

Eglash, R. (1999). *African fractals: Modern computing and indigenous design*. New Brunswick, NJ: Rutgers University Press.

Gleick, J. (1988). *Chaos making a new science*. New York: Penguin.

Goldstein, J. (1999). Emergence as a construct: History and issues. *Emergence: Complexity and Organization, 1*(1), 49–72.
Grabow, S. (1983). *Christopher Alexander and the search for a new paradigm in architecture*. London: Orial Press.
Kauffman, S. (1993). *The origins of order self-organization and selection in evolution*. New York: Oxford University Press.
Kuhn, T. (1962). *The structure of scientific revolution*. Chicago, IL: University of Chicago Press.
Lovelock, J. (2009). *The vanishing face of Gaia*. New York: Basic Books.
Lovelock, J., & Margulis, L. (1974). Atmospheric homeostatis by and for the biosphere: The Gaia hypothesis. *Tellus. Series A, 26*(1–2), 2–10.
Lyle, J. T. (1994). *Regenerative design for sustainable development*. New York: Wiley.
McCormack, J., Dorin, A., & Innocent, T. (2004). Generative design: A paradigm for design research. In J. E. Redmond (Ed.), *Proceedings of Futureground*. Melbourne, VIC: Design Research Society.
McHarg, I. (1992 [1969]). *Design with nature*. Washington, DC: Wiley.
Mehaffy, M. (2009a). Horizons of pattern languages. In H. Neis & G. A. Brown (Eds.), *Current challenges for patterns, pattern languages and sustainability* (pp. 103–111). Portland, OR: PUARL Press.
Mehaffy, M. (2009b). *From pattern languages to generative codes: A report on the work of Christopher Alexander and Colleagues, and its application to the Regeneration of Traditional Settlements. Centre for Environmental Structure – Europe*. Portland, OR: Centre for Environmental Structure.
Neis, H. J., Brown, G. A., Gurr, J. M., & Schmidt, J. A. (2012). Generative process, patterns, and the urban challenge. In *Fall 2011 International PUARL Conference*. Portland, OR: PUARL Press.
Nelson, J. (2005). Branch growth and sidebranching in Snow Cristals. *Crystal Growth & Design, 5*(4), 1509–1525.
Noble, D. (2006). *The music of life: Biology beyond the genome*. Oxford, NY: Oxford University Press.
Polkinghorne, J. J. (1998). *Belief in God in an age of Science (The Terry Lectures ed.)*. New Haven, CT: Yale University Press.
Roös, P., & Jones, D. (2017). Knowledge of making life: Design patterns for regenerative-adaptive design. In *DesTech 2016: Proceedings of the International Conference on Design and Technology* (pp. 203–210). Dubai, United Arab Emirates: Knowledge E. https://doi.org/10.18502/keg.v2i2.616
Salingaros, N. A., & Mehaffy, M. W. (2006). *A new theory of architecture*. Solingen, Germany: UMBAU-VERLAG Hareld Puschel.
Seamon, D. (2007). Christopher Alexander and a phenomenology of wholeness. In *EDRA – Environmental Design Research Association, Architecture Department*. Manhattan, KS: Kansas State University.
Solomon, E. P., Berg, L. R., & Martin, D. (2010). *Biology (9, Illustrated ed.)*. Belmont, CA: Brooks Cole Publishing Company.
Stark, W. (2012). Tacit knowledge and innovation patterns for communities and social systems. In H. B. Neis (Ed.), *Generative process, patterns, and the urban challenge* (pp. 89–95). Portland, OR: PUARL ARUS Press.
Stefanovic, I. (1991). Evolving sustainability: A rethinking of ontological foundations. *The Trumpeter, 8*, 194–200.
Stevens, P. (1974). *Patterns in nature*. Boston: Little, Brown & Co..
Stewart, I. (2001). *What shape is a snowflake? Magical numbers in nature*. London: Weidenfield & Nicolson.
Thompson, D. W. (1961). *On growth and form*. Cambridge, UK: Cambridge University Press.
Van der Ryn, S. (2005). *Design for life*. Layton, UT: Gibbs Smith.
Van der Ryn, S., & Cowan, S. (1996). *Ecological design*. Washington, DC: Island Press.
Wolfram, S. (2002). *A new kind of science*. Champaign, IL: Wofram Media, Inc..

Chapter 10
A Regenerative-Adaptive Pattern Language

Abstract Going beyond sustainability, which is mostly focused on a minimum or 'less harm' approach, the concept of *regenerative design and development* considers the improvement of a sustainable system as going beyond just being 'sustainable'. The central message from the *regenerative paradigm* is that not only should human activities that are degenerative to the environment be minimized, but rather we should move towards a concept of maximising human activities that restore and regenerate ecological systems.

The regenerative design approach, which includes regenerative development, engages communities and practitioners in a dialogue to develop higher orders of aspirations of sustainability through understanding and problem-solving by storytelling – in essence, developing a language for place that regenerates and improves the environment. This shifts the focus of a design problem from singular, separate parts (or mechanical approach) to an all-encompassing, *integral approach* that looks at the *whole system*. However, *forces of change* are always present, and regenerative systems need to have the ability of adaptation. To deal with this complexity, the *regenerative-adaptive pattern language* was developed, which considers the dynamics of both humanmade and natural environments and includes the formulation of the '*notion of regenerative-adaptive patterns*' equation. This chapter proceeds with the devising of the *regenerative-adaptive pattern language*, accumulating key aspects and messages from previous chapters, and charters a new knowledge system. The concluding section describes a *regenerative-adaptive design model* that can be used to apply the *pattern language* to the design of our human settlements, with the premise that this model provides an innovative solution that goes beyond the current practice of sustainable design and development. The chapter concludes with the *fundamental pattern* NOTION OF REGENERATIVE-ADAPTIVE PATTERNS [10], providing instructions how to navigate through the complexity of the pattern language.

Keywords Regenerative paradigm · Regenerative-adaptive patterns · Pattern language · Forces of change · Ecological worldview · Notion of patterns · Adaptation · Regenerative design

Upward links:

… in the TRANSFORMATION OF WHOLENESS [9] we have identified that the 15 properties of wholeness unfold specific core patterns and generative codes of a place following the steps of morphogenetic sequences. The generative processes of design are based on REGENERATIVE ECOSYSTEMS [8], and a generative process of design needs to follow the layers of place as indicated in NATURE'S DESIGN [7]. Regenerative-adaptive design deals with multiple layers of complexity at place, and requires the guidance of the NOTION OF REGENERATIVE-ADAPTIVE PATTERNS [10].

P. B. Roös, *Regenerative-Adaptive Design for Sustainable Development*, Sustainable Development Goals Series, https://doi.org/10.1007/978-3-030-53234-5_10

10.1 Introduction

> An environment of community will not come to life unless [each environment], each place, each building, each street, each room, becomes unique, as a result of careful and piecemeal processes of adaptation (Alexander, Neis, & Moore Alexander, 2012, p. 19).

Undeniably, the identification of the nature of planning and design problems sets the foundation for what Alexander calls a 'pattern language' and potential methods of a generative process that can create sustainable form. Logical structures formed the order of the theories of McHarg (1992 [1969]), Lyle (1994), Alexander (2001–2005a, 2005b, 2005c, 2005d, 1979), and Alexander et al. (1977). As we have seen in Chap. 10, the rationale behind Alexander's method is the use of logical order or logical structures, similar to what was used in the creation of form by unselfconscious cultures (Indigenous cultures) (Alexander, 1964, p. 31). Further, we need to consider the issues of climate change adaptation, the establishment of sustainable communities and the achievement of resilience for both natural and human systems. Expanding on this, in this chapter I further explore the principles and the evolution of a *Regenerative-Adaptive Pattern Language* theory. The proposed theory informs a *regenerative-adaptive design model*, which bring together the various parts of *adaptation*, *indigenous knowledge*, *pattern language*, and *regenerative design* as a *coherent whole*, tested in application to a specific place, and documented in the next chapter, Chap. 11. But first, let's unpack the fundamentals of the *regenerative-adaptive pattern language* theory (Fig. 10.1).

10.2 Regenerative-Adaptive Design

In the previous chapter, Chap. 9, the consideration of *the whole* was identified as an important factor in the theories explored, and the conclusion summarises *wholeness* as being fundamental to the ecological design and planning of our built environments. We have established that Alexander's *hypoth-*

Fig. 10.1 Complex mathematical repetition in the formations of shells and pebbles. (Photo by Author)

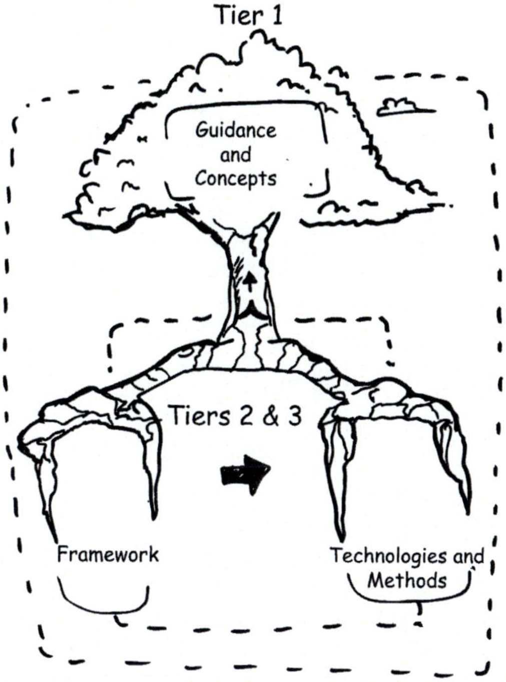

Fig. 10.2 Three-tier model of the regenerative development and design methodology. (Drawing by Jesse Delmo, synthesised by Roös, adapted from Mang & Reed, 2012)

esis of wholeness defines a *generative process* where there is always a sequence, an order, and a morphogenetic process that drives the organic unfolding of a place (Alexander, Schmidt, Alexander, Hanson, & Mehaffy, 2005, pp. 2–3). Linking to natural systems, this *hypothesis of wholeness* is also evident in both McHarg's (1992 [1969]) and Lyle's (1991) arguments that support a new living systems, ecological worldview.

According this premise of the ecological systems worldview, the radical changes required for the earth to remain fit for human and other species' habitation cannot happen without the transition from a 'mechanical' to an 'ecological' worldview (Du Plessis, 2009, 2012; Metzner, 1999). Du Plessis (2012) describes that the regenerative sustainability paradigm emerges from this transition, and takes hold as the new ecological systems worldview, which becomes embedded across fields of knowledge and across disciplines. Regenerative approaches started with descriptions, concepts and design theories such as Lyle's *Design for Human Ecosystems* (1991 [1985]), and *Regenerative Design* (Lyle, 1994), and through the adoption of this ecological worldview in the built environment professional practice early regenerative practitioners evolved this worldview complex. This was accepted by various practitioners in the design fields of architecture, landscape architecture and planning (Kuhn, 1962; Sanford & Mang, 1992).

Exploring a regenerative methodology as an example of how consciously a worldview shapes new practice in ecological design, Mang and Reed (2012) summarise a 'Regenerative Development and Design Methodology' based on the experience and application of this methodology in the professional practice of architecture, planning and design. The premises of the regenerative development and design methodology involves the role of humans and nature, working together in a developmental process that improves the value of the whole, and operates to take integrated systems to the next level, setting a higher order for regeneration (Mang & Reed, 2012, p. 34). Represented in a three-tier model, this order has been organized in six specific concepts: regeneration, development and design, place, pattern literacy, story, and potential (Fig. 10.2).

The regenerative design approach, which includes *regenerative development*, engages communities and practitioners in design teams in a dialogue to develop higher orders of aspirations of sustainability through understanding and problem-solving by storytelling. This shifts the focus of the problem to the context of the *whole system*. This process is an open dialogue growing out the emerging understanding of a place that comes from the knowledge of the community. The approach includes three phases of generation: understanding conceptualising the relationship to place, design for harmony, and the co-evolution of human and habitat to sustain its survival capacity (Mang & Reed, 2012, p. 31).

These concepts support the fundamental principles of the *Regenerative Design* aspect of the *Regenerative-Adaptive Pattern Language* theory, articulated in the next section.

10.3 Regenerative Design Fundamentals

Regenerative design is embedded in the acceptance of a *new ecological worldview,* and when actions and decisions by designers and planners are a result of consciously evaluating and expressing a solution based on their worldview complex, it can provide a pattern for the ongoing evolution of human beings (Mang & Reed, 2012). The role of humans is thus important in the regenerative process. The status quo of green or eco-efficient design minimising the impact on the environment is insufficient, and we as humans need to take our place as 'part of nature' and not as 'above nature'. From this perspective, regenerative design means the reconnection of human activities and aspirations in line with the evolution of natural systems; essentially, the co-evolution of both. This implies shifting human development activities back into alignment with nature's processes. It implies every human settlement needs to arrange itself around its watershed's capacity to support life (Mang & Reed, 2012, p. 26). This then defines the work of regeneration.

10.3.1 Regeneration

The central message from the *regenerative paradigm* is that not only should human activities that are degenerative to the environment be minimized, but rather we should move towards a concept of maximising human activities that *restore and regenerate ecological systems*. The co-evolutionary and partnered relationship between humans and the natural environment is thus embedded in a continually evolving world, and according to Kauffman (1993) the primary concern for sustainable futures is the process of evolution (Kauffman, 1993). This requires a process of regeneration where all levels of work produce a field within which the improvement of living systems can take place. Regeneration needs then to *address all levels of scale* that will reveal the essence and potential of the place, and all levels of scale become part of the *integrated whole* (Lyle, 1991 [1985]; Reed, 2007).

10.3.2 Place

A *specific place* is located in a hierarchy of relationships within the ecosystem structure according to a structural order; each place has a *unique character* and *value*, and as part of *morphogenetic sequences,* it has a specific combination of the properties of wholeness. Regeneration should return a place to its core position in *the whole*, and through regenerative design a place can help humans experience the intimacy and responsibility of the living world and find a meaningful identity in the context of wholeness. Place value is then based on the living system nested in its place, and to achieve connectedness is to understand the *mutuality* of the relationship of *human and nature*, and the understanding of *generating patterns*.

10.3.3 Patterns

To be able to assess a place, and to find meaningful and shared phenomena of the place (in essence understanding and connecting the generating patterns), the first task is to determine the reach of the place in terms of levels of scale. The *levels of scale* are formed naturally from the way the whole system brings itself together. In simple terms, the question is: *How big is here (the place)*? It is not always immediately apparent and visible; Mang and Reed quote Kelly about how to define a place, and the latter notes the following in the identification of patterns and place (in Mang & Reed, 2012, p. 32):

> Wherever you live, your tiny spot is deeply inter twined within a larger place, embedded fractal like into a whole system called a watershed, which is itself integrated with other watersheds into a tightly interdependent biome. At the ultimate level, your home is a cell in an organism called a planet. All these levels interconnect. (Kelly, 2005, in Mang & Reed, p. 32)

Seeing the complexity that surrounds a place can be overwhelming and complex if there is no way to understand place in its individual and collective connections. Emerging *core patterns* of a place can be identified through the order and structure of the whole (Alexander, 2001–2005b; Mehaffy, 2009). Alexander refers to these *core patterns* that pull everything together as strong centres. An essential step in regeneration is thus being able to identify the strong centres of a given place. Key to comprehending place as a living system or whole is to understand the ongoing and distinctive strong centres and interconnected patterns in which it organises the complex array of relationships that produce its activities, its growth and its evolution.

These strong centres and interconnected patterns form the framework for the story of place. Just as traditional storytelling provides the traditional knowledge of the nature of a place, the emergence of a *story of place* begins with an assessment of it based on a *whole systems approach*. This story of a place includes identifying and understanding the heritage, cultural, economic, geographic, demographic, climatic and ecological characteristics as part of – THE WHOLE [1].

The narrative structure for the story of place emerges by developing *a pattern language* to understand how geological, geographical, natural, and human history and culture have interwoven through time to create the unique character of a place (Mang & Reed, 2012). Out of that understanding a set of strong centres and organising patterns emerges that provides a science-based narrative through a *generative process*, named by Alexander, Neis, Anninau, & King (1987) and Alexander (2004, 2001–2005a, 2005b, 2005c, 2005d) as the '*generative code*'. Pattern literacy is key to this process, as described in Chap. 9. This allows for understanding the complex, dynamic relationships that constitute a place. The resulting narrative of the generative code has two dimensions: first the *strong centre* creating patterns that depict the working of a place, which reflect its unique character; and second, the *vocation of the place,* which has the unique potential to contribute to the larger ecological system which it is part of, interconnecting and evolving with.

10.3.4 The Generative Code and Regenerative Design

Regeneration in its pure form is

> …the process of renewal, restoration, and growth that makes genomes, cells, organisms, and ecosystems resilient to natural fluctuations or events that cause disturbance or damage. Every species is capable of regeneration, from bacteria to humans (Carlson, 2007, p. 400).

Every place is capable of regeneration, and humans can play a role in the regenerative process of a place by using a generative approach that envisions complex living systems, and the ecosystem of a specific place in particular, as dynamic wholes evolving and changing through time (Mang & Reed, 2012). This 'living systems thinking' demands a way of thinking that includes a world comprised of systems rather than a world of mechanical parts and building blocks. This is the process of storytelling, of creation; it is possible through generation, and can be categorised using a *generative code*.

Chapter 9 described the generative code in detail, and Neis, Brown, Gurr, and Schmidt (2012) explain in the context of an urban environment that the *generative code* is a process that include all the information to develop and build a neighbourhood in a way that creates living ones (Neis et al., 2012, p. 5). Of interest to the narrative of this book is the aspect of 'living neighbourhoods' and 'living structures', and the linkage of the *generative code* to the natural environment. In this process there is always a sequence and an order that follow the rules of the larger whole present in nature and living structures.

In this context of connecting to the natural environment, and in the regenerative design paradigm concerning the factors of regeneration as the living systems, place and patterns, I argue that the *generative code* can be explained as a process that includes all the information to design, develop, build and generate a settlement in a way that creates a 'whole place' that connects to its environment and is resilient in its aspects of both built and natural environments. This 'living systems thinking' as a *generative code* (Mang & Reed, 2012, p. 34) has a series of unfolding steps, identified through the patterns, that enable the people of a place to create a healthy, sustainable, and regenerative place. Alexander acknowledged the process of change through his hypothesis of morphogenetic sequences. In his Schumacher Lecture in Bristol (2004), entitle: *Sustainability and Morphogenesis: The Birth of a Living World,* he argued that the morphological growth of a place gives it character and allows the built environment to adapt to changes (Alexander, 2004). He identified that the *process of morphogenesis* applied to the growth of a place is through *patterns*, and the generation of form of what is coming, or what is about to be, is always drawn from the form of what was in the moment just before. I then conclude that this is closely linked to the *15 fundamental properties [of wholeness]* unfolding in a *generative sequence*, and thus supportive of the *notion of patterns* (Borchers, 2008). Adaptation is linked to the change of a place, noticeable in the patterns, and the *adaptive design* aspect of the *Regenerative-Adaptive Pattern Language* is articulated in the next section.

10.4 Adaptive Design Fundamentals

After a critical review of all the literature, I argue that the *component of change* is evident in the paradigms of regeneration, patterns, morphogenesis, generative process, and in the natural environment through evolution. It can also be identified in the built environment, and the growth of coastal settlements follows a pattern of change through time. Adaptation happens by necessity as a result of many influences, connections, and external factors. These changes also happen over long periods of time, as part of the whole. These indicate that patterns unfolding due to different forces that impact on a place result in consequences, with an established hierarchy (Borchers, 2000, 2001, 2008). This hierarchy of order, a structure that is present in the phenomena of evolution over a long period of time, is acknowledged and applied to the methods used in the works by Alexander (2001–2005a, 2005b, 2005c, 2005d, 1979), Alexander et al. (1977), Lyle (1994), McHarg (1992 [1969]), and as noted by Borchers (2000, 2001, 2008), in the notion of patterns.

However, anthropogenic climate change happens much faster than the slow process of evolution in nature, and the adaptive capacity of natural ecological systems cannot adapt in time for survival. The question is then: *Can the human-built environment adapt in time to the predicted climate changes, and consider the natural environment as an essential part in this adaptive process for the future resilience of both*? Borchers' equation (Borchers, 2008), as identified in Chap. 9, has been used and adapted to provide a new approach for adaptation to climate change *through a generative process*. The fundamental philosophy behind this approach is to include the *human*, *built* and *natural environments* as a *whole interconnected system*. This consideration of the whole interconnected system is represented within the new *regenerative-adaptive pattern language*, and analysed for application through the development of a '*notion of regenerative-adaptive patterns*' equation.

10.5 The Notion of Regenerative-Adaptive Patterns Equation

During my research and investigations for more than 9 years to develop the theory of a *Regenerative-Adaptive Pattern Language*, I have tested Borchers' equation of the *notion of patterns* through further research of the work of Weick (1979), Dell (2002) and Stark (2012). Stark explored the innovative and creative actions of communities and organisations. While some social systems manage to adapt to today's complex world and are able to re-design their structure and environment using the potential available, other social systems are not be able to do this. Exploring these phenomena, Stark used a methodological approach, which aims to identify the patterns of innovative cultures in communities and organisations, by using improvisation (Dell, 2002; Stark, 2012; Weick, 1979). His analyses were based on Alexander's *A Pattern Language* (Alexander et al., 1977), and for application to social structures and organisations, Stark adapted Borcher's *notion of patterns* and developed an equation that can be applied to the *complex phenomena* of *social* and *organisational interactions* (Stark, 2012, p. 92). Upon further investigation of Stark's equation, I found that this was more dynamic and adaptable than Borchers' '*notion of patterns*' one, which tends to be static. This is similar to what has been criticized about Alexander's *Pattern Language,*[1] focusing on the functional aspects of space and not on the geometry of unfolding sequences of patterns in it (Pontikis, 2012). A comparison of Borchers' '*notion of patterns*' equation and Stark's equation are described in Appendix 2.

To be part of a dynamic, regenerative and adaptive process, I adapted the structure of Stark's equation and adjusted the equation from a '*pattern*' to a '*regenerative-adaptive pattern*' to include regenerative considerations and the potential of design and adaptation challenges. The result is an equation with an embedded regenerative-adaptive characteristic, and I refer to this as '*the notion of regenerative-adaptive patterns*',[2] indicated as follow:

$$rgp = \{nda, f_1 \ldots f_i, std, tsp_{15}, e_1 \ldots e_i, r1, pot\}$$

Each *regenerative-adaptive pattern* (*rgp*) in this instance displays a function of:

- A name of typical *design or adaptation challenge* (*nda*);
- A set of *forces* which have an impact on (f_i);
- The specific place settings (*centres*) and *time dynamics* (*std*);
- The *transformations* [*of wholeness*] specific to place (tsp_{15});
- One or various examples of the specific *core pattern* (e_i);
- *Regenerative attributes* (*r*1); and

[1] Grabow (1983) noted that pattern language as a process alone offered little help in transforming a particular design vision and the construction of a place into actual wholeness. This led Alexander to develop it further to include a generative process and into the next level of patterns, *the Morphogenetic sequences* (*The nature of order: An essay on the art of building and the nature of the universe*, books 1 to 4; Alexander, 2001–2005a, 2005b, 2005c, 2005d)

[2] In my PhD Thesis, I have referred to the equation initially as the '*notion of regenerative patterns*', but due to further testing and analysis, I realised that the equation *inherently includes* both regenerative as well as adaptive qualities, and thus amended the description to better reflect the outcomes as the '*notion of regenerative-adaptive patterns*'.

- The *evolutionary potential* of the specific design or adaptation challenge (*pot*).

I concluded that the fundamental principles of the *Regenerative-Adaptive Pattern Language* are present in the '*notion of regenerative-adaptive patterns*' equation. These fundamental principles are *always* linked in one way or the other to the *fundamental patterns* of the language.

These are:

- Any place, built or natural environment always have the *challenge of adaptation* to forces that impacts its environment; these present itself in typical design, planning and development challenges (*nda*);
- *Forces of change* are continuously present that impact on the place, as well as on the adaptation patterns that occur. These forces induce change in the natural as well as the built environment (f_i);
- The place settings specific to location are interconnected to its *centres* (the fundamental primary entities[3]), and change of the *centres* happens in the *dynamics of time*; some are immediate, others in short periods, and others over long periods of time (*std*);
- *Transformations* [*of wholeness*] specific to place are always present. These transformations (referred to as the *15 fundamental properties [of wholeness]*[4]), at any given moment are part of an order structure (Alexander et al., 2012, p. 431), which we may identify as *the wholeness* of that place or system (tsp_{15});
- One or various examples of the specific *core patterns* occur in the place, and can be described and identified. These *core patterns* are never independent, or alone in existence. They are always interconnected to other patterns and entities (upwards or downwards), including to the *fundamental patterns*; they connect and form wholeness, and follow a set of rules part of a *generative code* (e_i);
- Essentially nature evolves and adapts to improve; regeneration works at different levels of scale and produces a field within which the improvement of living systems can take place. Design can assist in the healing of places that are damaged, and must be based on *regenerative attributes* (*r*1); and
- The potential of specific design or adaptation challenges can provide opportunities to assist a community (human and ecological) to evolve the value-generating capacity of their system as a whole, identifying their adaptive, *evolutionary potential*[5] (*pot*).

The principles of the *Regenerative-Adaptive Pattern Language* as described above and presented in the '*notion of regenerative-adaptive patterns*' equation demonstrate the premises of the regenerative paradigm to include the role of humans and nature, working together in a developmental process that improves the value of *the whole*. This includes the opportunities of co-evolution due to the impacts of changes. The six concepts, as defined by Mang and Reed (2012), of regeneration, development and design, place, pattern literacy, story, and potential are supported by the '*notion of regenerative-adaptive patterns*' as part of the *Regenerative-Adaptive Pattern Language*. Further, the forces of change to a place need to be taken into consideration, and the dynamics of adaptation are added to the *regenerative-adaptive pattern language* process.

To be able to test this new *regenerative-adaptive pattern language* theory at a specific place, the *Regenerative-Adaptive Design Model* was developed and used to combine the principles of regenerative-adaptive design and a pattern language with a risk-based adaptation process. The goal is that this

---

[3] Alexander and Neis explained the essence of centres of the whole as 'each centre is recursively dependant on other coherent centres for its own coherence. To understand this idea is to define all centres in this way, as the fundamental primary entities' [of a place] (Alexander et al., 2012, p. 430).

[4] See Chap. 9, Sect. 9.6 for a detailed explanation of the term '*15 Fundamental Properties of Wholeness.*'

[5] *Potential* is defined as the inherent capacity for coming into being, for growth and development. From the living systems theory perspective, all living systems are distinguished by a unique essence, and all have, based on that uniqueness, an inherent potential to which they are moving toward or away, depending on their state of integrity and vitality or health. Regenerative design identifies the patterns that reveal the underlying energy flows (Lyle, 1991), including the actual and the *potential* that shape that system (Capra, 1996; Von Bertalanffy, 1968).

model can be applied to coastal communities to better appreciate and understand how the landscape will change, in ecological and human contexts, and therefore inform their adaptive responses. This model can inform the community's *potential* and assist in establishing *resilient coastal settlements* in a *future changed climate*.

10.6 The Regenerative-Adaptive Design Model

As described in Chap. 3, according to the most recent definition of adaptation issued by the IPCC Working Group II, as well as Working Group III, 2014, the adaptation of natural systems is supported by human intervention. The definition states that:

> Adaptation is the process of adjustment to actual or expected climate and its effects. In human systems, adaptation seeks to moderate harm or exploit beneficial opportunities. In natural systems, human intervention may facilitate adjustment to expected climate and its effects (IPCC, 2014b, p. 40).

When adaptation planning uses hazard and risk management process to plan and manage risks due to climate change, the results will be more resilient if the process include social, economic, and environmental sustainability aspects. A prerequisite for sustainable development in the context of climate change is to address the underlying causes of vulnerability, including the structural inequalities that create and sustain access to both human and natural resources. This involves integrating adaptation planning and risk management into all social, economic, and environmental domains, including policy and the use of a bottom-up and top-down approach that involves the local community members (IPCC, 2014a).

10.6.1 Adaptation Planning

Adaptation planning is mostly based on the *hazard and risk management process* (GORCC, 2012). The principles of risk management are included in the risk assessment process; they are applied to adaptation planning by various governments, and are consistent with the broad approach to climate change adaptation promoted by various governments across the world, and by the Australian Government as well (AGO, 2006). In Australia, the risk framework for hazard management is based on the *Australian Standard for Risk Assessment and Management AS/NZS ISO31000:2009*. Adaptation planning considers various approaches and pathways to resilience; however, limits are faced when thresholds or tipping points associated with social and natural systems are exceeded, posing severe challenges for adaptation by the local community. Choices and outcomes for adaptive actions to climate events usually need to reflect divergent capacities and resources and multiple interacting processes (IPCC, 2012). Actions are framed by trade-offs between competing prioritized values and objectives, and the consideration of the adaptation process needs to look at different visions of development that can change over time.

The hazard risk management process indicates that multiple interacting processes with forces of change result in risk that needs to be managed, mitigated or avoided. The forces of change that increase the level of risk include climate aspects, natural variability, anthropogenic climate change, and socio-economic processes, which include governance. These factors influencing each other are a complex system, and potential solutions can be developed for identifying the *iterative processes*, which are similar to patterns in a pattern language.

Iterative approaches that evolve over time can allow development pathways to integrate risk management so that diverse solutions can be considered. Adaptive Planning as an *adaptive process* can be informed by design and planning options that include the consideration of ecological sustainability.

The regenerative aspects of nature, inclusive of patterns, is represented in the *Regenerative-Adaptive Design Model*. This adaptive process can be referred to as the *regenerative-adaptive system of processes*, as described in the next section.

10.6.2 Regenerative-Adaptive System of Processes

In the analysis of the '*Regenerative Development and Design Methodology*' (Mang & Reed, 2012), we have identified that the *regenerative process* is an open dialogue growing out of the emerging understanding of a place that comes from the knowledge of the community. The approach includes three phases of generation: understanding and conceptualising the relationship to place, design for harmony, and the co-evolution of human and habitat to sustain its survival capacity. These phases are encapsulated in the *Living systems thinking framework* and demonstrated in an evolutionary spiral, depicting a growing systemic capacity as it actualizes a specific project (Fig. 10.3).

The living systems thinking framework advocated by Mang and Reed (2012) depicts three phases for design and development. First is understanding and conceptualising the *right relationship to place*. Second is designing for *harmony with the place*, and third is the consideration of *co-evolution of all aspects of the place*. This framework highlights that the people who inhabit the place need to regenerate, manage, and sustain it over time, thus achieving a state of sustainability.

A framework that considers *change over various time dynamics and forces of impact* can be represented as the *regenerative-adaptive system of processes*. This system of processes considers all the relevant dynamics of change, even the impacts of direct and instant change, as well as a constant process of evolution adapting to internal and external forces. These phases of change, in addition to understanding, harmonizing and sustaining, are encapsulated in the *regenerative-adaptive system of processes*, demonstrated in an *enhanced evolutionary spiral* (Fig. 10.4).

The *regenerative-adaptive system of processes* depicted in Fig. 10.4 *includes four processes* for the design and development of a place:

- *Place* – Identify and define the place in the context of its unique dynamics and potential, the community's relationship to it and its values, and the understanding of the phenomena that make it up (its levels of scale);
- *Core patterns* – The story of the place; identifying the core patterns evident in the built and natural environments that depict the working of the place and that reflect its unique character;
- *Generative code* – An accumulation of the core patterns of a place, which include all the information to develop and generate it in a way that creates a whole place connected to its environment; and
- *Regenerative-Adaptive* – Concept for a systematic regenerative-adaptive design, planning and development process that can realise the co-evolution of both human and natural habitats of the place, depicting its potential for future resilience.

The *regenerative-adaptive system of processes* provides guidance for the process flow in developing the *Regenerative-Adaptive Design Model*. The model needs to be able to take all the principles of regenerative-adaptive theory and include these in an overall framework that combines them all into a system of processes (Roös & Jones, 2017).

10.6.3 Model Attributes

The *attributes of the model*, based on a *system of processes*, include the following:

- *Understanding and define place*: 'How big is here?' Findings of heritage, cultural, economic, geographic, demographic, climatic and ecological characteristics;

Fig. 10.3 Living systems thinking framework that structures the approach to regenerative development. (Drawing by Jesse Delmo, synthesised by Roös, adapted from Mang & Reed, 2012)

- *Wholeness of place*: The cohesion of the structures of the place includes built and natural attributes and identification of the 15 properties of wholeness;
- *Diagnosis of place*: The overall geometric structure of the place setting, identifying the attributes and values that make a positive or negative contribution to its existence;
- *Generative code*: The accumulation of the core patterns of the place, which include all the information to generate it, identifying interconnected patterns of harmony that form the whole connected to its environment;
- *Regenerative attributes*: The qualities that can restore, renew, or revitalize the sources of energy and materials of the place and that support its existence;

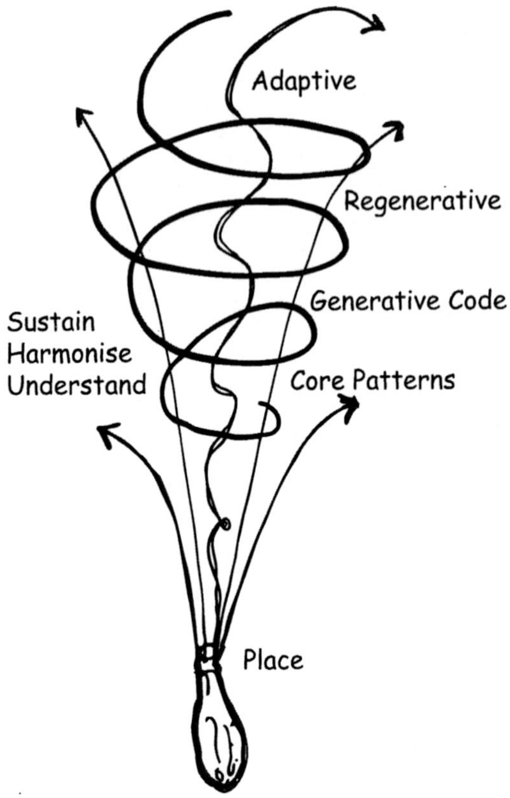

Fig. 10.4 Regenerative-adaptive system of processes. (Drawing by Jesse Delmo, synthesised by Roös)

- *Forces of change*: Identify forces of change that can impact the sustainable future of the place, including risk to social, economic, and environmental, cultural-heritage, and built and natural attributes;
- *Potential of place*: What is the vocation of the place, what is unique to it that has the potential to contribute to the wholeness of the larger system of which it forms part;
- *Regenerative patterns*: The combination of the various processes of the place, accumulating to support its morphological growth and regeneration, give it characteristics for resilience and allow it to adapt to changes;
- *Regenerative-adaptive pattern language*: The overall pattern language for the place connecting all attributes and the '*notion of regenerative-adaptive patterns*' with the *fundamental patterns*; and
- *Adaptation planning and design*: Use of the regenerative-adaptive pattern language and the regenerative-adaptive processes, principles and qualities of place as an input for an adaptation framework that can inform adaptation responses, contributing to the overall implementation of the *Regenerative-adaptive design model*.

Regenerative-adaptive design follows a *flow of patterns*. This begins with the recognition that a place is a dynamic entity with its own unique history and future, and thus is always *growing and evolving*, with constant forces that impact and change it. Place is constantly *influenced by the larger system* it is embedded in, and this concept informs the process flow of the application of the model,

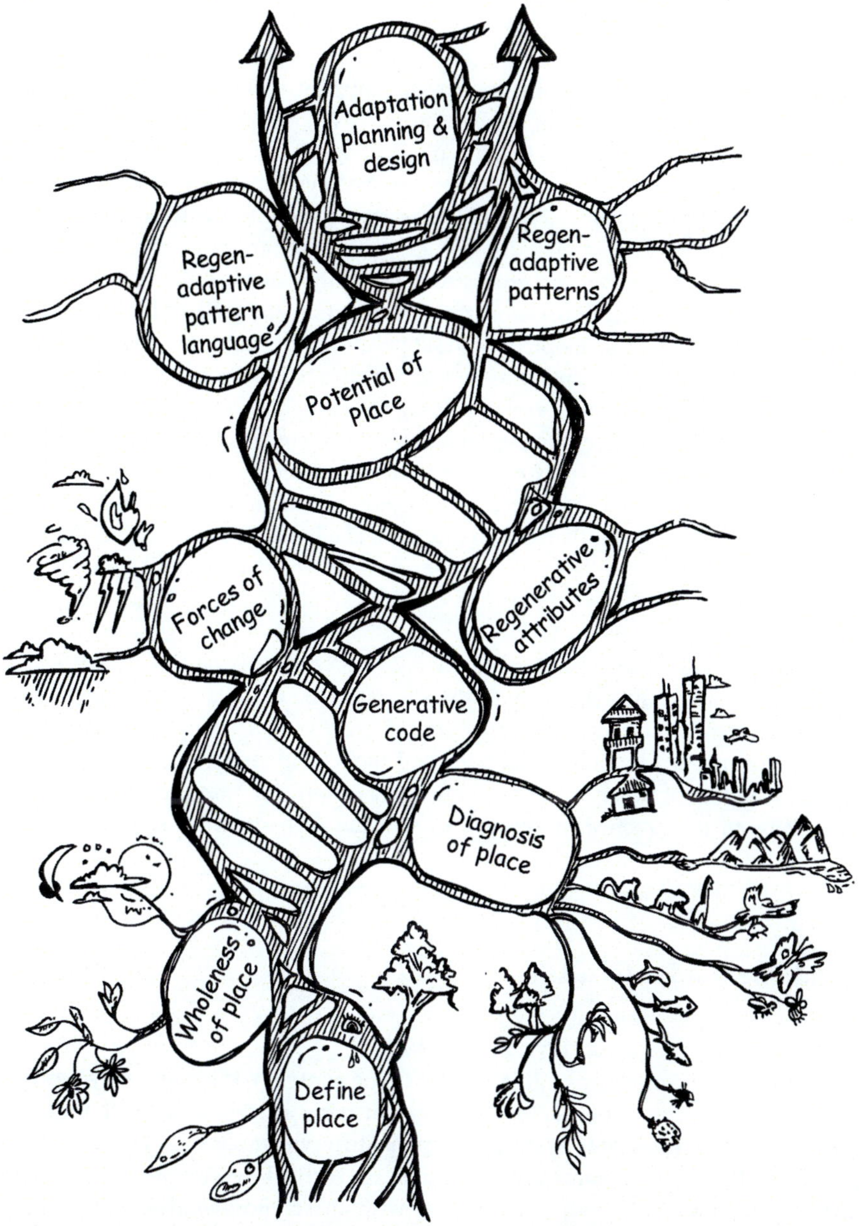

Fig. 10.5 Process flow of the regenerative-adaptive design model. (Drawing by Jesse Delmo, synthesised by Roös)

starting with *place* and concluding with local *regenerative-adaptive patterns*. The final step in the process is adaptation planning and design of the place. This process flow of the application of the model is indicated in Fig. 10.5. The application of this *regenerative-adaptive design model* is demonstrated in a case study, the coastal town of Anglesea in the state of Victoria, Australia, in the next chapter, Chap. 11.

10.7 Conclusion Statement

This chapter started with reconfirming that the consideration of – THE WHOLE [1] – is an important factor in the theories explored in this book, and the proposed *Regenerative-Adaptive Pattern Language* summarises 'wholeness' as fundamental to the ecological design and planning of built environments, which include the consideration of connections to natural systems. Based on Alexander's (2001–2005a, 2005b, 2005c, 2005d) hypothesis of wholeness, the development of the *Regenerative-Adaptive Pattern Language* evolved from the principles of the generative process, whereby there is always a sequence, an order, and a morphogenetic process that drives the organic unfolding of a place. Linking to natural systems, this *hypothesis of wholeness* is also evident in the work of Mang and Reed (2012), which encapsulates regenerative development and supports a *new living systems, ecological worldview*.

Exploring the fundamentals of adaptation, the force of change is identified as the factor that induces adaptive actions in the paradigms of regeneration, patterns, morphogenesis, generative process, and in the natural environment through evolution. Adaptation is linked to the change of a place, noticeable in its patterns. The fundamental principles of the *Regenerative-Adaptive Pattern Language* are represented in the '*notion of regenerative-adaptive patterns*' equation, and the *Regenerative-Adaptive Design Model* is used to apply the theory into practice. The collective narrative of the regenerative-adaptive pattern language and the application of the regenerative-adaptive design model makes up the *transformations of wholeness*.

Therefore:

To understand the *regenerative process* of a *whole system* that is part of THE WHOLE [1] is to understand the *patterns* and the generation of the system in the first place. Patterns provide a *place-based* understanding of the *integrated human and ecological systems*, which can only be appreciated by exploring the nature of generation processes and their relationship to patterns and pattern languages. It is thus important to apply the stipulations of *generative processes* from both human and ecological systems to the practice of *regenerative design* as part of the sustainable development discourse.

10.8 Fundamental Pattern 10

NOTION OF REGENERATIVE-ADAPTIVE PATTERNS [10]

In Nature there is no singular or climax state; to the contrary, it is always dynamic, with patterns of generative processes occurring and re-occurring in response to change forces. The regenerative process of the whole system occurs at different levels of scale, and can be either lateral-horizontal (increasing connection richness) or vertical/developmental (increasing complexity order). This pattern requires us to deal with these complexities specific to place and equips us to deal with regeneration and adaptation to significant change (Fig. 10.6).

Downward links:

Use this pattern NOTION OF REGENERATIVE-ADAPTIVE PATTERNS [10] to link all levels in the pattern language to a specific place – from the higher-level overarching *fundamental patterns*, down to the *core patterns* specific to a place. Always include all the attributes of the *Regenerative-Adaptive Design Model*. This will assist in dealing with *forces of change*, and through the application of an evolutionary process – EVOLUTIONARY ADAPTATION [11] – the outcomes will be dynamic in nature, resilient and part of the larger whole – THE WHOLE [1] – …

Fig. 10.6 Sketch of fundamental pattern – NOTION OF REGENERATIVE-ADAPTIVE PATTERNS [10]. (Drawing by Jesse Delmo, synthesised by Roös)

References

AGO. (2006). *Factors and methods workbook, December 2006: For use in Australian greenhouse gas emissions reporting*. Canberra: AGO – Australian Greenhouse Office.

Alexander, C. (1964). *Notes on the synthesis of form*. Cambridge, MA: Harvard University Press.

Alexander, C. (1979). *The timeless way of building* (Vol. 1). New York: Oxford University Press.

Alexander, C. (2004). *Sustainability and morphogenesis: The birth of a living world*. Berkeley, CA: Centre for Environmental Structure.

Alexander, C. (2001–2005a). The nature of order – An essay on the art of building and the nature of the universe, book one: The phenomenon of life. Berkeley, CA: The Center for Environmental Structure.

Alexander, C. (2001–2005b). The nature of order – An essay on the art of building and the nature of the universe, book two: The process of creating life. Berkeley, CA: The Center for Environmental Structure.

Alexander, C. (2001–2005c). The nature of order – An essay on the art of building and the nature of the universe, book three: A vision of a living world. Berkley, CA: The Center for Environmental Structure.

Alexander, C. (2001–2005d). The nature of order – An essay on the art of building and the nature of the universe, book four: The luminous ground. Berkeley, CA: The Center for Environmental Structure.

Alexander, C., Ishikawa, S., Silverstein, M., Jacobson, M., Fiksdahl-King, I., & Angel, S. (1977). *A pattern language: Towns, buildings, construction*. New York: Oxford University Press.

Alexander, C., Neis, H., Anninau, A., & King, I. (1987). *A new theory of urban design (Center for Environmental Structure)*. New York/Oxford: Oxford University Press.

Alexander, C., Neis, H. J., & Moore Alexander, M. (2012). *The battle for the life and beauty of the earth: A struggle between two world-systems*. New York: Oxford University Press.

Alexander, C., Schmidt, M., Alexander, M., Hanson, B., & Mehaffy, M. (2005). *Generative codes: The path to building welcoming, beautiful, sustainable neighbourhoods*. Berkley, CA: Centre for Environmental Structure.

Borchers, J. (2000, August 17–19). A pattern approach to interaction design. In Proceedings of the ACM DIS 2000 International Conference on Designing Interactive Systems, 1–10). New York: ACM.

Borchers, J. (2001). A pattern approach to interaction design. *AI & Society, 15*(4), 359–376; London: Springer-Verlag London Limited.
Borchers, J. (2008). A pattern approach to interaction design. In P. G. Satinder (Ed.), *Cognition, communication and interaction: Transdisciplinary perspectives on interactive technology* (pp. 114–129). London: Springer.
Capra, F. (1996). *The web of life: A new scientific understanding of living systems*. New York: Anchor Books.
Carlson, B. (2007).cBurlington, MA: Academic Press, Elsevier Inc.
Dell, C. (2002). *Prinzip improvisation (principles of improvisation)*. Koln, Germany: Walter Konig.
Du Plessis, C. (2009). *An approach to studying urban sustainability from within an ecological worldview*. Salford, UK: Research Institute for the Built and Human Environment, School of the Built Environment/University of Salford.
Du Plessis, C. (2012). Towards a regenerative paradigm for the built environment. *Building Research & Information, 40*(1), 7–22.
GORCC. (2012). *Coastal climate change vulnerability and adaptation*. Melbourne, VIC: Great Ocean Road Coast Committee – GORCC.
Grabow, S. (1983). *Christopher Alexander and the search for a new paradigm in architecture*. London: Orial Press.
IPCC. (2012). *Managing the risks of extreme events and disasters to advance climate change adaptation: A special report of Working Groups I and II*. Cambridge: IPCC, SREX/Cambridge University Press.
IPCC. (2014a). *Working Group III contribution to the IPCC 5th Assessment Report – Mitigation to climate change technical summary final draft*. Geneva: Intergovernmental Panel on Climate Change IPCC, Working Group III.
IPCC. (2014b). *Working Group II Contribution to the IPCC 5th Assessment Report: Climate change 2014: Impacts, adaptation, and vulnerability AR5 final draft report*. Geneva: Intergovernmental Panel on Climate Change, Working Group II.
Kauffman, S. (1993). *The origins of order self-organization and selection in evolution*. New York: Oxford University Press.
Kelly, K. (2005). *The Big Here Quiz*. (available at: http:// www.kk.org/cooltools/archives/000957.php) (accessed on 7 May 2011).
Kuhn, T. (1962). *The structure of scientific revolution*. Chicago, IL: University of Chicago Press.
Lyle, J. T. (1991 [1985]). *Design for human ecosystems – Landscape, land use, and natural resources*. California/Washington, DC: Island Press.
Lyle, J. T. (1994). *Regenerative design for sustainable development*. New York: Wiley.
Mang, P., & Reed, B. (2012). Designing from place: A regenerative framework and methodology. *Building Research and Information, 40*(1), 23–28.
McHarg, I. (1992 [1969]). *Design with nature*. Washington, DC: Wiley.
Mehaffy, M. (2009). Horizons of pattern languages. In H. Neis & G. Brown (Eds.), *Current challenges for patterns, pattern languages and sustainability* (pp. 103–111). Portland, OR: PUARL Press.
Metzner, R. (1999). *Green psychology: transforming our relationship to the earth*. South Paris, ME: Park Street Press.
Neis, H. J., Brown, G. A., Gurr, J. M., & Schmidt, J. A. (2012). *Generative process, patterns, and the urban challenge. Fall 2011 International PUARL conference*. Portland, OR: PUARL Press.
Pontikis, K. (2012). A green pattern language. In H. Neis, G. Brown, J. M. Gurr, & J. A. Schmidt (Eds.), *Generative process, patterns, and the urban challenge* (1st ed., pp. 103–113). Portland, OR: The PUARL Press.
Reed, B. (2007). Shifting from sustainability to regeneration. *Building Research & Information, 35*(6), 674–680.
Roös, P., & Jones, D. (2017). Knowledge of making life: Design patterns for regenerative-adaptive design. In *DesTech 2016: Proceedings of the international conference on design and technology* (pp. 203–210). Dubai, United Arab Emirates: Knowledge E. https://doi.org/10.18502/keg.v2i2.616
Sanford, C., & Mang, P. (1992). *A self-organizing leadership view of paradigms – traditions in business*. San Francisco, CA: Berrett-Koehler.
Stark, W. (2012). Tacit knowledge and innovation patterns for communities and social systems. In H. B. Neis (Ed.), *Generative process, patterns, and the urban challenge* (pp. 89–95). Portland, OR: PUARL ARUS Press.
Von Bertalanffy, L. (1968). *General systems theory: Foundations, development, applications*. New York: George Braziller.
Weick, K. E. (1979). *The social psychology of organizing* (2nd ed.). Columbus, OH: McGraw-Hill.

Check for updates

Chapter 11
Case Study: Application of the Regenerative-Adaptive Pattern Language

Abstract The fundamentals of the *regenerative-adaptive pattern language* have been described in previous chapters, including the formulation of the *regenerative-adaptive design model*. The need to apply the key principles of the *notion of regenerative-adaptive patterns equation* to the design and planning of our future built and natural environments has been highlighted. In this chapter the application of all the elements of the *regenerative-adaptive pattern language,* the *regenerative-adaptive design model,* and the *notion of regenerative-adaptive patterns equation* are tested on the case study of a coastal town called Anglesea, located on the Victorian coast in Australia.

This chapter start with an introduction to the case study area, the Great Ocean Road coastal region, and progresses with documenting the application of the *regenerative-adaptive design model* to the case study town Anglesea. The case study area has been analysed using the framework of the *regenerative-adaptive pattern language*. Similar to the process of *A Pattern Language* (Alexander et al., *A pattern language: Towns, buildings, construction.* Oxford University Press, New York, 1977) and the Ecological systems method, *the whole* is considered and the analysis starts with the *bigger regional context*, moving down in levels of scale to the *local context* of Anglesea. Findings resulted in the identification of many similarities between the settlements along the coast, concluding with detailed documentation of the application of the *regenerative-adaptive design model* to the town of Anglesea. Recommendations are provided how these findings can be applied to adaptive planning and design of other coastal settlements and larger cities around the world. The chapter concludes with the *fundamental pattern* EVOLUTIONARY ADAPTATION [11], which includes practical applications of the *regenerative-adaptive pattern language*.

Keywords Evolutionary adaptation · Regenerative-adaptive design model · Regenerative-adaptive pattern language · Generative codes · Core patterns · Wholeness of place

Upward links:

… in the process of adaptation this pattern – EVOLUTIONARY ADAPTATION [11] – helps to complete the process of applying the NOTION OF REGENERATIVE-ADAPTIVE PATTERNS [10]. The ground rules were set in NATURE'S DESIGN [7], where the ecological method of levels of scale reveal Nature itself as the primary influencer for decision-making in adaptation design and planning.

P. B. Roös, *Regenerative-Adaptive Design for Sustainable Development*, Sustainable Development Goals Series, https://doi.org/10.1007/978-3-030-53234-5_11

11.1 Introduction

In every walk with nature one receives far more than one seeks (John Muir, 1877)

Reading through all the chapters of this book starting from Chap. 1 will help form a narrative on how the overall *regenerative-adaptive pattern language* fits together. However, up to this point it is possible for the reader to raise the question of how all of this can be applied to a real place, and conclude that it is complex. Further, the enormous complexity of the topic can cause confusion, and indeed it is not uncommon that the understanding of *a pattern language theory* causes this puzzlement in many instances (Neis, Brown, Gurr, & Schmidt, 2012). Knowledge of *patterns* and the *pattern language theory* are useful when applied to a real problem, and further reading is recommended to grasp this understanding. Recommended readings include *A Pattern Language* (Alexander et al., 1977); *The Timeless Way of Building* (Alexander, 1979); Chaps. 1 and 5 in *The Nature of Order, Book One, The Phenomenon of Life* (Alexander, 2001–2005a); Chap. 2 in *The Nature of Order, Book Two, The Process of Creating Life* (Alexander, 2001–2005b); and Chap. 7 – Form Languages and Their Vocabulary in *Unified Architectural Theory: Form, Language, Complexity* (Salingaros, 2013) (Fig. 11.1).

Further, in Appendix 3 to 6, I have attempted to provide detailed information, explanations and descriptions of the various elements related to the *regenerative-adaptive pattern language*, the *regenerative-adaptive design model*, and the key principles of the *notion of regenerative-adaptive patterns equation,* which includes: *forces of change*, *centres*, *transformations of wholeness*, *core patterns*, *fundamental patterns*, *regenerative attributes*, and *regenerative-adaptive patterns*. To be able to implement this *new pattern language* in a place one must acknowledge that the underlying theory is the *regenerative-adaptive pattern language*, and the theory can be put to use by using the framework of the *regenerative-adaptive design model* and applying the *notion of regenerative-adaptive equation*

Fig. 11.1 Anglesea river and boardwalk (photo by Author)

to a specific place. This 'systems of processes' methodology is demonstrated as part of the *regenerative-adaptive design model* in Fig. 11.2.

This methodology was used to analyse 6 coastal towns along the Great Ocean Road Coast (the larger region), which informed the identified *similarities* of *generic generative codes* and the *core patterns* established for each town. For each town, further steps included the identification of *wholeness of place*; the *regenerative attributes* of place; forces of change; the *long-term history of human and landscape interaction*; and the development of *regenerative-adaptive patterns* of place. Due to the extensive data analysis, and mammoth undertaking of 6 years' research, it is beyond the scope of this book to report on each town's analysis. Only the town of Anglesea is reflected upon.

11.2 Application Methodology

The collection of and interpretation of data to understand the phenomena of patterns was conducted using an abductive reasoning strategy. The *nature of patterns,*[1] which are evident in the changing landscape, social and cultural perspectives, place character and climate change impacts, and the understanding of place by local residents in the coastal towns, justify the use of an abductive reasoning strategy. Further, methods and generative processes described in Alexander's *A Pattern Language* (1977), *The Timeless Way of Building* (1979), and Book 1 and 2 of *The Nature of Order* (2001–2005a, b) informed the methodology for data collection, interpretation, analysis and community engagement for this case study. The methods in these theories underpin the *regenerative-adaptive design model* (RADM); the 'system of processes' that informs the 8 stages of information collection and interpretation; the development of the regenerative-adaptive patterns; and the application of the RADM for identifying adaption planning and design options. The stages include:

- Stage 1: Understand and *define place*;
- Stage 2: Explore the *wholeness of place*;
- Stage 3: *Diagnosis of place* identifying the *core patterns*, developing a *generative code*, and identifying its *regenerative attributes*;
- Stage 4: Identify the *forces of change* that impact place;
- Stage 5: Identify the *long-term history of human and landscape interaction*;
- Stage 6: Develop the *regenerative-adaptive patterns* of place, including application of the *notion of regenerative-adaptive equation*;
- Stage 7: Describe the overall *regenerative-adaptive pattern language* of the place, confirming whether all the attributes support the *fundamental patterns*; and
- Stage 8: Identify options for future *adaptive planning and design.*

The stages and corresponding attributes of the *regenerative-adaptive design model* are described in Table 11.1.

11.3 The Great Ocean Road Coastal Region

As noted previously, it is important to understand the broader context of a settlement in its regional context and location within the levels of scale in an ecological framework. What is unique about the Great Ocean Road coastal region and its settlements? Understanding this context will help to develop

[1] The 'nature of patterns' is an abstract occurrence of physical relationships, which resolve a small system of interacting and conflicting forces that is independent of all other forces.

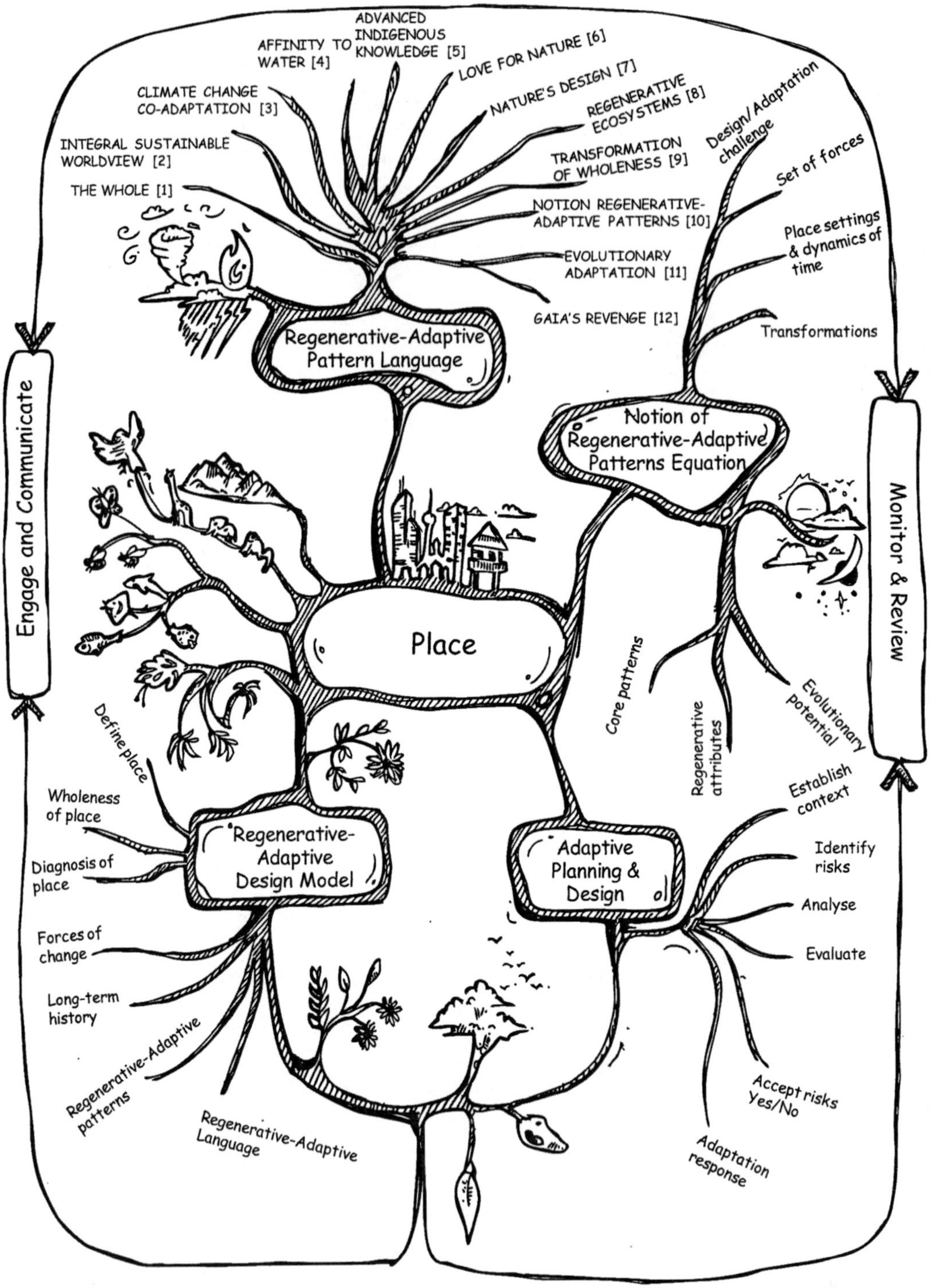

Fig. 11.2 Flow diagram indicating the method of implementing the regenerative-adaptive pattern language (drawing by Jesse Delmo, synthesised by Roös)

Table 11.1 Stages of the application of the *regenerative-adaptive design model* (RADM)

Stage	Description	Attributes of the RADM
Stage 1	**Understanding and define place:** Data collection and analysis of information to understand the historical context of the town as well as identifying its natural and built attributes and climatic context, including information on demographics.	How big is here? Understanding and defining place in its regional setting Place in its context of the unique dynamics and phenomena that constitute it
Stage 2	**Explore the wholeness of place:** Identify the structure of wholeness that exists in the settlement, including the existing physical configuration that represents the 'sense of place'. What are the deep patterns that exist which are unique to the town and its landscape? What is the human-nature connection and its contribution to sense of place? Walk the land, collect photographic data, on-site sketching, feel, write and map attributes that unfold from the built and natural aspects of the place.	Identification of the *fifteen fundamental properties of wholeness*, the transformations specific to the place
Stage 3	**Diagnosis of place:** Identify the overall *geometric structure* of the place. The town diagnosis allows the identification of *core patterns* of the place – An identification of interconnected patterns that are the cohesion of everything that forms the harmony of the whole, accumulating into a *generative code*. The town diagnosis helps to identify within its broader context the *regenerative attributes* that can assist in the restoration of resources and energy that supports its existence. *Unfolding of place* includes: Finding special places; Focal points, visual values, views; Place character, cultural places; Centres supporting a living structure, identification of the 'the heart' of the town; Places to heal, places that needs repair; Vulnerability of assets facing risks due to environmental impacts; Prospect, considering the context of sacred space, open space, and positive space; and Boundaries, natural and built boundaries of the town.	The overall *geometric structure of place*: Attributes that have a positive as well as a negative contribution to the town's existence, identification of the *core patterns* of place The *Generative code*, representing the collation of core patterns of the place, identifying interconnected patterns that result in harmony with the whole *Regenerative attributes*, including the identification of the qualities that can restore, renew, or revitalise the sources of energy and materials of the place that support its existence
Stage 4	**Forces of change that impacts place:** The identification of the forces of change that impact upon place. Place in its larger context has the *potential* to be regenerative-adaptive, using forces of change as an agent of improvement.	*Forces of change* including risk to social, economic, environmental, cultural-heritage, and built and natural attributes *Potential of place,* indicating what is unique to place that has the potential to contribute to the wholeness of the larger system of which it is part
Stage 5	**Long term human-nature interaction:** Identify the longer-term history of *human and landscape interaction* of the region from the Indigenous people's perspective, to understand the *deeper forms* of the land, the changing landscape over long periods of time, the climate patterns, and the human settlement patterns of the past.	Wholeness of place, the connection to *Country* (land and nature) over long periods of time, from the perspective of local Indigenous peoples Forces of change, the changes of landscape and human connection to *Country* over long periods of time, settlement patterns Potential of *Country*, what is unique from *Country* that has the potential to contribute to the wholeness of the larger system of which it is part.

(continued)

Table 11.1 (continued)

Stage	Description	Attributes of the RADM
Stage 6	**Regenerative-adaptive patterns of place:** Using the *regenerative-adaptive processes*, principles and qualities of a place as an input to a framework that can inform adaptation and design responses. This stage includes the *regenerative-adaptive patterns* as an input for the application of the *regenerative-adaptive design model*.	Development of the *regenerative-adaptive patterns* of place, using the '*notion of regenerative-adaptive patterns*' equation.
Stage 7	**A regenerative-adaptive pattern language:** Describe a *regenerative-adaptive pattern language* of the place, confirming whether all the attributes and principles of the language support the overall *fundamental patterns*.	Description of the *regenerative-adaptive pattern language* and alignment with the *fundamental patterns*.
Stage 8	**Adaptive planning and design:** Using the findings and outputs of the *regenerative-adaptive pattern language* process to inform potential adaptive planning and design solutions.	Adaptation planning and design, including the regenerative-adaptive processes, principles and qualities of a place as an input to an *adaptation framework* that can inform adaptation responses, contributing to the overall implementation of the *regenerative-adaptive design model*.

a *unified pattern language* for this area and its settlements with unique character. Living next to nature is the attraction for communities along the Great Ocean Road coast, a traveller's paradise with some of Australia's best surf beaches, bustling resort towns, spectacular scenery, and lush rainforests. Beyond visual attributes, it is the ecological, social and economic values that boost a shift in Australian culture from the bush and city. In Victoria, the migration to the coastal towns from the city of Melbourne resulted in the emergence of the third culture in Australia, that of 'the culture of the beach' (Salt, 2003). This is indeed the culture of communities along the Great Ocean Road coast. These communities survive today because of the high visual and natural attributes they have, which underpin their existence and economic survival. It is these same attributes that make coastal towns attractive places to live, which is now threatening their current existence (Roös & Jones, 2013). Change these landscape attributes and there is a dramatic effect upon their context, economic, social and environmental sustainability. The values that are important for residents and that make the Great Ocean Road coastal region an attractive tourism destination are evident in its natural, built and social/recreational attributes.

For the purposes of this investigation, the Great Ocean Road coastal region stretches from Torquay in the east to Peterborough in the west, just south of the town of Warrnambool. Six towns were selected for sampling, and included Port Campbell, Apollo Bay, Wye River, Lorne, Anglesea, and Torquay. The town of Anglesea was selected as a case study due to its unique character, as well its representativeness of similar small towns along this coastline (Fig. 11.3).

11.4 Case Study Town: Anglesea

Coordinates: 38.24,29 S / 144.11,07 E.

Postcode: Victoria, 3230.

The town is located next to the Anglesea River inlet, with a wetland, hilly geomorphology, large trees, bush and coastal scrub, and a limestone coastal cliff that butts against the sandy beach area. The vegetation is predominantly heath woodland, lowland forest, and coastal headland scrub. With a unique coastal village character (Fig. 11.4), and the setting of the town within the bush, river plain and the coast makes Anglesea a major attraction for holidaymakers that visit the Great Ocean Road coast. Nearby coastal settlements with similar characteristics are Lorne, Aireys Inlet, Bells Beach and Jan Juc.

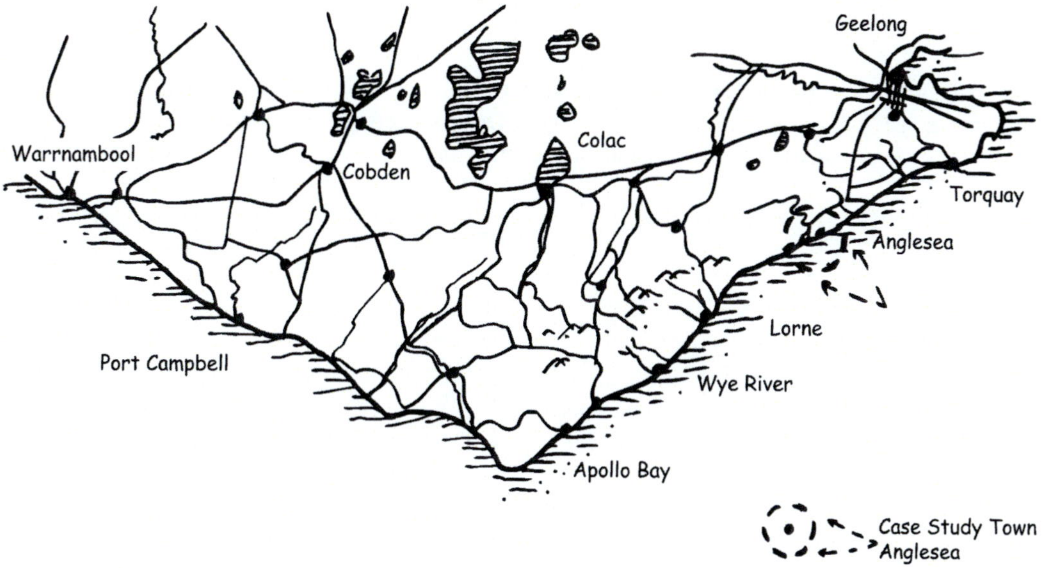

Fig. 11.3 The Great Ocean Road coastal region, indicating the location of Anglesea (drawing by Jesse Delmo)

Fig. 11.4 Unique town character of anglesea (photo by Author)

11.5 Understanding and Define Place

Understanding and defining place records the heritage, cultural, economic, geographic, demographic, climatic and ecological characteristics of the town. These aspects are part of the place's settings specific to location, and are interconnected with centres that have been influenced by dynamics of time (*std*).[2]

11.5.1 Brief History of Anglesea

Before European settlement, early explorers noted the *Wadawurrung* people occupied the land in the area close to the river (Richards, 1998; Clark, 1990; PIRG, 1977). The remoteness of Anglesea restricted any settlement or development, and growth of the town only started in the late 1800s. The earliest European settler of the region was John Airey who moved to Roadknight and Aireys Inlet south of Anglesea in 1839. The first land in Anglesea (known then as Swampy Creek) was occupied and settled by Michael Fallon in 1845. In 1884, the then shire of Barrabool gazetted Anglesea as a town and between 1894 and 1890 blocks of land were sold (Weber & Weber, 2010). To cross the River, the Anglesea Bridge was built in 1968 and allowed for the further growth of Anglesea, mostly due to its tourism attraction, activities on the river, and beach. Today, Anglesea is predominantly a tourist town. The recently closed Alcoa open cut coal mine was for many years a provider of employment for a small portion of local residents.

11.5.2 Natural Attributes

Anglesea is set within the Otway Ranges, located at the edge of the Great Otway National Park. The Great Otway National Park and the reserve – The Anglesea Heath[3] – surround the town. Various creeks and the Anglesea River carve through the landscape and coastal bush, with its inlet expanding through the beach and into the ocean, with long stretches of sections of sandy beaches on both sides. High limestone cliffs embrace the sandy beaches, and in various areas rock formations protrude into the sea. The hilly forests result in the dramatic landscape scenery of the town. Similar to other small settlements on this coast, the vegetation of Anglesea is a mixture of introduced and local native species.

11.5.3 Built Attributes

Various homes are simple in form and reflect the 'beach shack' and 'fisherman's cottage' style of buildings. Buildings predominantly sit within or below the tree canopy, and are single-storey mixed with double-storey in certain areas. In the commercial precinct of the town buildings are predomi-

[2] This aligns with the (*std*) component of the 'notion of regenerative-adaptive patterns' equation: 'The specific place settings and time dynamics'.

[3] The Anglesea Heath is a 7000 hectares parcel of land that nearly surrounds the whole of the town. This name was given to the title of the land, but it is not a typical heathland, open with low shrubs. Bush and trees with a diversified ecosystem cover the Anglesea Heath. The EVC classification of this area includes woodland, mixed lowland forest, swampy riparian woodland, sedgy wetland and shallow fresh water marshland (McKellar, 2009).

nately double-storey, with a higher density zoning, and buildings are located on the street front. Anglesea lost most of its early heritage buildings in the fires of 1891, and then later in the *Ash Wednesday* fires of 1983, which had the biggest impact on the landscape and the settlement of the town (Weber & Weber, 2010). As with most of the small towns along this coast, out of character buildings are present, and these are 'modern', bulky, boxy in style with large footprints on the land.

11.5.4 Demographics

In summary, the population of Anglesea was 2454 in 2011 (UCL/ Urban Centre Locality); of these, 48.5% were male and 51.5% were female. Aboriginal and Torres Strait Islander people made up 0.8% of the population (ABS, 2011). The town is about 110 kilometres from the state of Victoria's capital city of Melbourne. Anglesea is in the federal electorate of Corangamite. The median age of people in Anglesea (UCL) was 48 years. Children aged 0–14 years made up 16% of the population and people aged 65 years and over made up 23.8%. Of these, 47.3% were employed full-time, 41.4% were employed part-time, and 3.9% were unemployed (ABS, 2011; DTPLI, 2014). In Anglesea, 33.4% of private dwellings were occupied and 66.6% were unoccupied (ABS, 2011).

11.5.5 Local Climate

The climate of Anglesea is warm temperate and is considered to be Cfb according to the Köppen-Geiger climate classification. Similar to the other towns along the Great Ocean Road, Anglesea has a significant amount of rainfall during the year. This is true even for the driest month. The temperature here averages between 13.9 °C and 16.3 °C and the rainfall averages 619.4 mm annually (BOM, 2014).

11.6 Explore Wholeness of Place

Driving down towards Anglesea on the Great Ocean Road, views towards the town give the impression of a settlement within the coastal bushland, embraced by trees and vegetation. This character of the town is a strong element that influences how the *15 fundamental properties of wholeness* connect and strengthen each other. The setting amongst vegetated hills and valleys and its location next to the River are the source of its attraction for human settlement, and provide for a 'sense of stillness', an aspect that is strong in the most obvious occurrence of Fundamental Property 13 [The Void] and 14 [Simplicity and Inner Calm]. Additionally, the dense vegetation and coastal bush that embrace the town result in a feeling of the bush and town merged with each other, representing Property 15 [Not Separateness]. The coastal bush, setting of houses integrated within the trees, and the River re-occurs as strong elements in most of the *15 fundamental properties of wholeness* - (tsp_{15})[4] for Anglesea.

The *wholeness* of the location at Anglesea, according to the *15 properties,* is summarised in Table 11.2. Detailed information on the properties is listed in the *Wholeness of Place Evaluation Table*, in Appendix 3.

[4] This aligns with the (tsp_{15}) component of the '*notion of regenerative-adaptive patterns*' equation: 'The transformations specific to place'.

Table 11.2 Evidence of the fifteen fundamental properties of wholeness in anglesea

Location	Fifteen fundamental properties of wholeness	Town boundary	Catchment/regional boundary
Anglesea	Levels of scale	8	5
	Strong centres	11	5
	Boundaries	8	7
	Alternating repetition	8	9
	Positive space	9	7
	Good shape	4	5
	Local symmetries	5	6
	Deep interlock and ambiguity	4	6
	Contrast	6	4
	Gradients	3	3
	Roughness	5	5
	Echoes	4	6
	The void	3	2
	Simplicity and inner calm	1	2
	Not separateness	1	1
	Total	80	73

11.7 Diagnosis of Place

11.7.1 Geometric Structure

The geometric structure considering the whole context of Anglesea is a diagram that captures the large-scale structure of the town, indicated in Fig. 11.5.

Features from the diagram that determine the character and essence of the place:

1. The River is located in the middle of the settlement, dividing the town in two;
2. Surrounding hills are covered with trees and vegetation;
3. The Great Ocean Road crosses the River over a bridge in the centre of the town, which connects the residential areas on the east and west regions of the River;
4. Local shops and amenities are located along the Great Ocean Road;
5. Four places are indicated as access to the town from the north, east, west and southwest;
6. Combination of a formal street grid, and an informal street grid running along contour lines and the topography of the land;
7. Scattered houses between trees on the edges of the settlement;
8. Open grassy area next to and along the meandering River, including wetlands and saltmarsh;
9. On both sides of the River inlet there are long stretches of sandy beaches, edged with limestone cliffs in the north and rocky coast in the south;
10. Landscape pinnacle protruding into the ocean at the southern end of the town; and
11. A large scar in the landscape located behind the town, representing the open cast coal mine pit.

For a detailed list of the attributes identified in the *diagnosis of place* of Anglesea, refer to Appendix 4.

Fig. 11.5 Large scale geometric structure of anglesea. (diagram by Author)

11.7.2 Core Patterns

The interconnected *core patterns* from which place organises the complex relationships that produce its activities and existence result in the *story of the place*. These core patterns are never independent or alone in existence (e_1).[5] They are always interconnected to other patterns and entities, connecting and forming wholeness, following a set of rules, forming and being part of a *generative code*. The structure (built up) of a *core pattern* consists of:

- Core pattern title - the name of the core pattern;
- Relation to other core patterns - the connection of the pattern with other patterns in a higher scale or a lower scale;
- Description of the core pattern - the explanation of the core pattern, its occurrence and its purpose; and
- Instruction - Proposed physical actions or connections to implement or achieve the core pattern.

Following is a summary list of the *core patterns*. For a detail description of the core patterns including instructions, applications and relevance to the town of Anglesea refer to Appendix 5.

[5]This aligns with the (e_1) component of the '*notion of regenerative-adaptive patterns*' equation: 'One or various examples of the specific core pattern'.

- Our *Country* [CP1]. Deep cultural and spiritual connections with the land exists within the region, where the connection with *Country* is evident in the Dreamtime Stories of the Wadawurrung and Gadubanud peoples.
- *Hinterland to Coast Connection* [CP2]. The hinterland of valleys and ridges connects the coast with river and creek catchments flowing down to the sea.
- *The Natural Amphitheatre* [CP3]. Formations of hills, valleys and cliffs form at various locations a natural amphitheatre (a cove) in which the town is positioned.
- *Contrasting Landscapes* [CP4]. The steep uplifts of the Otway Ranges and deeply dissected valleys and cliffs on the coastline transferring into long stretches of sandy beaches provide for contrasting landscapes.
- *Forests and Woodlands* [CP5]. The trees of the Otway forest, as well as the woodlands of the coastal bush, create stature, and connect the landscape with the town in many ways.
- *The Heads and Cliffs* [CP6]. The cliffs at the river inlet provide for dramatic landscape features and the contrast between the soft water of the ocean, gentle sandy beaches, and hard heads and cliff formations.
- *River of Life* [CP7]. The river and creeks can be seen as the arteries that inject life into the town and surrounds, both in a natural landscape, environmental and human settlement context.
- *Wetland and Saltmarsh* [CP8]. Wetlands and saltmarsh are key to the functioning of ecological systems at the intersection of the ocean and the river inlets.
- *Beach, Rocks and Surf* [CP9]. Small secluded beaches enclosed with a rocky coastline and steep cliffs; in other instances, long stretches of sandy beaches.
- *The Great Ocean Road - Connections and Crossings* [CP10]. The Great Ocean Road connects the towns with each other along the coast, with connector roads to the main Great Ocean Road.
- *The Formal-Informal Street Grid* [CP11]. The layouts of various precincts within the town are arranged according to a formal and/or informal street grid alongside hills and valleys, rivers and inlets.
- *Nomad's Rest* [CP12]. The caravan park and cottages support the economic stability for each town, accommodating holidaymakers and tourists.
- *The Recreation Reserve* [CP13]. The Recreation Reserve is the place where community recreation activities occur, including sport fields, playgrounds, gardens, shelters with BBQ and picnic tables, and pathways.
- *The Shopping Strip* [CP14]. Small commercial zones in various locations, as centres to the town, are predominantly along the main road. It is a 'living centre' with local and tourists activities combined, including historical buildings to enhance town character.
- *The Esplanade* [CP15]. The Esplanade connects a cluster of walkways, foreshore reserves, open public spaces, as well as key community centres to enhance connectivity.
- *The Jetty and the Pier* [CP16]. They provide access to the rivers and sea for fishing and recreational activities.
- *Valued Community* [CP17]. Encourage activity and community gathering places that promote a cohesion of friendship, peacefulness and a sense of belonging.
- *Local Heritage and Culture* [CP18]. Acknowledge local heritage and historical built and natural assets, Aboriginal special and sacred sites, and historical town values.
- *Boundaries* [CP19]. Boundaries are formed between built form and natural elements, neighbourhoods and sub-cultures; strengthen these boundaries to enhance local character.
- *At the Heart* [CP20]. Strong centres support the overall functionality of the town. One or more centres need to be classified as being 'at the heart' of the town, the neighbourhood or community precinct.
- *Tread Lightly* [CP21]. Tread lightly on the land; ensure minimal disturbance to its physical, ecological and biophysical processes.

- *House Mix* [CP22]. Buildings to use similar materials and follow a *form language* that gives a specific, local and living character to the coastal settlements.
- *Pathways and Walks* [CP23]. Pedestrian pathways and walks provide for walkability and a connection between beach areas, river walks, nature walks and internal connections between commercial areas, houses and recreational parks.
- *Views* [CP24]. Advantage points and views in key areas in the town need to be celebrated and enhanced.
- *Sense of Danger* [CP25]. Acknowledge the dangers of natural hazards such as bushfires and flooding.
- *Meeting Place* [CP26]. Community clusters and subcultures come together at a communal meeting place. This is where boundaries between neighbourhoods and cultures fuse together to form a sense of belonging.

11.7.3 The Generative Code

The *generative code* is a process that includes all the information to design, develop, and regenerate a settlement in a way that creates adaptive and resilient built and natural environments. This code is an accumulation of the *core patterns of a place*, which include all the information to develop and generate a whole place that is connected to its environment. Additional to the *core patterns* of a place, the *key attributes of place* of the town needs to be considered when the overall *generative code* is applied. An overall generic *generative code* has been established for the towns of the Great Ocean Road Region and is listed in Table 11.3. Further, in addition to the overall *generative code* listed here, each town will have its own subset of attributes and assets representing the values of the community. These will represent unique characteristics, local attributes and unfoldings to be considered for the town.

The structure of a *generative code* is as follows:

- Name of the generative code;
- An explanation of what the generative code is to accomplish;
- The generative code statement listed in bold type; and
- The core patterns aligned to the generative code.

11.7.4 Regenerative Attributes

The *regenerative attributes* (*r*1)[6] of the coastal town include the qualities that can restore, renew, or revitalise the sources of energy and materials of the place that support its existence. As indicated by Lyle (1994), a regenerative system provides for continued replacement, and such a system generally has the qualities of integration of natural processes, such as minimum use of fossil fuels or net zero use, reuse of recycling materials, and the use of renewable resources (Lyle, 1994, p. 11). *Regenerative attributes* also include food sources, social structures, regeneration of ecological systems and the use of local energy flows.

The following *regenerative attributes* of Anglesea have been identified, as indicated in Table 11.4.

[6]This aligns with the (*r*1) component of the '*notion of regenerative-adaptive patterns*' equation: 'Regenerative attributes of place'.

Table 11.3 Generic generative codes for the settlements along the Great Ocean Road coast

Generative Code:	[GC1] Deep Connections to the Land
Explanation:	Deep cultural and spiritual connections with the land exist within the region of the Great Ocean Road coast, embedded in the longer-term history of human and landscape interaction of the place. The changing landscape over long periods of time, the climate patterns, and the human settlement patterns of the past and current are visible within the overall geometric structure of the place. The wholeness of the region connects the unique characteristics of each settlement together; it defines 'how big is here' and helps to put together the context of place dynamics and phenomena in relation to the deep connections to the land. This Generative Code seeks to identify the spirit of the place, and then to establish steps for the preservation of these connections. The connection to *Country* will be established by protecting special and sacred sites of the Aboriginal people.
Statement:	A continued effort shall be made to identify the deep spiritual and cultural connections to the land in order to preserve it and the wholeness of the region that is evident in its geometric structure and the spirit of the place.
Core Patterns Aligned:	Our Country [CP1], Hinterland to Coast Connection [CP2], Local Heritage and Culture [CP18]
Generative Code:	[GC2] Diagnosis of the Land
Explanation:	The land of the region of the Great Ocean Road coastal region is characterised by natural beauty and formations. Towns are settled within the beauty of the landscape, surrounded by the steep uplifts of the Otway Ranges and deeply dissected valleys and cliffs on the coastlines, extending into coastal grasslands and agricultural lands. Forests and woodlands provide for a variety of ecology and vegetation systems, enriching the soils and catchment areas surrounding the towns. Visiting these places, walking around on the land, one will encounter various places with natural beauty and special characteristics. This Generative Code seeks to protect them and enhance them by providing steps that will ensure the preservation of these special places within the land. The diagnosis shall include the recording of these special places, the geography of the land, its shape, the rivers and creeks, their fauna and flora, and their relation to the settlements.
Statement:	The natural beauty of the land shall be recorded on a diagnosis map of the land, and steps taken to protect these special places that will be listed in the core patterns related to the land. Its landscape attributes shall be respected, and work carried out to enhance existing habitats, both in vegetation, wildlife, and human occupation.
Core Patterns Aligned:	Hinterland to Coast Connection [CP2], The Natural Amphitheatre [CP3], Contrasting Landscapes [CP4], Forests and Woodlands [CP5], The Heads and Cliffs [CP6], River of Life [CP7], Wetland and Saltmarsh [CP8], Beach, Rocks and Surf [CP9]
Generative Code:	[GC3] Diagnosis of Settlement
Explanation:	Towns along the Great Ocean Road coast are settled within the beauty of the landscape, each settlement functioning within its own structure and setting, yet also being linked to the other settlements within the region. The existence of these towns and their social and economic survival is closely linked to the attributes and services that the landscape offers. Each town inherited attributes and values that contribute to its existence, underpinned by latent centres and core patterns. The interconnected core patterns are the cohesion of everything that forms the harmony of the whole. This Generative Code aims to establish those latent centres and core patterns, enhance them, and strengthen them.
Statement:	Establish the latent centres and core patterns of each settlement, enhance them and strengthen them. Conduct a detailed town diagnosis of each settlement, to help identify within its broader context the regenerative attributes that can assist in the restoration of sources and energy that support its existence.
Core Patterns Aligned:	The Great Ocean Road - Connections and Crossings [CP10], The Formal-Informal Street Grid [CP11], Nomad's Rest [CP12], the Recreation Reserve [CP13], The Shopping Strip [CP14], The Esplanade [CP15], The Jetty and Pier [CP16], Boundaries [CP19], At the Heart [CP20], Pathways and Walkways [CP23], Views [CP24], Meeting Place [CP26]

(continued)

Table 11.3 (continued)

Generative Code: [GC4] Potential of Settlement
Explanation: The settlements are exposed to future changes in climate as well as population growth. The growth of the towns will be more exposed to the risks of climate change impacts such as increased heat and more frequent bush fires, inundation by sea level rise, floods and coastal recession. The sustainable future of these towns are rooted in their connections to the environment, in a social, environmental and economic context. The potential of settlements can be identified and can be expressed in regenerative-adaptive design and planning outcomes. Potential of place indicates what is unique to it that can contribute to the wholeness of the larger system of which it is part for the future resilience that can be established through its core patterns, attributes and values. This Generative Code aims to identify those core patterns, values and attributes that will provide the place's potential for future resilience.
Statement: Identify the potential of place that indicates what is unique to the settlement that has the potential to contribute to the wholeness of the larger system of which it is part, and the sustainable future of place to secure resilience in a future changed environment.
Core Patterns Aligned: At the Heart [CP20], Meeting Place [CP26], Tread Lightly [CP21], House Mix [CP22], Sense of Danger [CP25]

Table 11.4 Regenerative Attributes of Anglesea

Attribute	**Regenerative Potential**
Water flow [AS-RA1]	Surface water runoff, standing water bodies, order of small streams, the river and landscape topography in the Anglesea River, Salt Creek and Marshall Creek watershed.
Wind [AS-RA2]	Optimisation of local climatic conditions and wind to generate energy. Wind energy can be used along the cliffs at the coastline, as well as at the quarry of the decommissioned coalmine.
Solar [AS-RA3]	Energy source for natural processes and human needs, inclusive of conversion and storage.
Woodlands [AS-RA4]	Biodiversity and wildlife habitat of the Anglesea Heath and woodlands. Recreational uses and natural services.
River course [AS-RA5]	Ecosystem and terrestrial environment linking of the inland Otway forests, the Anglesea Heath and woodlands with the ocean at the river inlet. Topographic carving of valleys in the hills provides a connection with land and sea. Wetlands along the river provide for dynamic food webs and ecosystem stabilisation.
Ocean [AS-RA6]	Wave energy for power generation. Marine environment for controlled recreational use and fishing. Waves for recreational use such as surfing.
Small scale local agriculture [AS-RA7]	Small-scale local agriculture in the form of vegetable gardens located in private properties, as well as in parklands for potential of community gardens.
Waste [AS-RA8]	Biodegradable waste, organic waste recycling, and the recycling of grey water for a closed-loop system.
Flora and Fauna [AS-RA9]	Native vegetation and animals, relationships of wildlife habitat and domestic habitats. Optimizing biological diversity between habitats.
Soils [AS-RA10]	Practices that enhance soils to support local biodiversity and ecological systems.
Small business [AS-RA11]	Improve local economy by providing a mixture of shops and amenities to provide local products for residents, as well as tourism.
Heritage and culture [AS-RA12]	Local heritage and culture as a mechanism to strengthen sense of belonging, community wellbeing, and tourism attraction.

11.8 Forces of Change that Impacts Place

It is evident that the dominant environmental challenges that each town is facing are the impacts of climate change and sea level rise, inundation and loss of assets due to coastal recession, fires due to increased temperatures during summer months, and changes due to population growth. These forces of change impact the sustainable future of each town, including risks to their social, economic, environmental, cultural-heritage, built and natural attributes (f_t).[7] Similar forces of change for each town and associated risks include:

- *Sea level rise* - Permanent inundation, and loss of beaches, vegetation, buildings and infrastructure;
- *Coastal recession and inundation* - Storm surges resulting in recession of beaches, cliff collapse, and temporary inundation;
- *Flooding by extreme rainfall events* - Overland flooding and river flooding of properties, infrastructure, road access;
- *Drought and extreme heat waves* - Loss of vegetation, water shortage, and potential loss of life. Increased risk and frequency of bush fires; and
- *Population growth* - Migration to coastal towns resulting in increased 'coastal sprawl', destroying natural habitats and ecosystems. Additionally, new housing and buildings are out-of-character and result in the loss of the coastal place character of these towns.

11.9 Long-Term Human-Nature Interaction of Place

To understand the deeper forms of the land, the changing landscape over long periods of time, the climate patterns, and the human settlement patterns of the past, knowledge of place from the Aboriginal people needs to explored. Knowledge of special and sacred sites, settlement ruins, and areas of interest that represent settlement patterns and migration patterns across *Country* are collected and narrated. The *Wadawurrung* knowledge of the area as described in Chap. 6 is applied to the *regenerative-adaptive design model* to integrate Indigenous knowledge with the overall pattern language. Key elements of Indigenous knowledge include:

- Wholeness of place, the connection to *Country* (the land and nature) over long periods of time;
- Forces of change, the changes of landscape and human connection to *Country*; and
- Potential of *Country*, what is unique from *Country* that has the potential to contribute to the wholeness of the larger system of which it is part.

11.10 Regenerative-Adaptive Patterns of Place

Regenerative-adaptive patterns are the combination of the various processes of a place, accumulating to support its morphological growth and regeneration that give it characteristics for resilience and allow it to adapt to changes. The *regenerative-adaptive patterns* (*nda*)[8] are key patterns which need to

[7] This aligns with the (f_t) component of the '*notion of regenerative-adaptive patterns*' equation: 'A set of forces which have an impact on'.

[8] This aligns with the (*nda*) component of the '*notion of regenerative-adaptive patterns*' equation.

extend and include all the functions of the *notion of regenerative-adaptive patterns* equation.[9] For the town of Anglesea, 12 *regenerative-adaptive patterns* have been identified:

- [RAP1]: *Adaptive Built Environments* - includes considerations of the adaptive potential of buildings and infrastructure;
- [RAP2]: *Protect Nature's Adaptive Capacity* - considers the adaptive capacity of local natural systems, assists in the ability to adapt;
- [RAP3]: *Nature's Work as Continuous and Reciprocal Interactions* - allows life support functions to be processed through conversion, distribution, filtration, assimilation and storage with interaction throughputs;
- [RAP4]: *Optimisation and Multiple Functions* - inclusion of multiple functions and outcomes for systems to optimise resilience ability;
- [RAP5]: *Aggregate not Isolate, Integrate rather than Segregate* - integrate all parts to fix connections, aggregate to assist the inclusion of symbiotic relationships to promote regeneration;
- [RAP6]: *Self-Regulation and Feedback Loops* - include self-regulation of positive and negative feedback loop systems in processes;
- [RAP7]: *Produce no Waste, Recycle and Assimilate* - make use of all inputs and outputs for a closed loop or net positive system;
- [RAP8]: *Conversion of the Solar Income* - include passive solar systems for energy, heating and cooling, thermal storage and conversion;
- [RAP9]: *Scale Linking to Facilitate Flow* - shaping the medium to facilitate flow, scale linking for support of maximum function at the smallest scale;
- [RAP10]: *Storage as a Key Resource* - for energy, water and materials - maintaining adequate storage with balancing the replenish rate with the rate of use;
- [RAP11]: *Valued Renewable Resources and Services* - use and value existing natural, renewable resources for energy and biological services; and
- [RAP12]: *Human - Nature Connections for Healthy and Prosperous Environments* - the application of Biophilia strategies and designs to create healthy environments for both humans and nature.

For a detailed description of the 12 *regenerative-adaptive patterns* of Anglesea, and examples of the application of the patterns according to the *notion of regenerative-adaptive patterns equation*, refer to Appendix 6.

11.11 Regenerative-Adaptive Pattern Language of Place

The narrative of how the town of Anglesea aligns with The Whole can be described in the pattern language of place, using the *fundamental patterns* as follow:

The town of Anglesea is located in a diverse natural environment and possesses a 'quality without a name', in essence representing a 'sense of spirit'. Therefore, to continue into the future with a town character that has got 'sense of place', apply the *15 fundamental properties of wholeness* that makes up The Whole [1].

When we identify all the attributes of the larger region and those of Anglesea through the *regenerative-adaptive design model*, wholeness of place is established where humans and nature can co-exist in harmony by achieving a resilient sustainable future. Therefore, adopt integral sustainable design practices at multiple levels of complexity through the Integral Sustainable Worldview [2].

---

[9] The *notion of regenerative-adaptive patterns equation*: $rgp = \{nda, f_1 \ldots f_i, std, tsp_{15}, e_1 \ldots e_i, r1, pot\}$ is explained in detail in Chap. 11.

Changes on Earth happen at multiple levels of intricacy and take place in dynamic, social, economic, technological, biophysical and political contexts. Climate and environmental systems change rapidly and through the application of CLIMATE CHANGE CO-ADAPTATION [3] the community of Anglesea can achieve resilience. Therefore, identify the key issues of a changing climate, its impacts on the built and natural environment, and implement actions to be able to adapt to these changes.

Anglesea is located alongside the Anglesea River and the seas of the Bass Straight, which form part of the attraction for settlement, in essence a result of the inherent attraction of humans to large bodies of water. This AFFINITY TO WATER [4] requires us to acknowledge the complexities of the dynamics of coastal and water systems. Therefore, investigate the biophysical characteristics and social-ecological systems of place, in order to understand the patterns of human-nature affiliations.

When developing coastal management and planning solutions for the town of Anglesea, there is a need for the knowledge of the longer-term history of the land, the Indigenous knowledge that stretches over thousands of years. Considerations for coastal settlement needs to embrace the ADVANCED INDIGENOUS KNBOWLEDGE [5] of the *Wadawurrung*. Therefore, collaborate with the *Wadawurrung* to record their long-term history of Anglesea to better understand the dynamics of the lands and waters of *Country* and *Sea Country*.

The positive enhancement of connections between the local habitants of Anglesea and their natural environments is fundamental to the health and wellbeing of the community. Therefore, employ *Biophilia* to connect people to nature at physical, psychological and consciousness levels. Further, consider *deep interaction patterns* and apply biophilic design to embrace LOVE FOR NATURE [6], which will result in a *nature language* for the wellbeing of both humans and nature.

Regenerative-Adaptive Design looks at nature for its inspiration and model. Any future development in Anglesea needs to question how the human systems can be ordered in ways analogous to, and integrated with, the order of nature's systems at location. Acknowledging that everything is interconnected using NATURE'S DESIGN [7] will reveal the consideration of a *set of layers* from a place's geology to the plant, animal and human habitats. Therefore, locate places that are suitable for *sustainable development* with the resources and absorptive capacity of human life, and identify places where sensitive and eco-productive systems are not harmed.

Moving beyond sustainability as part of a regenerative future, the town of Anglesea needs to engage an input-output cycle, considering systems that restore, renew or revitalize their own sources of energy and materials, thereby creating sustainable systems. Beyond this baseline approach, design *human-nature ecosystems* as co-spatial processes by applying the pattern REGENERATIVE ECOSYSTEMS [8]. Therefore, identify and record all the underlying living processes that contribute to the co-evolution of *one holistic regenerative and adaptive structure* of the town of Anglesea and larger regional environment.

The patterns and processes of Anglesea at the later stages of development need to be embedded in the patterns of process and structure at the earlier stages, going as far back as to the first Indigenous settlement. Therefore, apply the knowledge from the past to the future through the TRANSFORMATIONS OF WHOLENESS [9] to recognise the interconnection of all and this unfolding to create living structures.

Forces of change are constantly present, and for continuous resilience the community of Anglesea needs to deal with these complexities of change specific to place. Therefore, using the fundamental pattern of NOTION OF REGENERATIVE-ADAPTIVE PATTERNS [10] equip the community to facilitate the evolution of Anglesea to both build up its level of complexity to greater maturity and stability, and increase its richness and redundancy to enable agile adaptation.

Ecological and settlement succession of Anglesea is only achievable if the pattern EVOLUTIONARY ADAPTATION [11] is applied, resulting in integrative complexity as the collective action of residents over time, moving towards continuous improvement and self-transcendence. Therefore, the *regenerative-adaptive patterns* of place must be fully embraced, progressing to adaptive planning and design, based on the findings from the *regenerative-adaptive pattern language,* the *regenerative-adaptive design model,* and the *notion of regenerative-adaptive patterns equation.*

Table 11.5 Risk management adaptation process and actions

Risk management adaptation process	Actions and relevant sections in RADM
Establish the context: Determine the objectives for the risk assessment, its scope, stakeholders who need to participate or be aware of it and the climate and climate change scenarios being considered	Focus group workshop Identification of valued assets and attributes of Anglesea Risk mapping with considered climate change scenarios Patterns of nature applications as per the *Wadawurrung* migrations and settlement patterns
Identify risk: Describe how climate change affects key elements of the system which is the subject of the risk assessment	Risk mapping with climate change scenarios that result in risk exposure of assets due to a sea level rise between 0.8 m and 1.4 m from the year 2030 to 2100 List of risks identified
Analyse risk: Consequences and likelihood of each specific impact is assessed and from this the overall level of risk is determined	Overall value of the risk is determined in this instance according to the wholeness of place and relevant regenerative-adaptive design principles, reflected in: Valued assets and attributes of Anglesea Forces of change
Treat risk: Options for treating priority risks are identified and evaluated. More effective and practically implementable measures are incorporated into action plans	Planning and design concepts, provided by the local workshop participants Core patterns instructions of implementation Generative code with unfolding steps Regenerative-adaptive patterns of place indicating the regenerative-adaptive system and outcomes. Specific regenerative opportunities and the potential for regenerative and adaptive futures of Anglesea are captured and reflected in the 'notion of regenerative-adaptive patterns' equation: $rgp = \{nda, f_1 \ldots f_i, std, tsp_{15}, e_1 \ldots e_i, r1, pot\}$

The potential is there for Anglesea to be celebrated and to thrive long term, rather than just surviving in a less diverse and inhospitable world due to GAIA'S REVENGE [12]. Therefore, acknowledge that the Earth is regulating itself and providing conditions for life. Thus we need to consciously make decisions for human habitation that is supportive, embedded and part of THE WHOLE [1].

11.12 Adaptive Planning and Design of Place

Making use of the *regenerative-adaptive design model* results in an integrated process of *adaptation planning*, *regenerative design*, and the *application of the regenerative-adaptive patterns*. However, the impacts of climate change to Anglesea need to be identified, and vulnerabilities established first, then integrated with the *regenerative-adaptive design model*. The first step for climate change risk adaptation planning and design for Anglesea is to complete the step-by-step actions of risk assessment and adaption as detailed in Table 11.5:

11.12.1 Considered Adaptation Planning and Design Responses

The next step in the risk and adaptation process will be to identify and propose ways to treat the risks, and consider adaptation planning and design options. The regenerative-adaptive design and planning solutions considered for the town of Anglesea include the following:

1. Apply the understandings of spiritual and cultural landscapes of the Aboriginal *Wadawurrung* people of their *Sea Country* – the wetlands, estuaries, rivers, lakes, and the sea – to the future adaptation planning and design considerations of Anglesea;
2. Establish protection measures of sacred and cultural heritage sites to allow for spiritual connection to the land;
3. Relocate the existing Anglesea Beachfront Family Caravan Park to the open areas east of Ramsay Street and south of Inverlochy Street. This process will be a staged process with initial holiday units raised above ground levels to accommodate flood levels, and then progressively all structures will be relocated to the new location;
4. Allow for the natural progression of the River and estuary according to sea level changes, with new settlement boundaries being established by giving preference to protecting natural habitats;
5. Re-establish the wetlands and saltmarshes at the Lions Park Reserve to their pre-European settlement status, and allow for flood zones to the east of the Anglesea River inlet;
6. Rehabilitate the Alcoa mine site and reinstate the original Anglesea River course, including water storage and wetland systems to assist with the improvement of the hydrology of surface and ground water quality supporting the ecological functional systems of the River course[10];
7. Construct a sea wall along the Anglesea River and saltmarsh plains to protect built and natural assets from inundation. Sea wall to include hard and soft construction methods, including rock revetments and mangrove planting;
8. Establish land use planning controls to allow for a setback of 40 metres from the River and high-water mark to allow for coastal recession and consequently avoiding damage to built assets;
9. Main bridges and roads need to be constructed and moved to adapt to future scenarios of sea level rise;
10. Re-alignment of the Great Ocean Road as a loop road around Anglesea;
11. Include pathways, walking tracks and cycling tracks within new Anglesea River alignments, foreshore reserves and the coastal zone;
12. Strengthen the 'heart' of the town, including community-focused buildings such as the Anglesea Store and social gathering areas, and allow for one or more meeting places for the community. Gathering places to be modelled on the principle of centres and the circles of equality[11];
13. Establish a 'heritage strip' along the western embankment of the River, with key historical-related buildings and functions;
14. Progressively include changes in the built infrastructure to accommodate the changing environment, inclusive of building options with light frame constructions that can change over time;
15. Establish regenerative infrastructure to include renewable energy, solar and wind energy, water capture and storage, water-sensitive design and planning, zero waste recovery systems, and waste-to-energy systems;
16. Establish small-scale farming within the town boundaries including vegetable gardens and locally grown food;

[10] The Anglesea River catchment area is an important aspect to the ecological health of the area. The hydrology provides the patterns of rivers and flows, which support the abundance or the decline of plants, which in turn represent the climatic factors, and provide specific habitats for the animal species of the region. This is a key aspect to consider in ecological planning as per McHarg's ecological method (McHarg, 1992 [1969], pp. 127–136).

[11] Centres for gathering are based on the connection to *Country* and the Aboriginal belief that the position of human in nature is that of oneness. The engagement in conversation happens around the centre; the centre acts as the natural phenomena of gravity and wholeness.

17. Restructure of the Anglesea waste water works to include regenerative waste use systems, including aquaculture systems, surface-flow wetland systems, and root zone systems[12];
18. Restructure town public areas to include biodiversity corridors linking the River with various parklands and recreational areas;
19. Establish siting and building guidelines that allow future built forms to reflect the local character of the coastal architecture of Anglesea, respecting and celebrating the traditional and heritage architectural forms; and
20. Set in place a community adaptation response plan that includes requirements for applying the *regenerative-adaptive design model* for the sustainable and resilient future of Anglesea under a new future-changed climate scenario.

11.12.2 Adaptation Response Plan

The dynamic process of the *regenerative-adaptive design model* must be considered as the process for the implementation of actions and responses to impacts and risks, and for use in an adaptation response plan. This plan will provide guidance for actions to be taken based upon the proposed adaptation and design planning proposals for Anglesea, as well as the implementation of all of the features of the *regenerative-adaptive design model.* This is the last step in the process flow of the risk management framework.

11.13 Conclusion Statement

This chapter started with the documentation of the process of how to apply the *regenerative-adaptive design model* to the case study region of the Great Ocean Road coast and the town of Anglesea. The process included the development of the *regenerative-adaptive patterns of place*, using the '*notion of regenerative-adaptive patterns*' equation. Keeping within the unfolding principles of a generative system that follows a step-by-step process, the establishment of *core patterns* supported the further development and refinement of a *generative code*. The generative code with its core patterns was then used as an input for the development of the *regenerative-adaptive patterns based* on the '*notion of regenerative-adaptive patterns*' equation, which then evolved into the formal structure of the *regenerative-adaptive design model.*

Demonstrated in the results of the application of the *regenerative-adaptive design model,* this process indicates a more dynamic, adaptive outcome that includes a *generative code* with unfolding steps and specific regenerative-adaptive opportunities that consider *the whole*. Combined with adaptation planning and design principles that emerged during the application of the *regenerative-adaptive pattern language*, it is evident that the emerging phenomena of all the elements of the *15 fundamental properties of wholeness* are embedded in the adaptive capacity of the natural and the built environment of the case study area.

Therefore:

[12] Regenerative technology includes waste as a resource for regeneration, including the use of aquaculture systems, surface-flow wetland systems, and root zone systems. Examples as noted by Lyle included the Tijuana River Valley constructed wetland plan, the Lyle Centre for Regenerative Studies - aquaculture system and the waste treatment system, and the Arcata Marsh and Wildlife Sanctuary marsh treatment system, which is a combination of aeration oxidation ponds, treatment marshes and salmon hatchery in the lagoon (Lyle, 1994, pp. 225–243).

Fig. 11.6 Sketch of fundamental pattern – Evolutionary Adaptation [11] (drawing by Jesse Delmo, synthesised by Roös)

It is recommended to use the concept of *a regenerative-adaptive pattern language* in settlement planning and design, rather than a typical standard master planning and risk management process. The results and the outcomes are *dynamic* in nature, with the ability to deal with *forces of change*. This allow residents of a town to regenerate, adapt and apply lasting, *deep sustainable* practices for the ongoing function of that place for it to exist within the *larger whole*, and to create a *living structure* of *resilience* continuing into the future, adapting with circumstances as they are changing over time, following an *evolutionary process*.

11.14 Fundamental Pattern

Evolutionary Adaptation [11].

The Earth is in a constant state of change to calibrate conditions for life through evolutionary adaptation. Align human interventions and settlements in Nature with the evolutionary processes of a place, by applying the regenerative-adaptive pattern language, along with its related tools and methods. Ultimately this achieves resilience and continues improvement and self-transcendence (Fig. 11.6)**.**

Downward links:

Constant change is always present, and adaptation to change needs to be flexible and dynamic, following the same process of *creative fitting,* – Evolutionary Adaptation [11]. However, the Earth as a self-regulating system will rebalancing itself – Gaia's Revenge [12], responding to the impacts of Anthropogenic induced climate change …

References

ABS. (2011). *Australian bureau of statistics*. Retrieved 18 November 2014 from ABS Census Data: www.censusdata.abs.gov.au.

Alexander, C. (1979). *The timeless way of building* (Vol. 1). New York: Oxford University Press.

Alexander, C. (2001–2005a). *The nature of order - an essay on the art of building and the nature of the universe, book one: The phenomenon of life*. Berkeley, CA: The Center for Environmental Structure.

Alexander, C. (2001–2005b). The nature of order - An essay on the art of building and the nature of the universe, book two: The process of creating life. Berkeley, CA: The Center for Environmental Structure.

Alexander, C., Ishikawa, S., Silverstein, M., Jacobson, M., Fiksdahl-King, I., & Angel, S. (1977). *A pattern language: Towns, buildings, construction*. New York: Oxford University Press.

BOM. (2014). *Climate statistics for Australian locations*. (Bureau of Meteorology, Australian Government). Retrieved June 16, 2015 from Bureau of Meteorology: http://www.bom.gov.au/climate/averages/.

Clark, I. D. (1990). *Aboriginal languages and clans: An historical atlas of western and Central Victoria, 1800–1900*. Melbourne: Department of Geography and Environmental Science, Monash University.

DTPLI. (2014). *Regional Victoria population*. Department of Transport, Plannning and Local Infrastructure - Census 2011. Retrieved 02 February 2015 from DTPLI: www.dtpli.vic.gov.au.

Lyle, J. T. (1994). *Regenerative design for sustainable development*. New York: John Wiley & Sons, Inc.

McHarg, I. (1992 [1969]). *Design with nature*. Washington, DC: John Wiley & Sons Inc.

McKellar, B. (2009). The Anglesea Heath: Do you know the difference? [Blog post]. Retrieved 10 Dec 2015 from Angair Inc., http://angair.org.au/knowledge-bank/factsheets?id=251.

Muir, J. (1877). Mormon lilies. *San Francisco Daily Evening Bulletin*. Retrieved 25 Feb 2020 from https://scholarly-commons.pacific.edu/cgi/viewcontent.cgi?article=1052&context=jmb.

Neis, H. J., Brown, G. A., Gurr, J. M., & Schmidt, J. A. (2012). Generative process, patterns, and the urban challenge. In *Fall 2011 international PUARL conference*. Portland, OR: PUARL Press.

PIRG. (1977). *A Coastal Retreat*. Mulgrave, Australia: The Victorian Public Interest Research Group Limited.

Richards, T. (1998). *A predictive model of aboriginal archeological site distribution in the Otway range, occasional report No 49*. Aboriginal Affairs Victoria. Melbourne, Australia: Aboriginal Affairs Victoria.

Roös, P. & Jones, D. (2013). *Visions of the Surf Coast - Changing Landscapes Under Future Climate Effects*. UPE10 Symposium: International Urban Planning and Environmental Society. Sydney: ICMS Pty Ltd.

Salingaros, N. A. (2013). *Unified architectural theory: Form, language, complexity - a companion to Christopher Alexander's "the phenomenon of life: The nature of order, book 1"*. Portland, OR; Kathmandu, Nepal: Sustasis Press.

Salt, B. (2003). *The big shift: Welcome to the third Australian culture – The Bernard salt report*. South Yarra, Australia: Hardie Grant Books.

Weber, P., & Weber, G. (2010). *The grand challange: A history of the Anglesea recreation and sports Club 1911–2010: Including a brief history of Aires inlet and Anglesea*. Malvern, Australia: National Library of Australia.

Check for updates

Chapter 12
The Living Earth: Revenge or Celebration?

Abstract This final chapter brings together the cumulative knowledge built up throughout the book, and summarises how the *regenerative-adaptive pattern language* offers a holistic approach, informed by nature, to better craft and strengthen our built and natural environment for the future. The '*knowledge of making life*' is embedded in the fractal geometry and morphological sequences of nature, and if we can establish a process and method to reconnect the human-nature relationship and secure resilience through our actions – both in planning and design practice, and in sustainable development – we can potentially be liberated. However, the chapter highlights that the Living Earth system behaves as a single, self-regulating system to support life, as emphasised by the Gaia Hypothesis (Lovelock, *Atmospheric Environment* 6:579–580, 1972). Facing an unprecedented changing climate and environment due to anthropogenic causes, the question remains as to whether we as humans can achieve a symbiosis with the revenge or celebration of Gaia. The chapter concludes with the *fundamental pattern* GIAI'S REVENGE [12], reminding us of the ultimate reality that we as humans are only a small part of this larger self-regulating system of Life on Earth.

Keywords Living structure · Self-regulating system · Architecture for life · Living process · Regenerative system · Degenerative system · Gaia's revenge · Restoration · Regenerative-adaptive

Upward links:

... the fundamental process for dealing with forces of change at a specific place at the local community level, which are part of the regenerative progressions of place, is achievable through the application of the NOTION OF REGENERATIVE-ADAPTIVE PATTERNS [10]. Regenerative patterns allow for constant regeneration. Including adaptive capacity as part of the ongoing function of place, EVOLUTIONARY ADAPTATION [11] allows for establishing resilience for both human and natural systems. However, urgent action is required for a new INTEGRAL SUSTAINABLE WORLDVIEW [2] to align human's future with GAIA'S REVENGE [12].

12.1 Introduction

The care of the Earth is our most ancient and most worthy, and after all, our most pleasing responsibility. To cherish what remains of it, and to foster its renewal, is our only legitimate hope (Wendell Berry in Vaughan-Lee, 2016, 2–13, p. ix).

In the beginning, this book starts with the following statement:

P. B. Roös, *Regenerative-Adaptive Design for Sustainable Development*, Sustainable Development Goals Series, https://doi.org/10.1007/978-3-030-53234-5_12

> *"From the twentieth to the twenty first century humans have passed through a unique period, one in which the planning, design and construction of the built environment has been almost un-imaginably bad and unacceptable. The way we have designed and built our cities has resulted in a degeneration of the earth's natural systems, now eventuating in the unprecedented impacts of a changing climate. Undeniably a more volatile climate and rising sea levels threaten coastal settlements and will bring change to the coastal zone impacting nature as well as built environments. Humans need to mitigate the causes of, and adapt to climate change as well as the loss of natural systems"* (Chap. 1)

This is indeed a planetary issue.

When I started my in-depth research investigations back in October 2011, I believed that I would find 'the answer' to how to address this planetary issue. What was clearly missing from the sustainable design, planning and adaptation practice discourse is a more deeply integrated approach to the design and planning of human settlements that considers '*the whole*', which moves away from the current view that humans stand apart from nature, rather than participating, co-evolving and adapting with it. Consequently, in this book I proposed and tested a *regenerative-adaptive pattern language theory* towards investigating the possibilities of a holistic integrated design and planning method – one that incorporates the principles of *regenerative design* as well as an *adaptive pattern language*, attempts to re-establish our wholeness with nature, and considers the vulnerabilities of a changing landscape (Fig. 12.1).

I believed that the development of a model that could apply the proposed *regenerative-adaptive pattern language theory* to the design and adaptation planning of vulnerable coastal towns would provide an answer to this planetary issue. Even as I approached the end of this book, I still held tight to the belief that in its final pages I would sculpture the exact formula for the resilience of the coastal environments, human and nature combined, through the application of the 'regenerative-adaptive patterns equation'. Yet, the more I tried to write this concluding formula, the more evident it became that this proposed approach, the theory, the model and the equation are just starting to delve into the depths of this tremendously complex subject matter. That said, I believe that this book explored and established new foundations for the regenerative-adaptive design (and pattern language) narrative – hopefully to be further developed, refined and tested on real life projects by others to come.

12.2 The Regenerative-Adaptive Pattern Language Theory

The '*notion of a regenerative-adaptive patterns equation*' forms the foundation of the *regenerative-adaptive pattern language theory,* informing a design process that fundamentally supports a proposed new *integral sustainable worldview*. The application of the equation indicates a progression based on an ecological systems process that considers *the whole*, rather than an industrial or mechanical process that is a collection of individual parts. The '*notion of regenerative-adaptive patterns equation*' collectively mimics the wholeness of place in a complex, cyclic system where the entities of evolution of place (built and natural combined) become distributors and producers of materials, water, food, and energy as well as adaptors of change. This is in direct contrast to the current status and 'production' of the built environment. Inherently this process needs to include the features of *the living process that creates life* (as noted by Alexander, 2001–2005), a step-by-step adaptive procedure which includes the opportunity for feedback and correction at every incremental stage (Alexander, 2001–2005, p. 225).

The activity of regeneration – and of life, for that matter – happens gradually; as noted by Alexander the living structure of wholeness emerges slowly, step-by-step in an adaptive fashion (Alexander, 2001–2005, p. 230). This is fundamental in the adaptation of natural environments.

Applying the process of a *regenerative-adaptive pattern language* would require an integral sustainable design approach. This is where a planning, design and construction team needs to work closely in collaboration with the local community to be deeply rooted into place, and to have an understanding of the climatic, ecological, natural and cultural context of the site (Wahl, 2006). It will be necessary for the people of that specific place to be involved as part of the professional team, and an increased collaboration between professional fields that seldom work together – such as the archi-

Fig. 12.1 Waterfalls in the Otway forest. Walking in the forest, you sense the Living Earth (photo by Author)

tectural, engineering, planning, ecology, biology, geology, hydrology, agricultural and social-cultural, social-psychological sectors – would be required (Mang & Reed, 2012; Reed, 2007; Landry, 2006). Even with extensive collaboration, it will be necessary to apply a deep understanding of the principles of regenerative-adaptive patterns, nature's morphological sequences, and the systems of ecology to reach adaptive, resilient outcomes. I argue that only then perhaps will plans and designs based on this '*architecture for life*', rooted in nature's systems and the *conclusive whole,* transcend the metaphor into a deeper form of regenerative-adaptive design for sustainable development – one that will be able to function as a whole entity, continuously evolving to be resilient, becoming part of and supporting the self-regulating mother earth, Gaia.

12.3 Opportunities for the Future

As noted, this book put in place the fundamentals for the further continuation of the *regenerative-adaptive pattern language theory* discourse and applications of regenerative-adaptive design. While the philosophical basis of the theory appears to be a logical approach for the adaptation planning, redevelopment and regeneration of the built and natural environments, in practical terms it remains difficult to implement such diverse and large-scale solutions. However, I believe that we need to progress forward and test the theory as broadly as possible, as there is an urgent need to change the current unsuccessful methods of sustainable planning, design and development in the face of a rapid changing climate.

Recently, a small research team from the Live+Smart Research Laboratory at Deakin University took on the challenge, investigating how *regenerative-adaptive design* and *integral sustainable design* can collectively be applied to regenerate exhausted quarry and open cast mine sites. Working with their industry partner Grimshaw Architects, the team investigated the application of the principles of regenerative-adaptive design to conceptual and hypothetical propositions for the regeneration of the recently closed down Alcoa open-cast coal mine in Anglesea.

The project – "Heal the Scar" – looked at how we can best regenerate exhausted quarry and mine sites, and holistically (environmentally, economically, socially and culturally) enrich both the sites and their surrounding community. The research methodology followed a process of inquiry, critical literature and case study reviews, theory and model development, and a final pilot study of the open cast coal mine in Anglesea. The findings informed visionary, hypothetical-scenario propositions for the specific site. Fundamental to the methodology was the use of the *Integral Sustainable Design Framework* (DeKay, 2011) and the *Regenerative-Adaptive Design Model* (Roös, 2017) (Fig. 12.2).

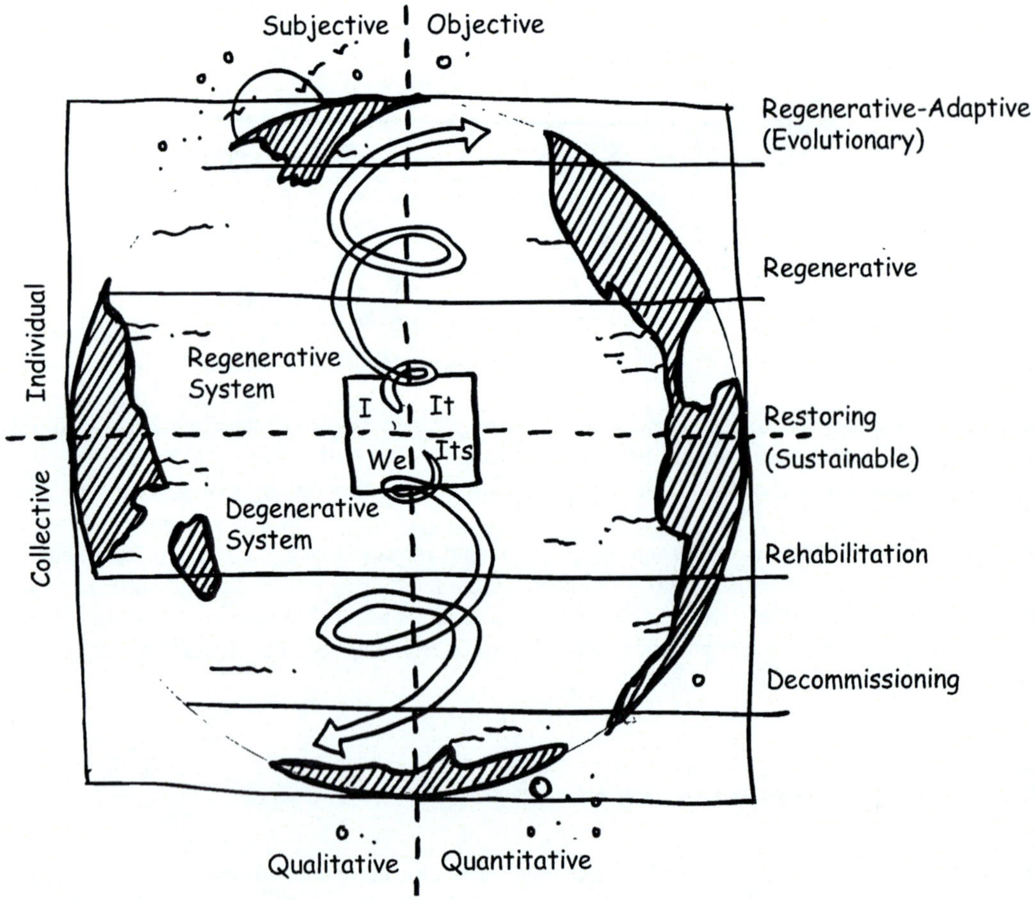

Fig. 12.2 Regenerative-adaptive pattern language and integral sustainable design theory working together for a new future, as demonstrated in the Heal the Scar Project (drawing by Jesse Delmo, synthesised by Roös)

The project team used the four-quadrant model of the Integral Sustainable Design framework as a guide to explore the individual and collective (societal) aspects of potential rehabilitation and regeneration propositions, as well as quantitative performance criteria (both at building/infrastructure performance and regional/town mapping scale) to investigate integrated social, economic, cultural and environmental sustainability outcomes for the chosen pilot study site (Roös, 2020). Guided by the Integral Sustainable Design framework and the Regenerative-Adaptive Design Model, three scenarios were developed (Potter, 2020):

1. *Restorative* – Reconnect the waterways by designing for the natural catchments and flows with new nature-inspired constructed water courses and linear park trail. This provides the foundation for the broader ecological corridors to connect through the site (adjacent to the mine area whilst it is undergoing restoration) and to provide an opportunity for people to access the site from day one via the waterway and to overlook the restoration of the mine.
2. *Regenerative* – Building from the baseline of restoration, the local community participates in the enhancement of the place, supporting continual habitat improvement through applied knowledge systems. The activities and uses on site are designed for deep research and learning, founded upon giving back to and living in harmony with nature. The site establishes productive and abundant sources of food, water and energy to serve the community who visits and sustains it.
3. *Regenerative-Adaptive* – Share the site's vitality, productivity and learnings with broader city and global networks. It becomes a 'living lab' and a place of continuous adaptation and advancement of synergistic systems.

The results of this conceptual project indicated that indeed it is possible to go beyond the current practice of sustainability, moving toward regeneration and adaptation, as presented in *Heal the Scar – Regenerative Futures of Damaged Landscapes* (Roös, 2020) (Fig. 12.3).

12.4 The Revenge or Celebration of Gaia

We have clearly identified in the early chapters of this book that the climate is rapidly changing and we are experiencing impacts such as coastal erosion, flooding, extreme heat, and bushfires. The living Earth is responding to the anthropogenic global warming; we have passed the tipping point, and as a self-regulating system, the Earth will change its climate to bring back the equilibrium necessary to maintain the conditions for life on this planet. In the Gaia Hypothesis, James Lovelock and Lynn Margulis (1974) put forward that the Earth is a complex self-regulating system which, as a whole, seeks a physical and chemical environment that is optimal for life (Lovelock, 2009; Lovelock & Margulis, 1974).

This is indeed the case, and as human beings we need to acknowledge that we are part of this larger natural system. Further, as argued by Wilson (2016), for the first time in the history of this planet the human species has become the architect of the environment, creating the Anthropocene epoch, bringing consequences to the Earth that will affect all life (both human and of the natural world) far into the geological future (Wilson, 2016, p. 2). According to Lovelock, our urban way of life demands more and more built-up areas, encroaching upon the domain of the living Earth, taking from it more resources than what it can sustain. Gaia is now changing according its own rules to regulate its intent to provide conditions for life – potentially to a state where human beings are no longer welcome (Lovelock, 2006, p. 9). However, if we return to our inherent deep connections to nature and embrace biophilia, where we use our love for nature as the initiator for our decisions, maybe we can re-establish a symbiosis with Gaia. The question remains of whether we, as humans, can achieve this symbiosis with the revenge or celebration of Gaia.

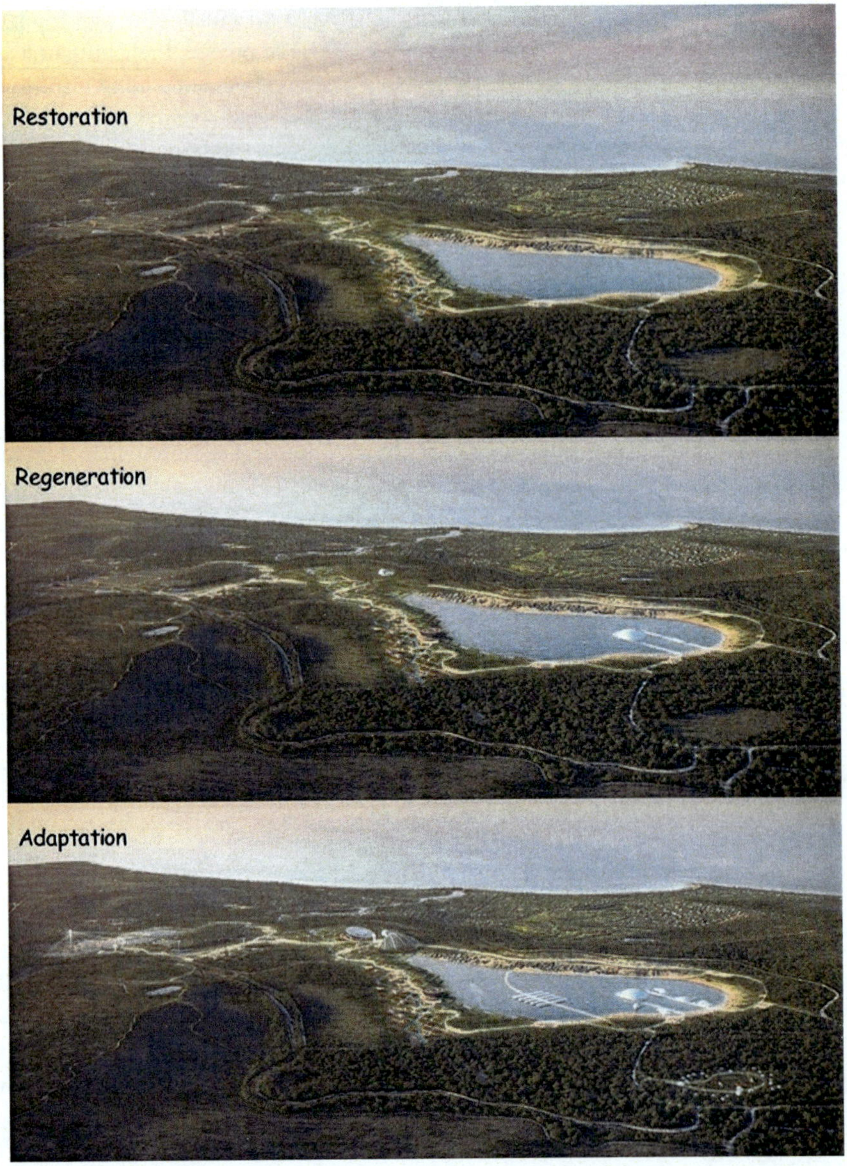

Fig. 12.3 Heal the scar proposition – restoration, regeneration and adaptation (courtesy from Roös, 2020)

12.5 Conclusion Statement

The *regenerative-adaptive pattern language* initiated a new perspective on the human-nature relationship and our need to realise our place in this world, as well as how we are absolutely part of *the whole*. The answers for adaptation to a future changing and dynamic climate is right in front of us, embedded within nature – in its patterns, fractal connections and formations. The narrative of this book hopefully results in connecting all the parts, and as a whole piece of literature, it is *in itself a pattern language* which demonstrates the context of *wholeness*. If the reading is to be meaningful, it is not just a matter of reading the words as separate parts in separate chapters, but rather the unfolding of *fundamental patterns* of each chapter as part of a *complete pattern language* that result in an act of initiating a new *integral sustainable worldview*.

Therefore:

If we want to achieve a more sustainable and resilient future for both human-built environments and the natural environment, we need to continue this conversation of *a regenerative-adaptive pattern language*. In this narrative, the regenerative dimension of 'Regenerative-Adaptive Design for Sustainable Development' begins with *Restoration* to a sustainable life-supporting state of affairs, and progresses in complexity to fundamental patterns that support *Regenerative* conditions in which biotic and cultural systems become self-healing and interactively self-maintaining. In the third stage, with greater complexity and differentiation and denser relationships, the *Regenerative-Adaptive* state of continuous improvement and self-transcendence can be attained (DeKay & Bennett, 2020). This sequence, in integral terms, is a vectorial progression of levels, which hopefully can help us as humans to achieve a symbiosis with the revenge or celebration of mother Gaia, our Living Earth.

12.6 Fundamental Pattern

GAIA'S REVENGE [12]

For more than four billion years the Earth has regulated itself, providing conditions for life. Gaia is now responding to anthropogenic-induced global climate change, and as a self-regulating system, will change her climate to restore equilibrium conditions for life. Therefore, embrace THE WHOLE and implement a regenerative-adaptive pattern language to celebrate the Living Earth and avoid impending consequences (Fig. 12.4)**.**

Downward links:

The *regenerative-adaptive pattern language* sets the scene for the future, providing us with guidance on a new approach to working with Nature, and highlighting that a new manner of thinking needs to be adopted if humanity wants to survive. The critical question we are asking ourselves now is: "Will we choose to endure the

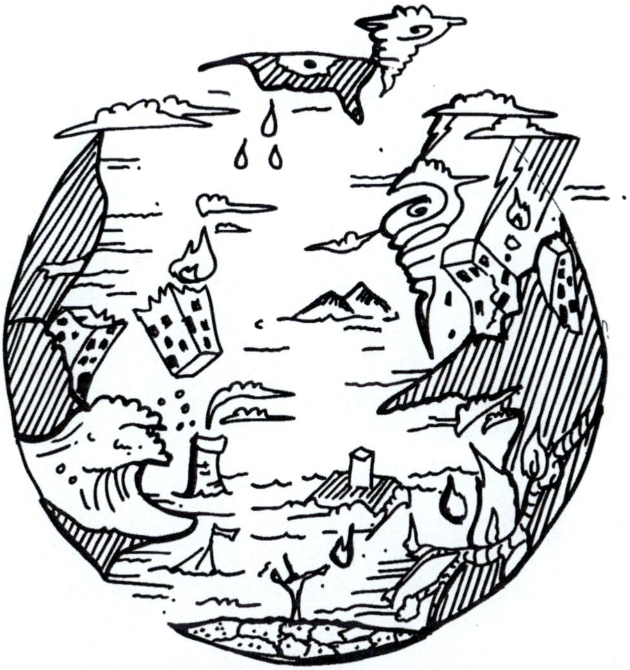

Fig. 12.4 Sketch of fundamental pattern – GAIA'S REVENGE [12] (drawing by Jesse Delmo, synthesised by Roös)

revenge of Gaia – GAIA'S REVENGE [12] – and survive in a less diverse and inhospitable world, or celebrate the good news that we can align ourselves with THE WHOLE [1] for long term thriving?" In a distant future, when humanity looks back to the Anthropocene, the 'Epoch of Man', what will be the answer to this question?

References

Alexander, C. (2001–2005). *The nature of order – An essay on the art of building and the nature of the universe, book one: The phenomenon of life*. Berkeley, CA: The Center for Environmental Structure.

DeKay, M. (2011). *Integral sustainable design: Transformative perspectives*. London, Washington: Earthscan.

DeKay, M., & Bennett, S. (2020). Foreword. In P. B. Roös (Ed.), *Heal the scar – Regenerative futures of damaged landscapes*. Geelong, Australia: The Live+Smart Research Lab, Deakin University.

Landry, C. (2006). *The art of city making*. London: Earthscan.

Lovelock, J. E. (1972). Gaia as seen through the atmosphere. *Atmospheric Environment, 6*(8), 579–580.

Lovelock, J. E. (2006). *The revenge of gaia*. London: Penguin Books.

Lovelock, J. E. (2009). *The vanishing face of gaia*. New York: Basic Books.

Lovelock, J. E., & Margulis, L. (1974). Atmospheric homeostatis by and for the biosphere: The Gaia hypothesis. *Tellus, 26*(1–2), 2–10.

Mang, P., & Reed, B. (2012). Designing from place: A regenerative framework and methodology. *Building Research and Information, 40*(1), 23–28.

Potter, A. (2020). Regenerative architecture. In P. B. Roös (Ed.), *Heal the scar – Regenerative futures of damaged landscapes*. Geelong, Australia: The Live+Smart Research Lab, Deakin University.

Reed, B. (2007). Shifting from sustainability to regeneration. *Building Research & Information, 35*(6), 674–680.

Roös, P.B. (2017). *Regenerative-adaptive design for coastal settlements: A pattern language approach to future resilience*. PhD Thesis, Deakin University, Geelong, Victoria, Australia.

Roös, P. B. (Ed.). (2020). *Heal the scar – Regenerative futures of damaged landscapes*. Geelong, Australia: The Live+Smart Research Lab, Deakin University.

Vaughan-Lee, L. (Ed.). (2016). *Spiritual ecology: The cry of the earth* (2nd ed.). California: The Golden Sufi Center.

Wahl, C. D. (2006). Bionics vs biomimicry: From control of nature to sustainable participation in nature. In C. Brebbia (Ed.), *Design and nature III: Comparing design in nature with science and engineering* (pp. 289–298). Wessex, UK: Wessex Institute of Technology.

Wilson, E. O. (2016). *Half earth: Our planet's fight. For life*. New York, London: Liveright Publishing Corporation.

Appendix 1
Alexander's Mathematical Definition of Wholeness

The Structure of Wholeness

The *notion of regenerative-adaptive patterns equation* described in Chaps. 10 and 11, developed by Roös and based on the fundamentals of the Borchers and Stark equations, further attempts to include the context of '*wholeness*'. Each element of the equation is interconnected with the others, and includes interconnected higher-level with lower-level patterns formed by the dynamics of change, and considers the process of adaptation (Chap. 11). The *regenerative-adaptive pattern language* accepts that the ecological systems process and all phenomena in the region of a specific place constitute in the context of '*the whole*': everything connected and interdependent on each other. Inherent to the *notion of regenerative-adaptive patterns equation* are the *15 properties of wholeness*. This mathematical equation of the regenerative-adaptive patterns demonstrated the *wholeness of the whole region* within which the case study town of Anglesea is located, and thus Alexander's mathematical explanation of *wholeness* is needed.

Mathematical Definition of *Wholeness* According to Alexander (2001–2005)

According to Alexander (2001–2005a, pp. 446–457),[1] *wholeness* (*W*) is a feature of a physical space that appears everywhere, in any part of matter or space. Characterised by a clear mathematical structure, consider any region of space or a place as (*R*). To include boundaries on this space, imposing a mesh or grid on the space will result in a number of points that are considered finite, not infinite. The result is that (*R*) contains a number (*n*) of points. In the real world there is usually some colouring or differentiation of type or character among the (*n*) points of (*R*), so that the region (*R*) has a visible and identifiable structure. The simplest colouring that produces a structure is black and white, and in the context of two-dimensional space, (*R*) would then be a drawing representing a particular object. In the case where the colouring of the region is not abstract, but concrete, the points may be assigned to physical materials, for example, solid and void, or various other attributes. The region (*R*) is thus intended to represent a part of the real world in its overall geometric form and organisation (Alexander, 2001–2005a, p. 446). The mathematical definition of wholeness is explained as follows by Alexander:

Wholeness (*W*) within a region (*R*) can be mathematically constructed as follows:

[1] Alexander, C. (2001-2005a). *The Nature of Order - An Essay on the Art of Building and the Nature of the Universe, Book One: The Phenomenon of Life.* Berkeley, California, USA: The Center for Environmental Structure.

P. B. Roös, *Regenerative-Adaptive Design for Sustainable Development*, Sustainable Development Goals Series, https://doi.org/10.1007/978-3-030-53234-5

> Within the region (R), that contains (n) points; there are (2^n) distinguishable subregions. A typical subregion is then named (S_i). What follows is that we construct (W) by recognising that there are different relative degrees of coherence that may be observed in the different subregions of (S_i). The regions of space have got coherence, and each subregion includes different relative degrees of coherence. The levels are called *life*, and the structure of the wholeness (W) relies on the distinctions of life that are explicit, and are used to erect structure. To make the different degrees of life explicit, the introduction of measures of life (c) is introduced on the subregions of (R). Each possible subregion is represented by (R), (S_i) with (i) ranges from (1) to (2^n). The life of the *i-th* subregion (S_i) is then to be (c_i). Each (c_i) is a number between 0 and 1, and every subregion of (R) is to be given a measure of life. The most coherent regions have a (c_i) that is close to 1 and the least coherent regions have a (c_i) that is close to 0 or close to 0^2 (Alexander, 2001–2005a, pp. 446–447).

Alexander noted that there is an objective measure of life that can be determined empirically for any given region within a given wholeness (Alexander, 2001–2005b, p. 446).[2] It is also possible to define various approximations to this empirical *life*, which can be attained as part of the functions of the internal structure of (R) and (W) by calculating the life of (S_i). The most coherent subregions of (R) are called *centres*. A region then will be considered more or less centred according to its level of life. The most coherent subregions (S_i), which have a (c_i) close to 1, will then be called centres of the region (R). Even among and within the centres there will be degrees of life, some that are more coherent than others, but all established through their life – a phenomenon of centredness in space.

In summary, Alexander thus defines wholeness as:

> I define wholeness (W) as the system, which is created by region (R), together with the measure (c), and all those subregions, which have measures more than some threshold and thus qualify as *centres*. For all practical purposes, the wholeness (W) is created by the interaction of the geometry of the region (R) and the rank order, which is created on the centres of (R) by (c) - (Alexander, 2001–2005b, pp. 447).

[2] Alexander, C. (2001-2005b). *The Nature of Order - An Essay on the Art of Building and the Nature of the Universe, Book Two: The Process of Creating Life.* Berkeley, California, USA: The Center for Environmental Structure.

Appendix 2
Patterns and Languages According to Borchers and Stark

As described in Chap. 10, the *'notion of regenerative-adaptive patterns'* equation was developed based on the fundamentals of Borchers' (2000, 2001, 2008) and Stark's (2012) equations for *patterns*. A comparison of Borchers' *'notion of patterns'* equation and Stark's *'Improvisional patterns'* follows:

Pattern Languages According to Borchers (2000, 2001, 2008)

According to Borchers (2000, 2001, 2008), the notion of patterns in a *pattern language*, used in fields of architecture, software design or sustainable innovation, will follow these formal principles:

> Each pattern is a set $p = \{n, f_1 \ldots f_i, s, e_1 \ldots e_i\}$ of a name n, forces $f_1 \ldots f_i$, the solution s, and examples $e_1 \ldots e_i$. It describes a commonly encountered design problem and suggests a solution that has been proven useful in this situation. A pattern language is a directed acyclic graph with each mode represents a pattern. There is a directed edge from pattern $p1$ to $p2$ if $p1$ recruits $p2$ to complete its solution. Edges pointing away from a pattern are its consequences, showing what lower levels of pattern need to be applied next. Edges pointing to a pattern are its context, the situations in which it can be applied. This relationship establishes a hierarchy within the pattern language. (Borchers, 2008; in Neis et al., 2012, p. 92).

To explain what *pattern languages* are made up of, Borchers (2000) noted that a pattern is a solution to a re-occurring design problem. The pattern pays special attention to the context within which it is relevant. It considers the competing 'forces' that it needs to balance, and also the positive and negative consequences as a result of its application. The pattern then references higher-level patterns in the context in which it is applied, and lower-level ones that could be considered below the current one to further assist in refining the solution (Borchers, 2001, p. 364). This hierarchical structure results in a comprehensive collection of patterns into *a pattern language*.

The following section is an extract from Borchers (2000), providing the explanation of a typical pattern language and a formal hypertext of a model of *A Pattern Language* (1977), as summarised in the previous paragraph.

According to Borchers (2000, 2001) a formal description of patterns makes it less ambiguous for the parties that participate in the design process or problem to decide what a pattern is supposed to look like, according to its structure and content. It also makes it possible to design methods that help designers in writing and readers in understanding the patterns. Explanation of the structure and content is as follows (Borchers, 2001, pp. 363–365):

P. B. Roös, *Regenerative-Adaptive Design for Sustainable Development*, Sustainable Development Goals Series, https://doi.org/10.1007/978-3-030-53234-5

The formal syntactic definition of a pattern language is:

- A pattern language is a directed acyclic graph (DAG) **PL** = (**P**,**R**) with nodes **P** = $\{P_1, \ldots, P_n\}$ and edges **R** = $\{R_1, \ldots, R_m\}$;
- Each node $P \in \mathbf{P}$ is called a pattern;
- For $P, Q \in \mathbf{P}$: P *references* $Q \Longleftrightarrow \exists R = (P,Q) \in \mathbf{R}$;
- The set of edges leaving a node P ∈ P is called its references. The set of edges entering it is called its context.
- Each node P ∈ **P** is itself a set $P = \{n, r, i, p, f_1 \ldots f_i, e_1 \ldots e_j, s, d,\}$ of a name *n*, ranking *r*, illustration *i*, problem *p* with forces $f_1 \ldots f_i$, examples $e_1 \ldots e_j$, the solution *s*, diagram *d*.
This definition is expanded with the following semantics:
- Each pattern of a language captures a reoccurring design problem, and suggests a proposed solution to it. The language is made up of a set of these patterns applicable to a specific design domain, for example such as urban design or architecture. Each pattern has a context represented by edges pointing to it from other interconnected higher-level patterns. They sketch the design situations in which it can be used;
- Similarly, its positions in the language shows what lower-level patterns can be applied after it has been used. This relationship creates a hierarchy within the pattern language. It leads the designer from patterns addressing large-scale design issues, to patterns about small design details, and helps the identification and location of related patterns;
- The name of a pattern helps to refer to its central idea and reasoning, and build a vocabulary for communication within a team or design community. The ranking of the pattern shows how universally valid the pattern the author believes this pattern is, and helps readers to distinguish early pattern ideas from truly timeless patterns that have been confirmed on countless occasions within the overall pattern language;
- The opening illustration of a pattern, gives readers a quick idea of a typical example situation for the pattern, even if they are not professionals. Media choice depends on the domain of the language, for example architecture can be represented by photos of buildings and sketches of built forms; landscape architecture may present photos and graphics of vegetation and landscapes; engineering my prefer coding, mathematical equations or engineering elements;
- The problem states what the major issue is that the pattern attempt to address. The forces further elaborate the problem statement. They are aspects of the design that need to be optimised. They usually come in pairs contradicting each other;
- The examples section is the largest of each pattern that describes existing situations in which the problem at hand can be (or has been) encountered, and how it has been solved in those situations through the application of the patterns;
- The solution generalises from the examples a proposed (or proven if already been applied before) way to balance the forces at hand optimally for the given context. The proposed solution is not simply prescriptive, but generic so that it can generate a solution (or alternative solutions) when it is applied to concrete problem situations of the specified context; and
- The diagram supports the solution by summarising its main idea in a graphical way, omitting any unnecessary details. For experts or specialists in a particular field, the diagram is quicker to grasp than the opening illustration. The choice of media to be used again depends on the domain; a graphical sketch for architecture or pseudo-code or diagram for engineering, etc. (Borchers, 2001, p. 364)

With these definitions, a formal model for *a pattern language* is in place. However, formalisation must not obstruct readability and the clarity of the material that has been described. The formal code should not hinder the process of writing the patterns within a language, and the results should still be accessible in various formats, including in electronic, linear, and printed documentation. Each part of a pattern, and its connections to other patterns, are usually presented as several paragraphs in the pattern description; other media, such as images, animations, audio recordings, etc., are used to strengthen it (Borchers, 2000, p. 2).

Improvisional Patterns According to Stark (2012)

Stark (2012) further expands on Borchers' equation (2000, 2001, 2008) using the 'pattern language of music' as a process of organisation in the field of improvisation. For improvisation in organisation and social systems, Stark developed a protracted version of Borchers' equation that includes an additional step in the process: '*con*' - the consequences from the challenge. Consequently Stark's explanation of each pattern (*p*) is as follows:

- A typical organisational or systemic challenge (*nc*);
- A set of forces (*f*) which have an impact on;
- The specific settings (*s*) and time dynamics (*t*);
- One or more solutions (*sol*);
- Examples (*e*) of the patterns; and
- The consequences for the challenge (con).

The patterns then represent and create a set of relationships that prove to be viable solutions for the challenges presented, but that also create new patterns and relationships while performed based on the forces they display. Stark's equation systematically put order to these relationships. The patterns therefore can be the key to understanding the organisation's deep levels, the 'unknown' of modern and complex organisational and social cultures. According to Stark, the 'improvisional patterns' therefore bear the built-in dynamics of repetitive principles combined with solutions, and includes extreme variable timeframes that are similar to fractals (Stark, 2012, p. 92).

Both Borchers and Stark's work is built on the foundation of pattern languages initiated by Alexander et al. (1977). Although *A Pattern Language* (1977) sets out the 253 patterns of building, and does not describe the mathematical thinking behind the patterns, the initial calculations in *The Synthesis of Form* (1964) and the later work *The Nature of Order* (2001–2005a, b) by Alexander include and explain the formulation of the patterns in *A Pattern Language*, and also formulate the concept of 'wholeness'.

References

Alexander, C. (1964). *Notes on the Synthesis of Form.* Cambridge, MA: Harvard University Press.

Alexander, C., Ishikawa, S., Silverstein, M., Jacobson, M., Fiksdahl-King, I., & Angel, S. (1977). *A pattern language: Towns, buildings, construction.* New York: Oxford University Press.

Alexander, C. (2001–2005a). *The nature of order - an essay on the art of building and the nature of the universe, book one: The phenomenon of Life.* Berkeley, CA: The Center for Environmental Structure.

Alexander, C. (2001–2005b). *The nature of order - An essay on the art of building and the nature of the universe, book two: The process of creating life.* Berkeley, CA: The Center for Environmental Structure.

Borchers, J. (2000). A pattern approach to interaction design. In *Proceedings of the ACM DIS 2000 international conference on designing interactive systems,* August 17–19 (pp. 1–10). New York: ACM.

Borchers, J. (2001). A pattern approach to interaction design. *AI & Society, 15 (4),* 359–376. London: Springer-Verlag London Limited.

Borchers, J. (2008). A pattern approach to interaction design. In P.G. Satinder (Ed.), *Cognition, communication and interaction: Transdisciplinary perspectives on interactive technology* (pp. 114–129) London: Springer-Verlag.

Stark, W. (2012). Tacit knowledge and innovation patterns for communities and social systems. In H. B. Neis (Ed.), *Generative process, patterns, and the urban challenge.* (pp. 89–95). Oregon: PUARL ARUS Press.

Appendix 3
Wholeness of Place Evaluation Table: Town of Anglesea

Description

To capture the 'wholeness of place' as described in Chap. 11, the *15 Properties of Wholeness* are recorded in the *Wholeness of Place Evaluation Table* of the sample town Anglesea. Attributes identified are listed within the town boundary, and within the catchment boundary (larger regional area) according to the 15 properties, which are interpreted and allocated a value score.

Scoring and Evaluation

Quality Score: 3 – Strong, 2 – Evident, 1 - Weak

Although attributes can be identified that represent the *15 properties of wholeness*, in various instances the qualities of the attribute (or combination of the attributes) can be strong or weak in relation to the substantiation of the properties of wholeness. To be able to indicate how strong or weak the quality of the attribute(s) is that represent that specific property of wholeness, each of the fifteen properties have been given an evaluated *Quality Score* between 3 and 1, where 3 indicates the highest quality, and 1 represents the weakest. The 15 properties have been identified and scored within the town boundary, and also within the larger region/catchment boundary.

The Fifteen Properties of Wholeness

The *15 Properties of Wholeness* (based on the *15 Fundamental Properties* by Alexander, 2001–2005, pp. 144–294),[1] are described in in Chap. 10. The 15 fundamental properties used for the evaluation are:

1. Levels of Scale
2. Strong Centres
3. Boundaries
4. Altering Repetition
5. Positive Space

[1] Alexander, C. (2001-2005a). *The Nature of Order - An Essay on the Art of Building and the Nature of the Universe, Book One: The Phenomenon of Life.* Berkeley, California, USA: The Center for Environmental Structure.

P. B. Roös, *Regenerative-Adaptive Design for Sustainable Development*, Sustainable Development Goals Series, https://doi.org/10.1007/978-3-030-53234-5

6. Good Shape
7. Local Symmetries
8. Deep Interlock and Ambiguity
9. Contrast
10. Gradients
11. Roughness
12. Echoes
13. The Void
14. Simplicity and Inner Calm
15. Not Separateness

Wholeness of Place Evaluation: Anglesea							
Location	Fifteen Properties	Town Boundary	Qty	Score	Catchment Boundary	Qty	Score
Anglesea	Levels of Scale	Town perimeter with the Anglesea Heathland, Angahook - Lorne State Park, and the Great Otway National Park The rivers and creeks - Anglesea river, Marshall Creek and Salt Creek Anglesea Coal Mine Residential area Anglesea Golf Club The Anglesea river inlet Foreshore park Anglesea shops and restaurants in commercial area	8	3	The greater Anglesea Heathland catchment area Great Otway National Park Marshall Creek and Salt Creek catchment area, the Anglesea river Small creeks and valleys Beach and coastline	5	3
	Strong Centres	Anglesea river estuary and inlet Riverfront park, boardwalk Coastal Reserve Beach Anglesea Surf Life Saving Club Anglesea Golf Club Great Ocean Road shops Beachfront Family Caravan Park Soapy Rock Point Roadknight Ellimatta reserve sport field and oval	11	3	Cliffs and beach Anglesea valley Anglesea River Anglesea river wetlands and estuary Great Ocean Road bridge over Anglesea River	5	3
	Boundaries	Town perimeter with the Anglesea Heathland, Angahook - Lorne State Park, and the Great Otway National Park Marshall Creek Salt Creek Anglesea River Anglesea Coal Mine Residential area edge Anglesea Golf Club The beach and foreshore, the ocean	8	3	Forest and heathland Farmland Anglesea River Wetlands Great Ocean road Coastline The ocean	7	2

(continued)

Alternating Repetition	Trees and built form Coastline and beach Meandering Great Ocean Road Cliff formations at main beach Bends in the Anglesea river Street grid Shopfronts Single and double story houses	8	2	Coastal shrub vegetation Cliffs and beach Rock formations and sandy beach Inlet and river Otway forest trees and vegetation River and estuary Wetland Cliff erosions Patches of farm land Great Ocean Road	9	2
Positive Space	Beach Anglesea river Open areas between houses Open areas between shops River boardwalk Walking tracks Coastal bush and trees Front yards Back yards	9	3	Larger Anglesea heathland basin Otway forest Coastal bushland and scrubs Foreshore Wetland and estuary Anglesea river Ocean	7	3
Good Shape	Roads following the contours and topography Beach and cliffs Coastal architectural forms and shapes - beach cottage homes Trees	4	3	The Anglesea river meander Cliff form Rock formations Tree shapes and structures Wetland vegetation	5	3
Local Symmetries	Street grid and meandering streets following topography Materials of building structures Predominantly single storey buildings Buildings set within bush and vegetation Lay-out of Anglesea Beachfront Caravan Park units	5	3	Cliff form, beach and river inlet River banks Coastal scrub and flowers Trees and vegetation Creeks and rivers Ocean waves	6	3
Deep Interlock and Ambiguity	Alleys and spaces between shops in town centre Natural bush, open spaces houses Holiday park grassed areas, natural bush and wetland / riverbank vegetation Beach and cliff forms	4	3	River, creeks and wetland Coastal scrub and wetland vegetation Cliffs and beach Beach and ocean Coastline curves, rock formations and sandy beaches Dense bush, trees and homes with roads	6	3
Contrast	Building and bush vegetation River inlet and beach, with cliffs Inlet and Lions Park Reserve Anglesea river, beach and the beachfront caravan park Built form typology Materials and built form styles (or lack off)	6	3	Cliffs and beach, ocean Natural bush, trees and town built environment Coastal bush and farm land Ocean and beach inlet	4	3

(continued)

	Gradients	Cliff form at beach lower in height towards river and towards Point Roadknight Creeks and river networks collect in Salt Creek, Marshall Creek, and Anglesea River Beach shack homes integration with bush and vegetation	3	3	Inland grass land transfers into Forest trees, then coastal and scrubs closer to coastline Creeks and river networks collect in Salt Creek, Marshall Creek, and Anglesea River High dense forest cover decline in density closer to rivers, valleys, creeks and the coastline	3	3
	Roughness	Buildings in town centre differ in architectural form and style Open spaces in lots, between, lots, public areas and parks Forest edge and homes scattered in between River bank and beach front Cliff and beach	5	3	Cliffs and beach Beach, rock formations and ocean River estuary, wetland, coastal scrubs Natural coastal bushland and scrubs Forest trees and undergrowth vegetation	5	3
	Echoes	Cliff face and erosion of cliffs, and the beach Trees and houses Roof forms Trees and coastal scrub	4	3	Cliffs and beach Beach and ocean Sand and rock formations River meander Coastal scrub Ocean waves	6	3
	The Void	Anglesea beach and cliff Point Roadknight Cliff lookout	3	3	Anglesea river mouth Otway forest mountains and basin	2	3
	Simplicity and Inner Calm	Siting of houses within bush and forest	1	3	Anglesea river mouth Otway forest mountains and basin	2	3
	Not Separateness	Setting of houses, buildings and streets integrated within the trees and bush along the river and inlet, beach and ocean	1	3	Trees and coastal bush of the Otway forest and the Angahook-Lorne State Park connect with coastal vegetation and the cliff scrubs, which then connect to the beach and the ocean	1	3

Appendix 4
Diagnosis of Place Evaluation: Town of Anglesea

Summary of Town Diagnosis for Anglesea

Finding Special Places	**Views and Focal Points**
The Beach and River inlet	Anglesea Lookout
The Beach Cliffs	Anglesea River bank and wetlands
Anglesea Heath	River mouth
Coogoorah Park	Beach and cliffs
Surf Life Saving Club	Point Roadknight
Caravan park and camping grounds	Cliff top
Anglesea River and wetlands	
Place Character	**Centres**
Bush meets Sea	The shopping strip
Australian beach shack	Surf Life Saving Club
Setting within the Anglesea Heathland	The Beach and River inlet
Small coastal setting	The info centre
School camps in the bush	Caravan park
Bush walking trails	The golf club
	Coogoorah Park
Places to Heal	**Vulnerability**
The coal mine power station	Beach and River inlet
Air quality (pollution due to power station)	Caravan park
The Anglesea Heath	Homes in the bush at boundary of settlement
Coastal foreshore	Surf Life Saving Club
River and wetlands	Great Ocean Road bridge
	Natural bush and coastal heath
Prospect	**Boundaries**
Caravan park area	Anglesea River, inlet, beach and Cliffs
Anglesea Heath	Salt Creek and Marshall Creek
Foreshore Cliffs	Coastline and Ocean
Main beach and River inlet	Anglesea heath and bush
Coogoorah Park	Otway forest nature reserve
Bush and beach setting	Great Ocean Road

P. B. Roös, *Regenerative-Adaptive Design for Sustainable Development*, Sustainable Development Goals Series, https://doi.org/10.1007/978-3-030-53234-5

Appendix 5
Core Patterns

Summary of ***core patterns*** of the Great Ocean Road coast region, and relevance to the town of Anglesea

P. B. Roös, *Regenerative-Adaptive Design for Sustainable Development*, Sustainable Development Goals Series, https://doi.org/10.1007/978-3-030-53234-5

*** Figure illustrations by Jesse Delmo, synthesised by Phillip B Roös**

Core Pattern	Description
Our *Country* [CP1] 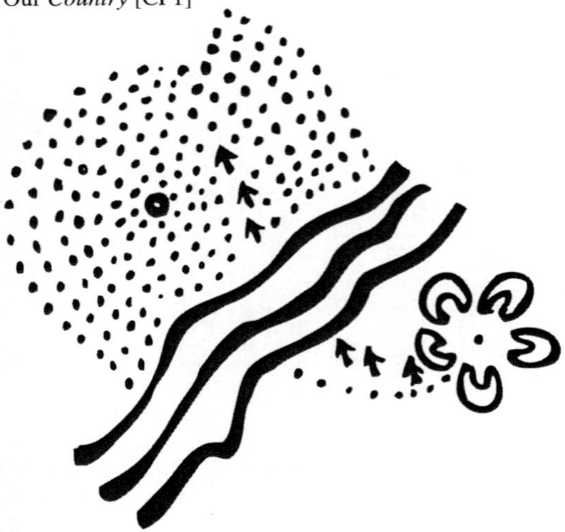 Relation to other Core Patterns: CP2, CP3, CP4, CP5, CP6, CP7, CP8, CP9, CP18 Alignment to Generative code: [GC1] Deep Connections to the Land	**Description:** Deep cultural and spiritual connections with the land exists within the bioregion of the Great Ocean Road coast; the connection with *Country* is evident in the Dreamtime Stories of the *Wadawurrung* and *Gadubanud* peoples. Not necessarily obvious and evident to European occupants, but certainly engraved within the living structures of the land, constantly present in the evolutionary process of the land. **Instruction:** The spirit of the place, the sacred sites, needs to be acknowledged, preserved and respected to allow for the evolution of harmony between humans, nature and spirit. Identify the connection to *Country*, establish ordinances that will protect the sacred sites, and build around the sites and places that people can use to connect to it, including spaces that preserve it. **Relevance to Anglesea:** Aboriginal sacred sites exist along the Anglesea River and the beach (middens). These sites must be protected.

Hinterland to Coast Connection [CP2]  Relation to other Core Patterns: CP1, CP3, CP4, CP5, CP6, CP7, CP8, CP9 Alignment to Generative code: [GC2] Diagnosis of the Land	**Description:** The hinterland to coast landscape is predominantly influenced by the Otway Range mountains, and is a prominent feature rising some 500 - 700 m above sea level. It comprises Lower Cretaceous sandstones, limestone and mudstones and, except for some of the higher parts, it has been deeply dissected into a rugged series of valleys and ridges, flowing down to the coast with surrounding foothills and plains to the east and west. This hinterland connects the coast with river and creek catchments flowing down to the sea. Consequently, this connection provides ecological, social, cultural, and economic viability to the towns of the region. **Instruction:** The development of town assets needs to be aligned with the attributes of the Hinterland to Coast Connection, and must not be located in places that will obstruct the natural connection between the coast and hinterland. **Relevance to Anglesea:** The Anglesea Heath surround the town, with a setting in an amphitheatre of the landscape linking the Anglesea River from the catchment area to the sea. The River is key to the functioning of the ecological system of the wetlands as well as the larger watershed of the Anglesea region, and its function and existence must be protected.
The Natural Amphitheatre [CP3]  Relation to other Core Patterns: CP1, CP2, CP5, CP6, CP7, CP8, CP9 Alignment to Generative code: [GC2] Diagnosis of the Land	At the larger scale the natural topography of a bay at each coastal town results in a topographical natural setting of an amphitheatre. At a more local scale formations of hills, valleys and cliffs form at various locations a natural amphitheatre (a cove) in which the towns are sited. These local natural amphitheatre settings results in dramatic landscapes and settlement locations that are protected to a certain extend against the harsh coastal weather patterns and conditions. Additionally, the half-circle shapes of the landscape result in 2 or more bays in the vicinity of the towns, with landscape pinnacles jutting out to sea. **Instruction:** Respect the land and the placement of any built or natural features to ensure integration and siting to support the paftery and to enhance the natural topography of each place. **Relevance to Anglesea:** Anglesea town is set within a naturally formed amphitheatre of hills and valleys, with direct connections to the sea, with coastal cliffs and wetlands and saltmarshes as part of the landscape. These natural attributes are the core attraction for settlement, and need to be respected.

(continued)

Contrasting Landscapes [CP4]  Relation to other Core Patterns: CP1, CP2, CP3, CP5, CP6, CP7, CP8, CP9 Alignment to Generative code: [GC2] Diagnosis of the Land	The steep uplifts of the Otway Ranges and deeply dissected valleys, and cliffs at the coastlines transferring into coastal grasslands and agricultural lands provide for the contrasting landscapes of the region. The various soil formations of the region result in a diverse vegetation regime, from grass plains and hills, coastal scrub, coastal heath and bushland, to dry and wet forests. These varieties in the landscape result in the natural beauty of the area, a source of attraction to human settlement both in permanent and short periods of stay. Additionally, varieties in landscape support a diversity of habitats and interconnected ecological systems that can function in a sustainable manner. **Instruction:** Acknowledge and respect the landscape attributes, and work with and enhance the existing habitats in vegetation, wildlife, and human occupation. **Relevance to Anglesea:** Wetlands, salt marshes, rivers, hills, beaches and cliffs with the Anglesea Heath forest provide for extraordinary contrasting landscapes. Working with, respecting and living with nature is fundamental in keeping the existing place character in place.

Forests and Woodlands [CP5]  Relation to other Core Patterns: CP1, CP2, CP3, CP4, CP6, CP7, CP8, CP9 Alignment to Generative code: [GC2] Diagnosis of the Land	The trees of the Otway forest, as well as the woodlands of the coastal bush, create stature and connect the landscape with each town in many ways. Forests and woodlands provide for extensive, varied ecology and vegetation systems, enriching the soils and catchment areas surrounding the towns. **Instruction:** Planting of trees and vegetation within the town need to include the local native species to enhance and to connect the larger forest and woodlands with the inner landscapes of the town. **Relevance to Anglesea:** The Anglesea Heath is surrounding the town, covered with natural woodland bush, set among hills and valleys. The local residents refer to the most noticeable character of Anglesea as living at the 'Beach and Bush'. This natural attribute must be conserved.

(continued)

The Heads and Cliffs [CP6]  Relation to other Core Patterns: CP1, CP2, CP3, CP4, CP5, CP7, CP8, CP9 Alignment to Generative code: [GC2] Diagnosis of the Land	Formations of sandstone, limestone and mudstone along the coast results in the heads and cliffs at various river inlets, providing for dramatic landscape features and the contrast between the soft water of the ocean, gentle sandy beaches, and hard heads and cliff formations. These heads and cliffs in variety of heights and formations act as landmarks and provide for lookouts and view points up and down the coast. The heads and cliffs provide for shelter against the elements of weather. **Instruction:** Where possible and without interference with natural attributes, site the built environment in areas of shelter. Preserve natural habitats and ecosystems, and integrate uses that enhance the landscape features. **Relevance to Anglesea:** The high cliffs on the northern beach of Anglesea undoubtedly create stature. The contrast between the water of the ocean, soft sand of the beach and the hard limestone formations visible in the façade of the cliff echoes formations of patterns of the environment. However, avoid buildings at or on the cliff, but rather provide look out points and walking tracks.

River of Life [CP7]  Relation to other Core Patterns: CP1, CP2, CP3, CP4, CP5, CP6, CP8, CP9 Alignment to Generative code: [GC2] Diagnosis of the Land	Rivers and creeks can be seen as the arteries that inject life into the towns and their surrounds, both in a natural landscape, environmental and a human settlement context. Rivers, wetlands and vegetation are keys to the functioning of the ecological system of the larger watershed. The water catchment areas along rivers provide for local water runoff capture and use. River inlets provide for access to the sea and various key recreational activities along rivers and creeks to support the social-economic functioning of the towns. **Instruction:** Leave the rivers and associated wetlands in their natural state, improve and repair areas that have been disturbed, and enhance the open corridors of biodiversity and vegetation along the riverbanks. **Relevance to Anglesea:** The Anglesea River can be seen as the 'heart' of the town. The River was and is the main attraction for holidaymakers and residents alike. The River connects the land with the sea, the beach surroundings and the inlet. The River is key to the functioning of the ecological system of the wetlands as well as the larger watershed of Anglesea region. The River as a whole system needs to be protected.

(continued)

Wetland and Saltmarsh [CP8] 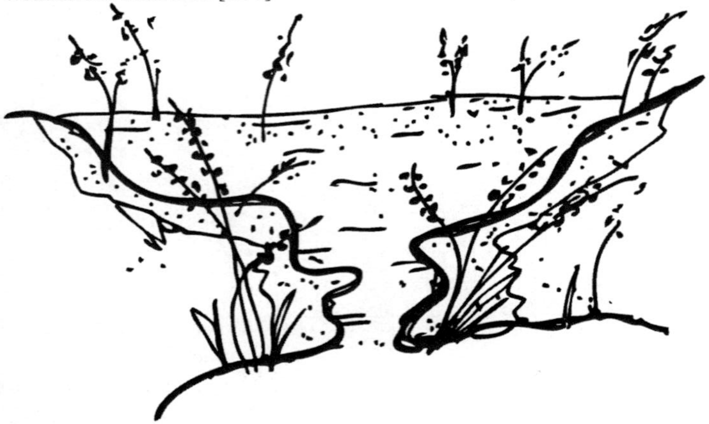 Relation to other Core Patterns: CP1, CP2, CP3, CP4, CP5, CP6, CP7, CP9 Alignment to Generative code: [GC2] Diagnosis of the Land	Wetlands and saltmarsh are keys to the functioning of ecological systems at the intersection of the ocean and the river inlets. Wetlands and saltmarsh are buffers for rising tides, king tides and flood events. **Instruction:** Where needed, improve and increase the extent of the wetlands and saltmarsh, keep existing ones intact, and avoid any development on or near the wetlands and saltmarsh. **Relevance to Anglesea:** The wetlands along the Anglesea River are a diverse ecological landscape. Vegetation ranges from damp herb rich woodlands, swampy riparian woodlands, coastal saltmarshes, mangrove strips and estuarine wetlands. These wetlands and saltmarshes must be protected against any development.
Beach, Rocks and Surf [CP9]  Relation to other Core Patterns: CP1, CP2, CP3, CP4, CP5, CP6, CP7, CP8 Alignment to Generative code: [GC2] Diagnosis of the Land	At each town a sandy beach on both sides of the river inlet occurs; in some instances small beaches enclosed with a rocky coastline and steep cliffs, in other instances long stretches of sandy beaches and rolling grasslands. Beach and surf recreational activities are important aspects for each town's economic and social existence. **Instruction:** Provide assets and infrastructure that can support the activities of the beach and the sea, without impacting upon the natural environment. Where necessary rehabilitate the foreshore and coastal primary dunes, and relocate existing built infrastructure. **Relevance to Anglesea:** The stretches of beach south and north of the River inlet are lined with a rocky coastline, as well as cliffs. The bay forms long stretch of sandy shores and surf that are used for surfing and other recreational uses. Beach and surf recreational activities are important assets for the town's economic and social existence. However, overuse and any development in this area must be avoided.

The Great Ocean Road - Connections and Crossings [CP10]  Relation to other Core Patterns: CP2, CP11, CP14, CP23 Alignment to Generative code: [GC3] Diagnosis of Settlement	The Great Ocean Road connects the towns with each other along the coast. In many instances the only accessible entry to the towns is the Great Ocean Road, with bridges crossing various rivers and steep valleys, cliffs and ridges. Connector roads to the main Great Ocean Road next to or from towns into the hinterland provide accessibility to the greater natural landscapes surrounding the towns. Accessibility for towns is critical in supporting the provision of goods, transport of people, and connection of the towns and cities to the larger region. **Instruction:** Enhance accessibility to towns by providing connector roads to the Great Ocean Road, bridges across rivers and creeks, and relocate main roads away from risk areas to negate coastal recession and inundation. **Relevance to Anglesea:** The Great Ocean Road is the main access into and out of the town of Anglesea. The Road forms a meander around the floodplains and along the riverbanks, along the coast and out into the Anglesea Heath and Otway Forest. The Great Ocean Road Bridge crosses the Anglesea River close to the centre of the town, and is the only road connection between the east and west areas of the town. However, the bridge and the road are located in low lying areas, and will be susceptible to sea level rise and storm surge inundation due to climate change impacts. Adaptation strategies need to be in place.

(continued)

The Formal-Informal Street Grid [CP11] 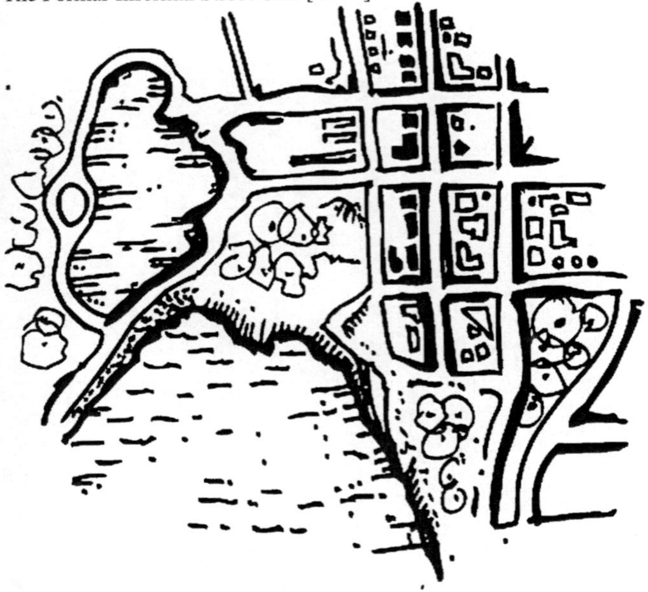 Relation to other Core Patterns: CP10, CP14, CP23 Alignment to Generative code: [GC3] Diagnosis of Settlement	Town layouts are arranged within a formal and/or informal street grid alongside rivers and inlets. Central to commercial nodes of the towns, the grid with same size blocks provides for a context of formality, expanding into informal street grids that meander along the contour lines of the topography. This phenomenon results in various instances, creating interest and the informal coastal character of town settlement. A combination of formal and informal street grids provides for interest, avoids fast through traffic, and close loops that connect and support walkable neighbourhoods. **Instruction:** The layout of roads need to be a combination of formal and informal grid patterns, following the topography of the site, and to include loop roads and dead ends to support connectivity for walking and bicycle tracks. **Relevance to Anglesea:** Noticeable of Anglesea is that some parts of the town are aligned in formal street patterns, where closer to edges of the town the informal street patterns follow the topography of the land. The contrast of formal and informal structures of the street layouts provides for a context of complexity. This character of complexity needs to be enhanced.

Nomad's Rest [CP12]  Relation to other Core Patterns: CP13, CP15, CP17, CP23 Alignment to Generative code: [GC3] Diagnosis of Settlement	The caravan park and cottages support the economic stability for each town, accommodating holidaymakers and tourists. **Instruction:** Provide facilities, amenities and a camping infrastructure that acknowledges the camping culture, which deliver adequate facilities for campers. **Relevance to Anglesea:** The holiday caravan park located at the inlet of the Anglesea River is a main tourist attraction. However, the location results in risks to future inundation, and an alternative location for camping needs to be considered.

(continued)

The Recreation Reserve [CP13]  Relation to other Core Patterns: CP12, CP15, CP17, CP23, CP26 Alignment to Generative code: [GC2] Diagnosis of the Land [GC3] Diagnosis of Settlement	The Recreation Reserve in most towns is the place where community recreation activities occur. The reserve includes an oval or rectangular shaped sports field with additional amenities that support local sports activities, community gatherings, and events, and it also provides for a safe venue in the event of bushfire emergencies. Activities hosted in facilities at the reserve range from football, cricket, tennis, playgrounds, gardens, shelters with BBQ and picnic tables, and pathways. **Instruction:** A mixture of facilities that respond to the local community needs must be provided within the Recreation Reserve. **Relevance to Anglesea:** The Coogoorah Park in Anglesea butts against the Anglesea River, and provides recreational activities for residents. Additionally, next to the Caravan park, bowling facilities, tennis courts and park grounds are used for recreational purposes.

The Shopping Strip [CP14]  Relation to other Core Patterns: CP10, CP11, CP15, CP20, CP26 Alignment to Generative code: [GC3] Diagnosis of Settlement	Small commercial zones in various locations as centres to the towns are predominantly along the main roads, or the Great Ocean Road that goes through. The Shopping strips entertain restaurants, boutique shops, art and craft shops, local businesses, and holiday accommodation. It is a ‘living centre’ with local and tourists activities combined, resulting in travellers and tourists visiting the region. In most of the towns the Shopping Strip features a few historical buildings and provide for a coastal town character, and as a result creates a sense of place. The buildings emphasise local heritage and culture. **Instruction:** The Shopping Strip must consist of a mixture of small businesses, providing different local products and services. Large chain stores or shopping centres must not be allowed. **Relevance to Anglesea:** The shopping street of Anglesea hosts a few restaurants, as well as businesses, offices, and additional amenities that support local food needs, community activities and gatherings, and is a stopping place for travellers and tourists on the Great Ocean Road. The character of the shopping strip needs to be kept in place, and out of character developments must be avoided at all times.

(continued)

The Esplanade [CP15]  Relation to other Core Patterns: CP9, CP14, CP16 Alignment to Generative code: [GC3] Diagnosis of Settlement	The beach Esplanade connects a cluster of walkways, foreshore reserves, d anopen public spaces, as well as key community centres including surf lifesaving clubs, information centres, community halls, holiday park and access to the beaches. **Instruction:** The Esplanade needs to connect a mixture of functional spaces and uses. **Relevance to Anglesea:** In Anglesea the natural setting and connection along the river, beach and caravan park and shopping strip can be defined as an 'esplanade'. However, to enhance this core pattern the integration with the core pattern of 'pathways and walks' will improve its use and functions.

The Jetty and the Pier [CP16]  Relation to other Core Patterns: CP9, CP14, CP15 Alignment to Generative code: [GC2] Diagnosis of the Land [GC3] Diagnosis of Settlement	The Jetty and the Pier provide access to the sea for fishing and recreational vessels. Jutting out into the sea, the Jetty and the Pier are favoured lookout posts for many visitors and local residents alike, and are an extension of the landscape pointing into the sea. Heritage values are associated with past historical uses of the Jetty and the Pier. **Instruction:** Provide for commercial and recreational uses connected to the Jetty and the Pier. **Relevance to Anglesea:** The Anglesea River doesn't have the capacity and depth to accommodate a Jetty and Pier for seagoing fishing and recreational vessels. However, the same principle can be applied for access to the waters of the river, for smaller craft such as small rowing boats, canoes and kayaks. In this instance extend the boardwalk into the river to allow for the mooring of these small boats and canoes.

(continued)

Valued Community [CP17]  Relation to other Core Patterns: CP12, CP13, CP18, CP19, CP26 Alignment to Generative code: [GC3] Diagnosis of Settlement [GC4] Potential of Settlement	Community collaboration is enriched by the many subcultures within the community. The mosaic of subcultures connects dispersed cultures amongst the neighbourhood, with activity and community gathering places that promote a cohesion of friendship, peacefulness and a sense of belonging. **Instruction:** Provide for opportunities to encourage and build on a 'friendly community'. All developments need to encourage community integration. **Relevance to Anglesea:** Due to the size of population the Anglesea community is close-knit. Values of the environment and the town's 'sense of place' provide for a valued community. The sense of place attribute must be understood and kept in place.

Local Heritage and Culture [CP18]  Relation to other Core Patterns: CP1, CP12, CP17, CP20, CP22 Alignment to Generative code: [GC2] Diagnosis of the Land [GC3] Diagnosis of Settlement	Acknowledge local heritage and historical built and natural assets, Aboriginal special and sacred sites, and historical town values. **Instruction:** Provide access to sites, and information on rich history. Protect historical buildings, repair, improve or re-establish historical buildings and sites to reinstate the heritage values of the area. **Relevance to Anglesea:** The neighbourhood character of Anglesea reflects a Coastal Hamlet, with homes and buildings of 'beach shack' style. The heritage elements and character of the town needs to be preserved and new modernist style buildings must not be allowed, as to preserve heritage and culture.

(continued)

Boundaries [CP19]  Relation to other Core Patterns: CP2, CP3, CP5, CP11, CP23 Alignment to Generative code: [GC2] Diagnosis of the Land [GC3] Diagnosis of Settlement [GC4] Potential of Settlement	The location of towns within an extensive natural environment, by default establishes a boundary at the edge of the towns between the buildings and the natural forests, woodlands or grasslands. The catchment areas around the town, as part of the rivers and creeks, have natural boundaries within the larger region. Within the town are neighbourhood boundaries, as well as subculture boundaries that are formed through natural processes of social-cultural growth. The strength of boundaries is essential to establish and maintain the town's own identifiable character. **Instruction:** Set regulations in place to restrict further growth of town boundaries into surrounding natural landscapes. Evaluate and initiate future growth within existing town boundaries, and clearly define the edges that cannot be extended. Create 'in-fill' developments to increase density, and support subcultural and neighbourhood boundaries. **Relevance to Anglesea:** The natural setting of Anglesea is within the Anglesea Heath that surrounds the town, and is covered with natural woodland bush, set among hills and valleys. The existing town boundary must be kept as is, to avoid the destruction of natural bush and environmental attributes.

At the Heart [CP20]  Relation to other Core Patterns: CP1, CP13, CP14, CP15, CP17, CP26 Alignment to Generative code: [GC4] Potential of Settlement	In each town, strong centres occur that give life to the place. These centres support the overall functionality of each town. One or more centres can be classified as 'at the heart' of the town, the neighbourhood or community precinct. These centres help the wholeness of place to come together by reinforcing its connections to other sub-centres or to a larger centre. **Instruction:** Identify the existing 'heart' of the town, neighbourhood or community, linking this centre with smaller sub-centres and a larger centre of the town and subregion. Identify the additional properties of wholeness, establishing the symmetries and boundaries to create a living structure for the town. **Relevance to Anglesea:** Various centres of function and attraction are located along the Anglesea River, which meanders right through the middle of Anglesea. The Anglesea River can be seen as the 'heart' of the town, supporting various centres of function.

(continued)

Tread Lightly [CP21]  Relation to other Core Patterns: CP12, CP13, CP22, CP25 Alignment to Generative code: [GC2] Diagnosis of the Land [GC3] Diagnosis of Settlement [GC4] Potential of Settlement	The development of the buildings and precincts in the towns will tread lightly on the land, ensure minimal disturbance to its physical, ecological and biophysical processes. Coastal character of the towns historically includes appropriate scale and built form. **Instruction:** Provide for siting and building guidelines that require the consideration of the natural environment, ensure minimal disturbance to the environment, and where applicable provide for reinstatement of the landscape where damaged. Keep buildings at a small scale and integrate with the natural environment. Buildings to respect nature and need to be lightweight, adaptable construction. **Relevance to Anglesea:** The current character of Anglesea is supporting mostly in-scale homes integrated within their surroundings. Implement regulations to enforce new buildings that enhance the town character and which will tread lightly on the land.

House Mix [CP22]  Relation to other Core Patterns: CP14, CP17, CP18, CP19, CP21 Alignment to Generative code: [GC3] Diagnosis of Settlement [GC4] Potential of Settlement	The form, geometry and style of the historical Victorian coastal built form has a deep coherence and substance that links with coastal settings. The buildings follow a similar form, use similar materials, and follow a form language that gives a specific, local and living character to the coastal settlements. The house mix in each town needs to be considered as a derivative of this form language, and the inclusion of the coastal character form, shape, volume, levels of scale, and coherent geometry results in its sense of place, local character and timeless architecture. **Instruction:** Establish a form language that will inform the house mix, types, settings, scale and footprint of buildings to enhance local character and sense of place. Bulky, monolithic 'box type' buildings with large footprints shall not be allowed. Buildings need to be shaped by local characteristics, use and needs. Buildings need to be different in size and position, but similar in different versions of local character type or form. **Relevance to Anglesea:** The neighbourhood character of Anglesea reflects a Coastal Hamlet, with homes and buildings of 'beach shack' style. However new homes are built with out-of-character styles, and it is required that a form language be established with design guidelines to preserve character.

(continued)

Pathways and Walks [CP23]  Relation to other Core Patterns: CP2, CP7, CP10, CP11, CP13 Alignment to Generative code: [GC3] Diagnosis of Settlement	Pedestrian pathways and walks provide for walkability and a connection between beach areas, river walks, nature walks and internal connections between commercial areas, houses and recreational parks. **Instruction:** Priority needs to be given to pedestrian routes, walks and paths. Integrate paths between the residential and commercial areas, connecting roads at right angles. Crossings of roads need to be raised, with different textures as roads finish. Paths need to follow the topography of the site, and meander through the neighbourhoods connecting clusters of communities. Create a network of walkable paths that are safe, have good signage, include resting areas (such as benches), include materiality and textures as way-finding, and include landscaping to match local vegetation types and habitats. **Relevance to Anglesea:** Link existing paths and walkways between the beach, the river side and the shopping strip to provide an interlinked network of walking tracks.

Views [CP24]  Relation to other Core Patterns: CP2, CP3, CP4, CP5, CP6, CP9, CP11, CP16, CP23, CP25 Alignment to Generative code: [GC2] Diagnosis of the Land	Advantage points and views at key areas in the town need to be celebrated and enhanced. Views down roads and from buildings with a good visual connection to nature and the surrounding environment stimulate calmness and assist in the identification of vocal points for orientation. Open areas provide spatial preferences for visual stimulation and a sense of belonging, safety and security. **Instruction:** Identify and conserve existing view lines of the forests and woodlands, rivers and beach, the natural amphitheatre of the area and the sea. Protect and enhance views down roads, lanes and walking paths. In open areas create prospect, conditions that allow for visual survey and contemplation of the surrounding area for both visual stimulation and potential hazards. **Relevance to Anglesea:** The Anglesea lookout provides for views and vistas of the area, and is a major attraction for residents and visitors alike. The point of the bay (Point Roadknight) protrudes into the ocean, with views and a lookout up and down the coast. Value these attributes and put in place measures to avoid overuse and damage to the natural environment due to high numbers of visitors in the area.

(continued)

Sense of Danger [CP25]  Relation to other Core Patterns: CP5, CP6, CP7, CP8, CP9 Alignment to Generative code: [GC2] Diagnosis of the Land	The location of the town is within a setting of forests and woodlands, as well as low-lying areas close to the sea and rivers. Acknowledge the dangers of natural hazards such as bushfires and flooding. **Instruction:** Provide for an awareness of hazards and put in place response measures as well as provide for controllable risk outcomes. Locate CFA and SCLC at key locations to make these services an integral part of the community fabric. **Relevance to Anglesea:** One of the highest risks for Anglesea is bushfires. The nature of the Anglesea Heath woodland bush is highly sensitive to fire ignition. Living with this hazard needs to be at the forefront of the awareness, culture and decisions of the community.

Meeting Place [CP26]  Relation to other Core Patterns: CP17, CP18, CP20, CP25 Alignment to Generative code: [GC3] Diagnosis of Settlement	In towns, various neighbourhoods, community clusters and subcultures come together at a communal meeting place. This is a place where boundaries between neighbourhoods and cultures fuse together to form a sense of belonging. **Instruction:** Create a meeting place for gatherings, social interaction, art and craft, social and community events. This place must be at least one big place within the town that can accommodate a gathering of a large group of people. **Relevance to Anglesea:** Various key locations within Anglesea provide meeting places for community members and visitors. The Anglesea SLSC building located on the foreshore, at the main beach, is used as a community gathering place. Link the SLSC with other key gathering places through walking tracks, and vehicle access paths.

Appendix 6
Regenerative-Adaptive Patterns: Application to the Town Anglesea

Twelve Regenerative-Adaptive Patterns of Anglesea

The 12 regenerative-adaptive patterns of Anglesea include the following:

- Regenerative-Adaptive Pattern [1]: Adaptive Built Environments - includes considerations of adaptive potential of buildings and infrastructure;
- Regenerative-Adaptive Pattern [2]: Protect Nature's Adaptive Capacity - considers the adaptive capacity of local natural systems, assists in the ability to adapt;
- Regenerative-Adaptive Pattern [3]: Nature's Work as a Continuous and Reciprocal Interactions - allows life support functions to be processed through conversion, distribution, filtration, assimilation and storage with interaction throughputs;
- Regenerative-Adaptive Pattern [4]: Optimisation and Multiple Functions - inclusion of multiple functions and outcomes for systems to optimise resilience ability;
- Regenerative-Adaptive Pattern [5]: Aggregate not Isolate, Integrate rather than Segregate - integrate all parts to fix connections, aggregate to assist the inclusion of symbiotic relationships to promote regeneration;
- Regenerative-Adaptive Pattern [6]: Self-Regulation and Feedback Loops - include self-regulation of positive and negative feedback loop systems in processes;
- Regenerative-Adaptive Pattern [7]: Produce no Waste, Recycle and Assimilate - make use of all inputs and outputs for a closed loop or net positive system;
- Regenerative-Adaptive Pattern [8]: Conversion of the Solar Income - include passive solar systems for energy, heating and cooling, thermal storage and conversion;
- Regenerative-Adaptive Pattern [9]: Scale Linking to Facilitate Flow - shaping the medium to facilitate flow, scale linking for support of maximum function at the smallest scale;
- Regenerative-Adaptive Pattern [10]: Storage as a Key Resource; for Energy, Water and Materials - maintaining adequate storage by balancing the replenish rate with the rate of use;
- Regenerative-Adaptive Pattern [11]: Valued Renewable Resources and Services - use and value existing natural, renewable resources for energy and biological services; and
- Regenerative-Adaptive Pattern [12]: Human - Nature Connections for Healthy and Prosperous Environments - the application of Biophilia strategies and designs to create healthy environments for both humans and nature.

P. B. Roös, *Regenerative-Adaptive Design for Sustainable Development*, Sustainable Development Goals Series, https://doi.org/10.1007/978-3-030-53234-5

Example of Application of the Regenerative-Adaptive Patterns: Anglesea

Regenerative-Adaptive Pattern [1] and [2] are used to demonstrate the application of the 'notion of regenerative-adaptive patterns' equation, where each element and its functions $\{nda, f_1 \ldots f_i, std, tsp_{15}, e_1 \ldots e_i, r1, pot\}$ are analysed and applied to the Regenerative-Adaptive Pattern *(rgp)*.

The Regeneration-Adaptive Patterns of Anglesea follows the structure below:

- The name and number of the Regenerative-Adaptive Pattern listed in bold type;
- A short statement of what the Regenerative-Adaptive Pattern is to represent in italics;
- The description of what the Regenerative-Adaptive Pattern is to achieve or accomplish; and
- Explanations of each element and its functions - $\{nda, f_1 \ldots f_i, std, tsp_{15}, e_1 \ldots e_i, r1, pot\}$ of the relevant Regenerative-Adaptive Pattern (*rgp*) equation.

Anglesea Regenerative-Adaptive Pattern [1]: Adaptive Built Environments

Statement:

Includes the considerations of the adaptive potential of buildings and infrastructure.

Description:

Although some buildings can be refurbished to be reused beyond their initial intended function, by its nature the built environment created by humans consists of static, fixed entities at their location. In most instances buildings are designed and built as completed, unchanged entities, and as a result abandoned or demolished after their useful functions have been exhausted. With the changing environment due to impacts of climate change, the built environment needs to be able to be adaptable according to future scenarios of change, and the future functional needs of the place where it is located. Adaptive built environments needs to be 'ecomorphic',[1] that is when the internal structures mimic and integrate with the natural systems within which they are located and interconnected (Van der Ryn, 2005, p. 156).

Explanation of elements and functions:

nda

Built attributes at risk due to sea level rise, inundation and coastal recession include the Anglesea Beachfront Family Caravan Park, the Anglesea Visitor and Information Centre, Anglesea Store, Anglesea Surf Centre and various buildings along the Great Ocean Road next to the Anglesea River. These built attributes need to be able to adapt in the future to the impacts of sea level rise, both in function and structure.

$$f_1 \supset f_i$$

Forces of change that have an impact upon the built attributes include:

- f_1 - Sea level rise between 0.8 m - 1.4 m at incremental levels between 2030 and 2100, that will result in permanent inundation;
- f_2 - Storm surge with increased wave heights and overland flooding due to extreme weather events will result in periodic temporary inundation; and

[1]According to Turner et al. (2004) an ecomorphic design should act and react according to its occupation, so that the environment and its inhabitants form an ecological whole (Turner, Mottram & Penn, 2004).

- f_3 - Recession of the coastline will result in beach, foreshore and cliff erosion, as well as instability of the soils.

std

The place settings specific to location are interconnected to its centres (the fundamental primary entities, the latent centres), and change of the centres happens in the dynamics of time; some immediate, others over short periods, and others over long periods of time. The key latent centres for Anglesea include the Anglesea Beachfront Family Caravan Park and the Anglesea Visitor and Information Centre as part of the boardwalk and promenade precinct along the Anglesea River, connecting walking tracks to the Coogoorah Park, the Anglesea Store, Anglesea Surf Centre and various small businesses and buildings along the Great Ocean Road. During holiday periods, these centres become very busy and are core attractions for visitors and holidaymakers.

$\boldsymbol{tsp_{15}}$

Transformations specific to place that include the 15 properties and form the wholeness applicable to the built assets for this regenerative-adaptive pattern are:

- tsp_1 - Levels of Scale [AS-FP1]; includes the inherent connection of the buildings to their hierarchy of built and natural environs, consisting of the building's purpose and function, connected to the foreshore park, beach, Anglesea River inlet, Anglesea River, the networks of creeks and rivers including Marshall Creek and Salt Creek, the wetlands, and the town perimeter, which is surrounded by the adjacent natural environments of the Anglesea Heathland, Angahook Lorne State Park, and the Great Otway National Park. The buildings of this regenerative-adaptive pattern are part of the larger whole that steps up in hierarchy of scale to the commercial precinct of the shopping strip, residential buildings of the surrounding residential area, public buildings as gathering places, the community hall and the Anglesea Golf Club;
- tsp_2 - Strong Centres [AS-FP2] that forms part of the whole. Deep connections to function include the Anglesea River estuary and inlet, the riverfront and boardwalk, the Coastal Reserve, and the Great Ocean Road shops linked to one of these strong centres, which include the Anglesea Beachfront Family Caravan Park;
- tsp_3 - Boundaries [AS-FP3] that strengthen the wholeness of the centres; they are the boundary of the Anglesea River estuary and inlet, the Anglesea River and the riverfront that borders the Anglesea Beachfront Family Caravan Park, the Great Ocean Road and the bridge across the River, and the cliffs at the beach front;
- tsp_4 - Positive Space [AS-FP5] that forms the space in-between the building structures and the landscape features, including the beach, open grass areas and walking tracks, front yards and back yards of cottages and camping sites, the foreshore and the play area for kids of the Anglesea Beachfront Family Caravan Park; and
- tsp_5 - Contrast [AS-FP9] that creates the existence of place from the essence of distinctiveness, noticeable between the built form of the holiday cottages and amenity buildings of the campsites, and the organic forms of the surrounding landscape, vegetation, trees and the river.

$\boldsymbol{e_1 \supset e_i}$

Core patterns are interconnected and organise the complex relationships that produce activities and existence. They include the following:

- e_1 - Nomad's Rest [CP12] that results in the existence of place, providing accommodation for holidaymakers and tourists in the caravan park and cottages; this stimulates and supports the economic stability of the town;

- e_2 - At the Heart [CP20]; the Anglesea Beachfront Family Caravan Park [AS-CP9] is a strong centre that supports the overall functionality of the town; it helps the wholeness of place to come together by reinforcing its connections to other sub-centres such as the River [AS-CP2], the Anglesea River boardwalk, and a larger strong centre - the Shopping Street [AS-CP8];
- e_3 - Tread Lightly [CP21]; most of the cottages and buildings at the Anglesea Beachfront Family Caravan Park are made of light construction materials including timber structures and weatherboard cladding. These light structures can be relocated to adapt to the changes of the environment. More permanent structures, such as the shops at the shopping street, the Anglesea General Store, the Motor Inn, Great Ocean Road Resort, and Anglesea Information Centre are made of permanent materials such as concrete and brick. These latter buildings need to be regenerated with more semi-permanent materials to support the sustainable outcomes of this core pattern;
- e_4 - Pathways and Walks [CP23]; the Anglesea River boardwalk connects with other pathways and walks, connecting the commercial street, the beach area, River bank, the Anglesea Coogoorah park, nature park and houses. Pathways from the Anglesea Beachfront Family Caravan Park link the boardwalk with the opposite side of the River across the Anglesea River bridge; and
- e_5 - Meeting Place [CP25]; the strength and resilience of a community are fostered and improved by the possibility and allowance for one or more meeting places. These include the Surf Life Saving Club at the beach, Anglesea Bowling Club at the Lions Park Reserve located next to the Caravan Park, the Anglesea Art House and the Anglesea Community House.

*r*1

Regenerative attributes can assist in the healing of places that are damaged and let them evolve, adapt to improve and regenerate to be resilient. The built environment of Anglesea can be resilient if ecological sustainability principles are considered as part of development in the context of adaptation and regeneration. The regenerative attributes for the built environment (human environment) include:

- *r*1 - Wind [AS-RA2]; the optimisation of wind energy, inclusive of passive and active wind uses. Passive responses include the optimised orientation of buildings; include planting and other methods for windbreaks, placement of public gathering areas in and around buildings at the lee side of the wind pressures. For passive ventilation in buildings, include wind catchers. Active wind optimisation includes the use of small-scale wind turbines to provide supplementary power supply to buildings in Anglesea;
- *r*2 - Solar [AS-RA3]; the optimisation of solar energy inclusive of passive and active solar. Passive solar includes the optimised orientation of buildings, allowance of solar penetration in indoor and outdoor spaces, as well as collection of solar radiant heat in structures and objects. Active solar includes the use of solar photovoltaic panels and solar hot water heating systems, to be applied to all buildings in Anglesea;
- *r*3 - Small Scale Local Agriculture [AS-RA7]; the inclusion of vegetable gardens, local grown food dispersed within the gardens and public areas of the Caravan Park, restaurants, and community buildings will provide the opportunity to reduce the demand of produce to be sourced and delivered from outside of the Anglesea town/catchment boundary;
- *r*4 - Waste [AS-RA8]; the collection and processing of waste in a closed loop system, biodegradable waste processing to use for composting or capturing of the methane, organic waste recycling and processing, and the recycling of grey water and black water for potential uses of garden watering and transfer to potable water. Localised processing and storage at the source will be feasible due to the nature of the holiday town population fluctuations;
- *r*5 - Small Business [AS-RA11]; the provision of a mixture of shops and amenities with a variety of services and products that complement each other for securing a sustainable local economy. Establish complimentary services and product supply between shops and restaurants on the shopping street, the Caravan Park and other accommodation providers such as bed and breakfasts and guesthouses; and

- *r6* - Heritage and Culture [AS-RA12]; the identification and inclusion of heritage places strengthens the sense of belonging for the local community, and includes both Aboriginal and European attributes. The Anglesea River area includes 2 registered Aboriginal Cultural Heritage Places (ACHP) of significance, and at Point Roadnight there are places of high sensitivity to Aboriginal heritage and culture with major and well-developed stratified midden deposits.[2] Celebrate the sites with view posts, but also construct barriers to protect the middens. European heritage buildings need to be acknowledged with re-introduced functions and architecture, including the Angahook General Store (built in 1929, it was the local community hub before the Memorial Hall was built in 1954[3]). Future development of and next to the Angahook General Store needs to include typical heritage architecture buildings and functions to create a 'heritage strip' that can support both local economic and tourist activities.

pot

Potential includes specific opportunities to assist the Anglesea community to evolve the value-generating capacity of the place as a whole. Potential needs to include built attributes that will be able to adapt to, or be future-proofed against the impacts of sea level rise, storm surges and flooding events, both in function and structure of place. The adaptive, evolutionary potential opportunities for the built environment include:

- Establish and provide shoreline protection at the Anglesea River inlet as well as along the River, according to the topography and contours, to include soft engineering approaches such as mangrove planting and the construction of rock revetments along the River;
- Identify assets with key heritage and cultural values to Anglesea and protect these against the impacts of change through the construction of barriers, flood control systems and the raising of floor levels. Plan for future sea level rises by identifying new locations to relocate and rebuild heritage structures and buildings, reinforcing at locations that can represent the existing strong centres of place;
- Protect and strengthen key infrastructure in the short term. Plan relocation through a staged process over time, including the re-routing of the Great Ocean Road;
- Planned retreat of key buildings, roads and other infrastructure, houses and amenities to higher levels where coastal recession and inundation will not have an impact upon these assets; and
- Evolutionary potential for the built environment of Anglesea includes opportunities for buildings and the whole of town to operate as a net positive energy, water, material, food and resources consumptive environment.

Anglesea Regenerative-Adaptive Pattern [2]: Protect Nature's Adaptive Capacity

Statement:

Considers the adaptive capacity of local natural systems, assist to the ability to adapt.

Description:

As part of the larger ecological and human system, the local natural systems must have the adaptive capacity to adapt to changes of the environment they exist within, when it undergoes a change after disturbance. Ecological succession allows the species to change structure of natural community overtime, usually in timescales over decades (Sahney and Benton, 2008, p.762). With Anthropogenic cli-

[2]A total of 6 known Aboriginal cultural heritage places occur within the Anglesea Crown Land Reserves and have been acknowledged to have high scientific significance with radiocarbon dating which resulted that these middens to be at least 1385 ± 20 years old (Marshall, 1997, p. 21)

[3]The General Store continues to be recognised by the Anglesea locals as a reflection of early twentieth century community life, being a hub for township meetings, Red Cross meeting and other important community events prior to the construction of the Memorial Hall in 1954. The historical and social significance of the place is embodied in the surviving (and to an extend altered) building fabric and architectural form of its era (Rowe, 2008, p. 2)

mate change impacts that could be happening at a rapid pace, such as rising sea levels, measures needs to be put in place to allow local natural systems to have the capacity to adapt to these changes.

Explanation of elements and functions:

nda

Natural attributes at risk due to sea level rise, inundation and coastal recession include the Anglesea beach, Anglesea River and inlet, the cliffs, the wetlands and saltmarsh, Coogoorah Park, coastal vegetation, and parts of the Anglesea Heath. These natural attributes need to be protected to allow the natural systems to be able to adapt to the impacts of sea level rise.

$$f_1 \supset f_i$$

Forces of change that have an impact upon the natural attributes include:

- f_i - Sea level rise between 0.8 m - 1.4 m at incremental levels between 2030 and 2100, that will result in permanent inundations;
- f_2 - Storm surges with increased wave heights, and overland flooding due to extreme weather events that will result in periodic temporary inundations;
- f_3 - Recession of the coastline resulting in beach, foreshore and cliff erosion, as well as instability of the soils; and
- f_4 - Saltwater intrusions in the freshwater ecosystems higher up in the Anglesea River catchment.

std

Key latent centres of the natural environment for Anglesea include the Anglesea River estuary and inlet, the Coastal Reserve, Point Roadknight, cliffs and the beach, the Anglesea River, the Anglesea Heath, the Point Addis Marine National Park, and the Coogoorah Park. These latent centres of natural environment evolved over many thousands of years, and the values of these centres are at the heart of the environmental and ecological existence of nature in the Anglesea region. Additionally, these natural attributes are key values that are the initiators for human settlement, visitors to the area, and destinations for holidaymakers.

tsp_{15}

Transformations specific to place, which include the 15 properties and form the wholeness applicable to the natural attributes for this regenerative-adaptive pattern, are:

- tsp_1 - Levels of Scale [AS-FP1]; includes the inherent connection of the habitats in the Anglesea catchment to their hierarchy of natural environs. These consist of ecological systems commencing from the beach and foreshore to the Anglesea River inlet, Anglesea River itself, the networks of creeks and rivers including Marshall Creek and Salt Creek, the wetlands, the Anglesea Heathland, Angahook Lorne State Park, and the Great Otway National Park;
- tsp_2 - Strong Centres [AS-FP2] that form part of the whole, supporting the larger River catchment area, include deep connections of ecological functions comprising the Anglesea River estuary and inlet, the Coastal Reserve, Marshall Creek and Salt Creek, the Anglesea Heathland, Angahook Lorne State Park, and the Great Otway National Park;
- tsp_3 - Boundaries [AS-FP3] that strengthen the wholeness of centres; they include the boundary of the Anglesea River, the wetland system at Coogoorah Park, the surrounding forests and heathland, the Great Otway Ranges, the coastline and the ocean;

- tsp_4 - Altering Repetition [AS-FP4]; that forms the systematic fractal patterns of the Anglesea area. These include the coastal scrub and vegetation, repetition of tree species, rock formations in the coastal cliffs, scale linking the high cliffs down to the River estuary vegetation, the repetition of the meandering curves of the River, and patches of dispersed open farmland;
- tsp_5 - Positive Space [AS-FP5] that supports the coherent shapes and organic forms of the surrounding landscape; vegetation, trees and the River are dispersed between these forms and result in an emptiness with an inherent strength of wholeness;
- tsp_6 - Good Shape [AS-FP6] and Self Symmetries [AS-FP7]; the patterns in nature that form good shapes and local symmetries, such as the self-symmetry forms in the structure of Stringybark (*Eucalyptus sp.*) and Banksia (*Banksia sp.)* trees, the patterns of geological layers in the exposed coastal cliff, the meanders of the Anglesea River, the Coast Beard Heath (*Leucopogon parviflorus*) flowers, the roots of the Mangroves in the wetland, and the sand formations of the windswept Anglesea beach;
- tsp_7 - Contrast [AS-FP9] that results due to the existence of visual contrast within the landscape, noticeable between the ocean with the cliffs at the beach, the large trees of the Anglesea Heath with small coastal scrubs, the flowing River with the embankments and vegetation, and the dense Otway Forest with scattered grassland;
- tsp_8 - Gradients [AS-FP10]; noticeable areas of the transition from one space in nature to another, gradually forming levels of intensity in gradients. The trees of the Anglesea Heath gradually formulate gradients in scale from the tree trunk to branches, to smaller twigs and finally the leaves. The coastal grass and scrubs at the beach and cliff transition to forest trees, and the network of the creeks collects in the Anglesea River, flowing down into the ocean with various levels of scale; and
- tsp_9 - Roughness [AS-FP11]; the irregularities in formations created by growth and environmental changes in nature result in the distinctiveness of place. Anglesea is noticeably distinct from other coastal towns due to its surrounding vegetation (Anglesea Heath) and the roughness as a character of the coastal bushland.

$$\boldsymbol{e}_1 \supset \boldsymbol{e}_i$$

Core patterns are interconnected and organise the complex relationships that produce activities and the existence of place, and include the following:

- e_1 - Our *Country* [CP1]; connects the land with its inhabitants, both human and fauna and flora. Connections through energy flows and the spirit of the land that are frequently formed by the living structures of the land, constantly present in the evolutionary process of the land. The evidence rests in the ocean, the beach, Anglesea River and Anglesea Heath and Otway Forest trees, connected with the Aboriginal season occurrences[4];
- e_2 - Hinterland to Coast Connection [CP2]; the Anglesea beach with its cliffs gradually link with the hills and valleys of the Otway Ranges along with the River and creeks into the larger catchments of the hinterland, providing ecological and social-cultural viability to the region;
- e_3 - The Natural Amphitheatre [CP3] and The Heads and Cliffs [CP6]; at the larger scale, the natural topography of the Anglesea Bay results in a topographical natural setting of an amphitheatre within which the town is settled. The landscape forms a pinnacle jutting out to sea, resulting in a landmark with views up and down the coastline, known as Point Roadknight. The high cliffs at Anglesea River are part of the Demons Bluff Formation that underlies the coastal hinterland proper on either side of the Anglesea River, connecting the coast to inland in a geomorphological and geological context;
- e_4 - Forests and Woodlands [CP5]; the large trees of the neighbouring Otway Forest and the Anglesea Heath create stature, connecting the landscape with the town of Anglesea in many ways.

[4]The Aboriginal seasons and the cycles of nature provide for a close connection between human, habitat and all other creatures (Roös, 2014).

Trees and the vegetation in these woodlands provide for extensive and a variety of ecological systems, as well as wildlife habitats; and

- e_5 - River of Life [CP7]; the artery that injects life into the surrounds of the larger water catchment where the Anglesea River, wetlands and vegetation are key to the functioning of the ecological system. The River is the largest local drainage line in the immediate region. The River flows south-east towards the coast and has its headwaters in the hills and valleys to the north and northwest. Numerous tributaries that flow eastwards include Salt Creek, Marshy Creek and Edwards Creek, which all feed into the Anglesea River.

***r*1**

Regenerative attributes can assist in the healing of places that are damaged in the Anglesea catchment area, by allowing them to evolve, adapt and regenerate to be resilient. The natural environment of Anglesea can be resilient if the natural systems of the place are kept intact to allow adaptation to changes in climate. Additionally, where accelerated changes occur, the adaptive capacity of current ecosystems will not be able to adapt, and therefore putting in place ecological resilient strategies as part of future development will assist in the regenerative and adaptive capacity of place. These regenerative attributes include:

- *r*1 - Water Flow [AS-RA1]; the consideration of local water flows and storage systems as part of the hydrologic cycle will support the regenerative attributes of the local water systems for Anglesea, taking into account the landform, soil, plants and biological processes of the watershed.[5] Fitted into the topography of place, using landform alterations such as swales, berms, and depressions, they can aid water to flow and be regenerative for Anglesea, both in supply and as an agent for improving ecological systems and biodiversity. Water storages to be constructed and located according to the land and topography;
- *r*2 - Solar [AS-RA3]; the allowance of nature's optimisation of solar energy to regenerate natural systems including photosynthesis, absorption and reflection, providing heat sinks and evaporation for conversion and dissipation. Energy concentrations through the natural energy flow of place, can be used to support a low-entropy society for Anglesea;
- *r*3 - Woodlands [AS-RA4]; the protection and sustainable management to assist regeneration of woodlands that support the integrity of the overall watershed, with roots holding soils in place and thereby slowing down and absorbing surface runoff water. The support of an extensive biodiversity system in the Anglesea Heath, which includes about a quarter of the state of Victoria's recorded plant species and over 80 different species of orchids, is undoubtedly integral to the importance to the ecological system of the Anglesea area;
- *r*4 - River Course [AS-RA5]; the Anglesea River, wetlands and surrounding vegetation are key to the functioning of the larger ecological system in the Anglesea River catchment area. Because the Anglesea River is the largest local drainage line in the immediate region, it is key to the functioning of hydrology in the catchment area. Controlled water runoff, catchment and controlled flooding through the inclusion of wetlands and saltmarshes next to the River could assist in reinstating the balance of water chemistry and water flow that is current due to local extraction activities despite the recent closure of the Alcoa open cut Coal Mine[6] (the mine site is directly located upstream in the River course), and to reinstate the natural water flow of the River;
- *r*5 - Ocean [AS-RA6]; the ocean is a key attribute to the coastal communities along the Victorian coast both in services and the marine environment. The marine environment, just offshore from

[5]Regenerative technology for water systems needs to include the conversion, distribution, storage, assimilation and filtration process of the full hydrologic cycle. The regenerative water system includes ecosystem principles of the ecosystem structure, ecosystem function and locational patterns (Lyle, 1994, p.146).

[6]The water flow in the Anglesea River will drastically change due to River levels that will drop up to one meter when the pumps of the mine shut down (Noonan, 2016).

Anglesea, is an integrated system connecting the coastal waters, estuaries and wetlands with a diverse marine ecosystem. Shallow waters between 10 and 20 metres deep from the Ingoldsby Reef up to the Point Addis Reef system include weedy seagrass, kelp and brightly coloured sponge gardens that provide habitats for various fish species and other marine life. These environments are inherently connected to the onshore coastal environments. These environments can be protected and planned and managed to promote the larger interconnected ecosystem to support the goods and services they provide;
- *r*6- Flora and Fauna [AS-RA9]; the onshore coastal environment of the Anglesea area contains a wide range of habitats that support a diversity of plants and animals, including migratory animals and shorebirds (such as the Eastern Curlew, *Numenius madagascariensis*, and the Pectoral Sandpiper, *Charadrius melanotos*);
- Allow and enhance the natural horizontal migration of terrestrial fauna and flora communities by removing barriers (such as built infrastructure) that may prevent migration, especially having regard to the future temporary and permanent inundation of specific areas along the Anglesea River;
- *r*7 - Soils [AS-RA10]; the healthy soils of the Anglesea catchment are important to the natural environment of the area, including the biotic and abiotic elements. Soils influence the distribution of plant species and support the habitat of organisms, which in turn are key to the ecological functioning of the Anglesea River. Practices to protect the soil from erosion, acidification due to the close proximity to the ocean, and soil saturation will assist in regeneration of native vegetation resilience, as well as supportive habitats for wildlife and river fish species.

pot

Potential for the natural environment of Anglesea includes specific opportunities to assist the environment to evolve its value-generating capacity as an ecological place, including:

- Establish and provide natural methods for shoreline protection at the Anglesea River inlet as well as along the River, and include mangrove planting, saltmarsh vegetation, and wetland constructions along the River to mitigate flooding and tidal events as well as storm surges;
- Identify ecosystem processes specific to place that are adaptable and resilient against current climatic conditions. Apply these aspects of resilience to mitigation strategies for the Anglesea natural environment for future scenarios of climate change;
- Identify vulnerable ecosystems and natural habitats at risk from climate change impacts, and assist in migration options by providing biodiversity corridors;
- Set habitat provisions in place to allow for nutrient cycling, purification and water flows;
- Allow for the natural retreat of plants and animals to higher levels where coastal recession and inundation will not have an impact; and
- Establish conditions that can potentially allow organisms, plants and animal species of the Anglesea area to adapt to conditions that can enhance their evolutionary potential with novel functions to thrive in alternative environments for the future.[7]

[7] According to Hall the genome of each organism contains the potential to evolve novel functions to adapt to changing environments, embedded in its structure is the evolutionary potential to allow it to thrive in alternative and different environments (Hall, 1999).

References

Hall, B. (1999). Toward an understanding of evolutionary potential. *FEMS Microbiology Letters* 178, 1–6. Elsevier.

Lyle, J. T. (1994). *Regenerative design for sustainable development.* New York: John Wiley & Sons, Inc.

Marshall, B. (1997). *Archaelogical investigations of an aboriginal shell midden at point roadknight.* Anglesea Foreshore Committee of Management. Anglesea: Anglesea Foreshore Committee of Management.

Noonan, A. (2016). *Anglesea river greatest priority as community consultation over former Alcoa land begins.* March 28. Retrieved May 20, 2016 from ABC News: www.abc.net.au/news/206–03-28/.

Roös, P. B. (2014). Indigenous Knowledge and Climate Change: Settlement Patterns of the Past to Adaptation of the Future, in the *International Journal of Climate Change: Impacts and Responses, Common Ground Publishing, University of Illinois, Illinois, USA.*

Rowe, D. (2008). *Surf coast shire heritage study stage 2B.* Surf Coast Shire. Torquay: Authentic Heritage Services Pty Ltd.

Turner, A., Mottram, C., & Penn, A. (2004). An ecological approach to generative design. *Design Cognition Computing 04*, 259–274.

Index

P. B. Roös, *Regenerative-Adaptive Design for Sustainable Development*, Sustainable Development Goals Series, https://doi.org/10.1007/978-3-030-53234-5